Ansys Discovery 2025
Black Book

By
Gaurav Verma
Matt Weber
(CADCAMCAE Works)

Edited by
Kristen

ISBN # 978-1-77459-168-0

NOTICE TO THE READER

DEDICATION

To teachers, who make it possible to disseminate knowledge
to enlighten the young and curious minds
of our future generations

To students, who are the future of the world

THANKS

To my friends and colleagues

To my family for their love and support

Table of Contents

Chapter 2 : Ansys Discovery-Design II

Chapter 3 : Ansys Discovery-Design III

Chapter 4 : Ansys Discovery-Preparing Model

Chapter 5 : Ansys Discovery-Design Practical and Practice

Chapter 6
Ansys Discovery-Introduction to Analyses

Chapter 7 : Ansys Discovery-Structural Analyses

Chapter 8 : Ansys Discovery-Thermal, Electromagnetic, and Fluid Flow Analyses

Chapter 9 : Ansys Discovery - Multi-physics Analysis

Preface

ANSYS Discovery is an easy-to-use simulation software that helps engineers and designers test their ideas quickly. It allows you to see how a design will perform under different conditions, like heat, force, or fluid flow, without building a physical model. The software gives real-time results, so you can make changes and instantly see their effects. This makes it great for beginners who want to learn simulation without complex setup. It helps in improving designs faster and reducing mistakes before making a final product.

Ansys Discovery 2025 Black Book, 1st edition, is a comprehensive guide designed for engineers, designers, and analysts seeking to master Ansys Discovery. Covering essential topics from sketching and modeling to advanced simulations, this book provides step-by-step instructions with practical examples. It explores geometry creation, repair tools, design modifications, and simulation techniques, including structural, thermal, electromagnetic, and fluid flow analyses. With detailed explanations of user interface tools, constraints, meshing, and topology optimization, this book ensures a strong foundation for both beginners and professionals. Self-assessment sections help reinforce learning, making it a valuable resource for anyone looking to enhance their simulation and design skills with Ansys Discovery. Some of the salient features of this book are:

In-Depth explanation of concepts

Every new topic of this book starts with the explanation of the basic concepts. In this way, the user becomes capable of relating the things with real world.

Topics Covered

Every chapter starts with a list of topics being covered in that chapter. In this way, the user can easy find the topic of his/her interest easily.

Instruction through illustration

The instructions to perform any action are provided by maximum number of illustrations so that the user can perform the actions discussed in the book easily and effectively. There are about 600 illustrations that make the learning process effective.

Tutorial point of view

The book explains the concepts through the tutorial to make the understanding of users firm and long lasting. Each chapter of the book has tutorials that are real world projects.

Project
Projects and exercises are provided to students for practicing.

For Faculty
If you are a faculty member, then you can ask for video tutorials on any of the topic, exercise, tutorial, or concept. As faculty, you can register on our website to get electronic desk copies of our latest books, self-assessment, and solution of practical. Faculty resources are available in the `Faculty Member` page of our website (`www.cadcamcaeworks.com`) once you login. Note that faculty registration approval is manual and it may take two days for approval before you can access the faculty website.

Formatting Conventions Used in the Text
All the key terms like name of button, tool, drop-down etc. are kept bold.

Free Resources
Link to the resources used in this book are provided to the users via email. To get the resources, mail us at ***cadcamcaeworks@gmail.com*** or ***info@cadcamcaeworks.com*** with your contact information. With your contact record with us, you will be provided latest updates and informations regarding various technologies. The format to write us e-mail for resources is as follows:

Subject of E-mail as ***Application for resources of .... Black Book***.
You can give your information below to get updates on the book.

Name:
Course pursuing/Profession:
Contact Address:
E-mail ID:

About Author

The author of this book, Matt Weber, has written more than 16 books on CAD/CAM/CAE available in market. He has coauthored SolidWorks Simulation, SolidWorks Electrical, SolidWorks Flow Simulation, and SolidWorks CAM Black Books. The author has hands on experience on almost all the CAD/CAM/CAE packages. If you have any query/doubt in any CAD/CAM/CAE package, then you can contact the author by writing at cadcamcaeworks@gmail.com

The author of this book, Gaurav Verma, has written and assisted in more than 17 titles in CAD/CAM/CAE which are already available in market. He has authored Autodesk Fusion 360 Black Book, AutoCAD Electrical Black Book, Autodesk Revit Black Books, and so on. He has provided consultant services to many industries in US, Greece, Canada, and UK. He has assisted in preparing many Government aided skill development programs. He has been speaker for Autodesk University, Russia 2014. He has assisted in preparing AutoCAD Electrical course for Autodesk Design Academy. He has worked on Sheetmetal, Forging, Machining, and Casting designs in Design and Development departments of various manufacturing firms.

For Any query or suggestion

If you have any query or suggestion please let us know by mailing us on *cadcamcaeworks@gmail.com* or *info@cadcamcaeworks.com*. Your valuable constructive suggestions will be incorporated in our books and your name will be addressed in special thanks area of our books.

Page left blank intentionally

Chapter 1

Ansys Discovery-Design

Topics Covered

The major topics covered in this chapter are:

- *Introduction to Ansys Discovery*
- *User Interface of Ansys Discovery*
- *File Menu Options*
- *Sketching*
- *Performing Pull Operations*
- *Moving Objects*
- *View Arc Tools*

INTRODUCTION TO ANSYS DISCOVERY

Ansys Discovery is the application from Ansys which is used for rapid designing and simulation. It combines geometry modeling with instant physics simulation and high fidelity analysis. The software can work in collaboration with Ansys Workbench as well as standalone application for designing and analysis. To access this software, type **Discovery** in search bar of **Start** menu; refer to Figure-1 and click on the **Discovery** application option from the menu. The first page of application will be displayed; refer to Figure-2.

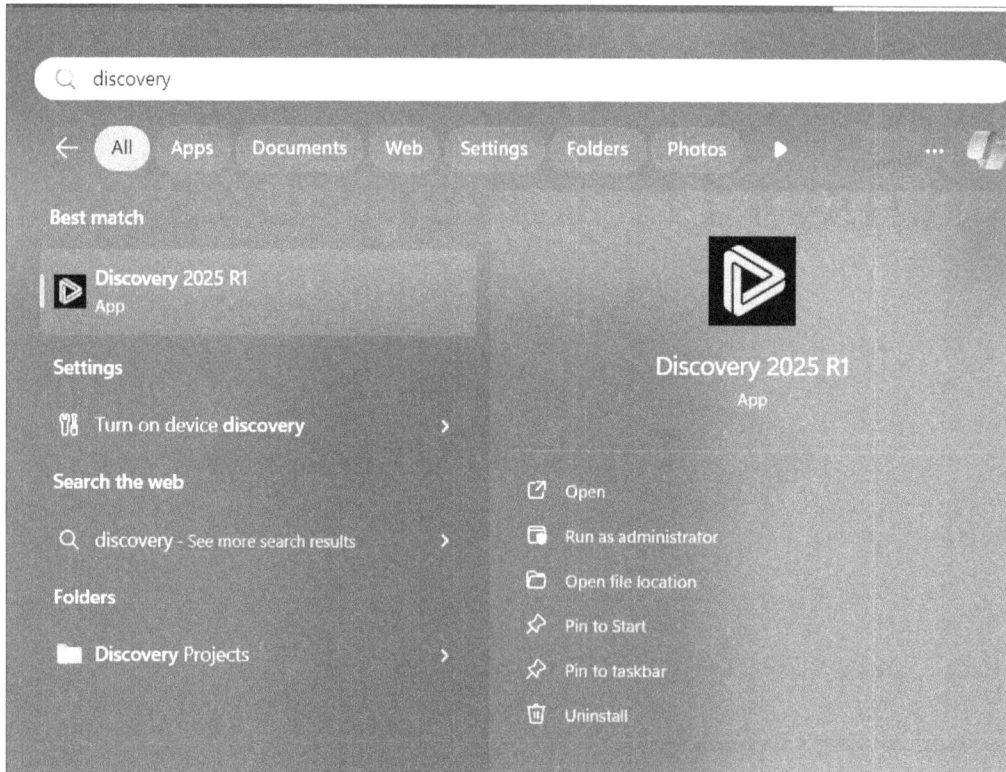

Figure-1. Discovery option in Start menu

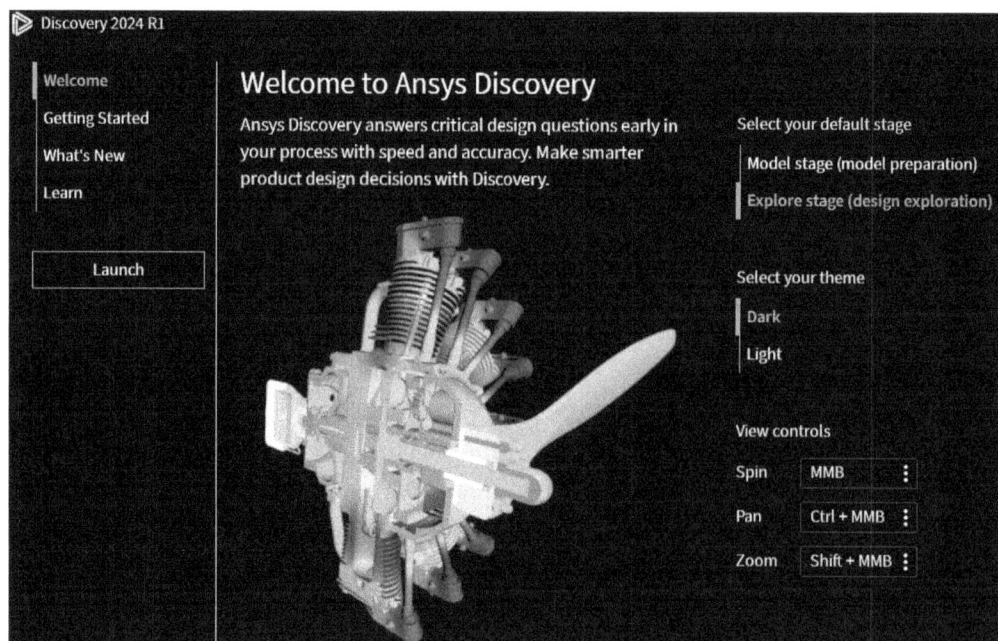

Figure-2. First page of Discovery

By default, **Welcome** option is selected at the left in First page so basic application settings are displayed for the software. Select the **Model stage (model preparation)** option from **Select your default stage** section to define that you are at stage of creating/preparing model. Select the **Light** option from the **Select your theme** section of page to use light background of application for working as it is easier for us to print this book. In the **View controls** section of page, you can set shortcut keys for various orientation operations like spin, pan, and zoom.

Select the **Getting Started** option from left area in the First page to check basic operation information about the software like how to switch between different stages of software, basic user interface, selection shortcuts, and so on. Similarly, select the **What's New** option from left area to check what are latest enhancements in the software and select the **Learn** option to access help documentation of software. Click on the **Launch** button from the First page. The **Ansys Product Improvement Program** information box will be displayed. Click on the **Continue** button. The Recent Files access dialog box will be displayed; refer to Figure-3. Click on the **New** button from the dialog box. The user interface of software will be displayed; refer to Figure-4.

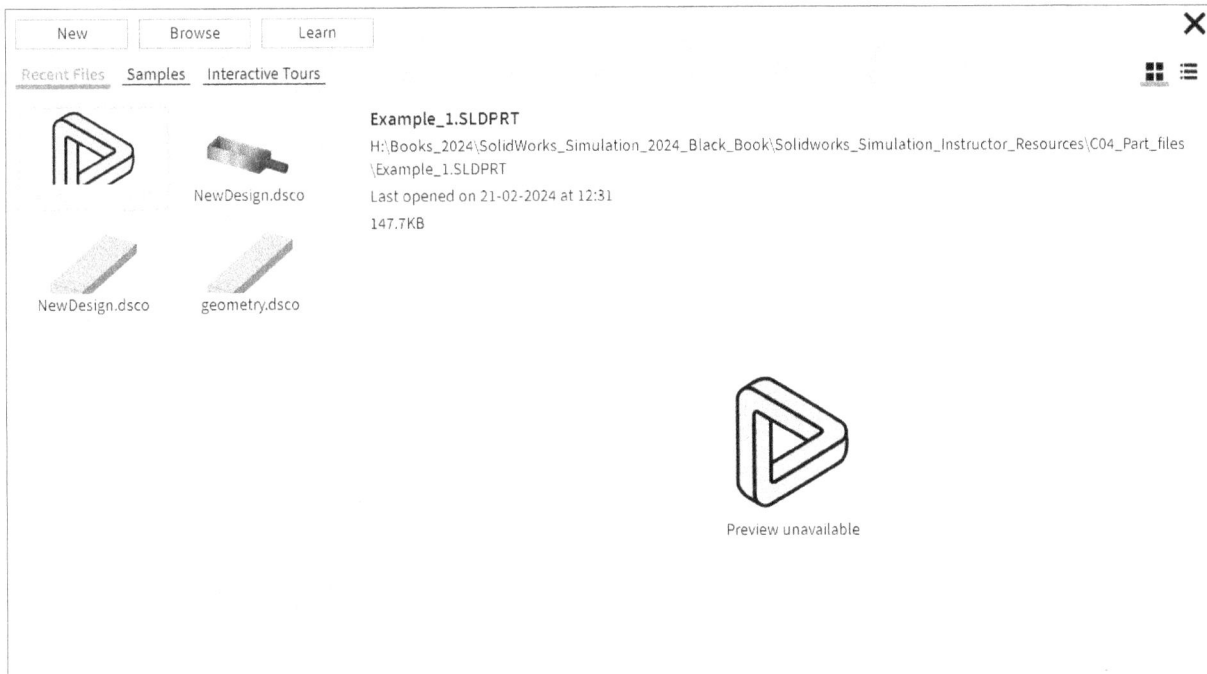

Figure-3. Recent Files access box

USER INTERFACE OF ANSYS DISCOVERY

Once you have started Ansys Discovery application then the user interface of software is displayed as shown in Figure-4. Various elements of user interface are discussed next.

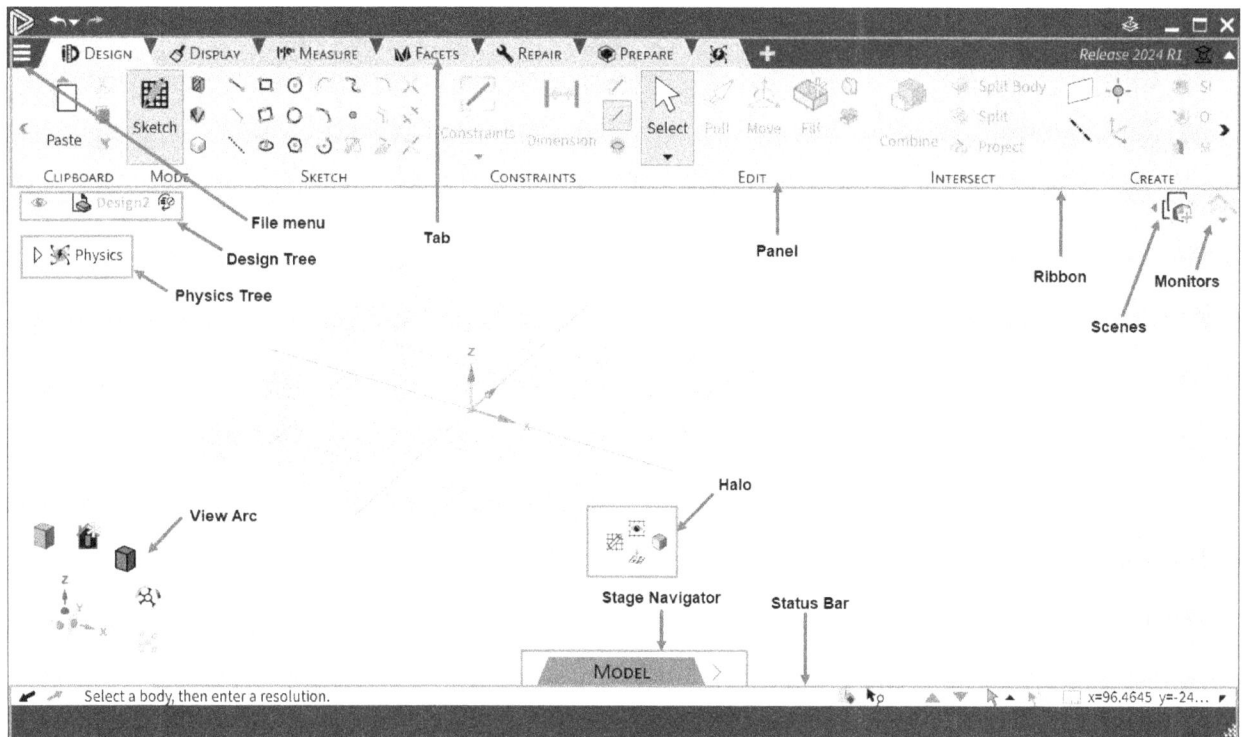

Figure-4. User interface of Ansys Discovery application

- **File** menu: The options in the **File** menu are used to perform file handling related operations and define settings for software.
- **Ribbon**: The **Ribbon** is a palette of tools grouped in **Panel** based on their similar functioning. The **Tab** in **Ribbon** is collection of these panels belonging to same field of operation like **DESIGN** tab will have panels related to tools used designing objects.
- **Design Tree**: The options in **Design Tree** are used to manage elements of current model (design). When you create an object like sketch, solid, surface, facets, and so on; they are added in the **Design Tree**.
- **Physics Tree**: The options in the **Physics Tree** area used to manage parameters and properties of analysis. When you apply simulation parameters like load, material data, contacts, and so on; they are added in the **Physics Tree**.
- **View Arc**: The options in the **View Arc** are used to change view style and view orientation.
- **Halo**: The options in **Halo** are used to access functions related to current active object/tool.
- **Stage Navigator**: The options in **Stage Navigator** are used to switch between various workspaces of Ansys Discovery. Select the **Model** stage option to perform model creation. Select the **Explore** stage option to perform quick design simulations. Select the **Refine** stage option to further analyze the model with refined inputs.
- **Status Bar**: The options in the status bar are used to check tool tips, selection tools, and quick measurement parameters.
- **Scenes**: The **SAVE CURRENT SCENE** tool is used to save current display in graphics area of application to reorient the model.
- **Monitors**: The options in the **Monitors** drop-down are used to check results of analysis in different strategies. The options are active after performing the analysis.

FILE MENU OPTIONS

The options in the **File** menu are used to handle file related operations like opening a file, starting a new file, saving the file, and so on. Various options of this menu are discussed next.

Starting New File

The **New** tool in the **File** menu is used to start a new document. You can also use **CTRL+N** shortcut key to perform the same operation.

Opening File

The **Open** tool in the **File** menu is used to open an already existing file. You can also use **CTRL+O** shortcut key to perform the same operation. The procedure to use this tool is given next.

- Click on the **Open** tool from the **File** menu. The **Open** dialog box will be displayed; refer to Figure-5.

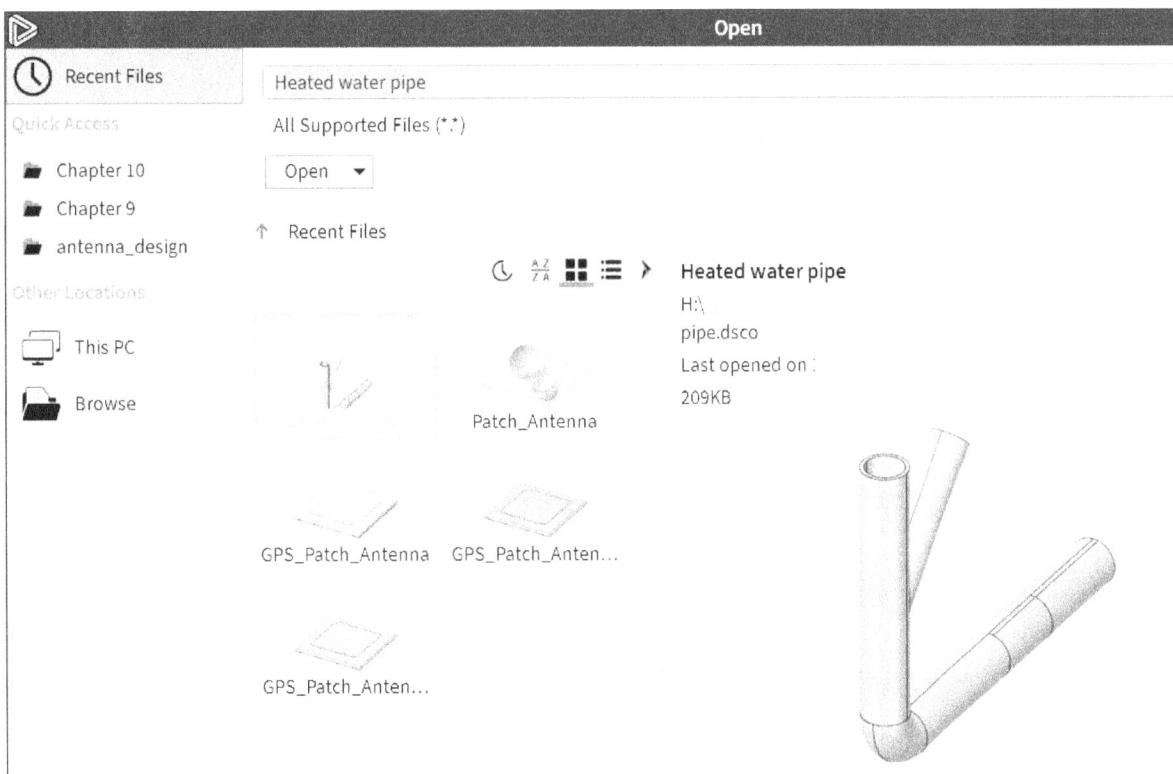

Figure-5. Open dialog box

- Set desired location in the dialog box and double-click on the file to be opened. The related project will open in Ansys Discovery.

Saving File

The **Save** tool in **File** menu is used to save current file at desired location in local drive. The procedure to use this tool is given next.

- After starting a new file if you click on the **Save** tool in **File** menu for first time then **Save As** dialog box will be displayed; refer to Figure-6.

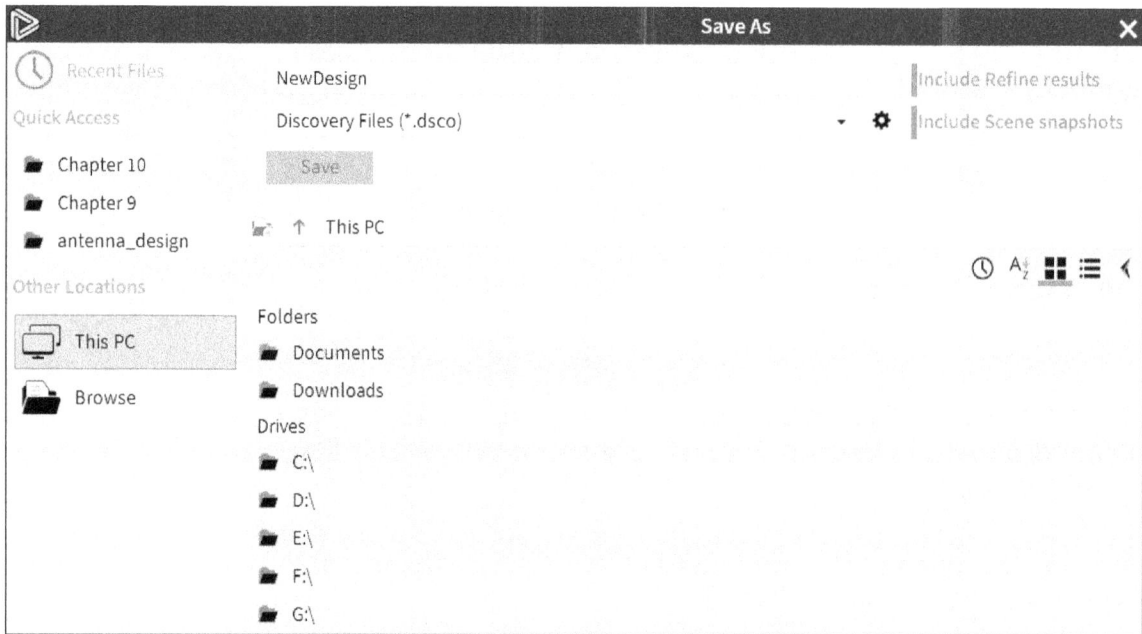

Figure-6. Save As dialog box

- Select desired location in the local drive and specify desired name of file in the **File name** edit box of the dialog box.
- Select the **Include Refine results** toggle button to include results data generated when performed analysis. Note that selecting this toggle button will increase the file size.
- Select the **Include Scene snapshots** toggle button to include result snapshots data generated after performing analysis.
- After setting desired parameters, click on the **Save** button from the dialog box. The file will be saved at specified location. Note that once you have saved the file then this dialog box will not be displayed again on saving the changes in project.

Save As

The **Save As** tool in **File** menu is used to save a copy of current file with desired name and at desired location in the local drive. The procedure is similar to **Save** tool discussed earlier.

Creating Reports

The **Create Report** tool is used to generate HTML file report of analysis parameters and results. The procedure to use this tool is given next.

- After performing analysis, click on the **Create Report** tool from the **File** menu. The **Create Report** dialog box will be displayed.
- Set desired name and location of the report file in respective edit boxes.
- Select desired option from the **Scene image resolution** drop-down to define size of image files generated from result as scene.
- Select desired option from the **Scene inclusion, 2D** drop-down to define which scenes are to be included in the report.
- After setting desired parameters, click on the **Create Report** button from the dialog box to generate html file report.

Importing Geometry

The **Insert Geometry** tool is used to import models created in supported CAD software like AutoCAD, CATIA, NX, Creo, Inventor, Fusion, and so on. The procedure to use this tool is given next.

- Click on the **Insert Geometry** tool from the **File** menu. The **Insert Geometry** dialog box will be displayed; refer to Figure-7.

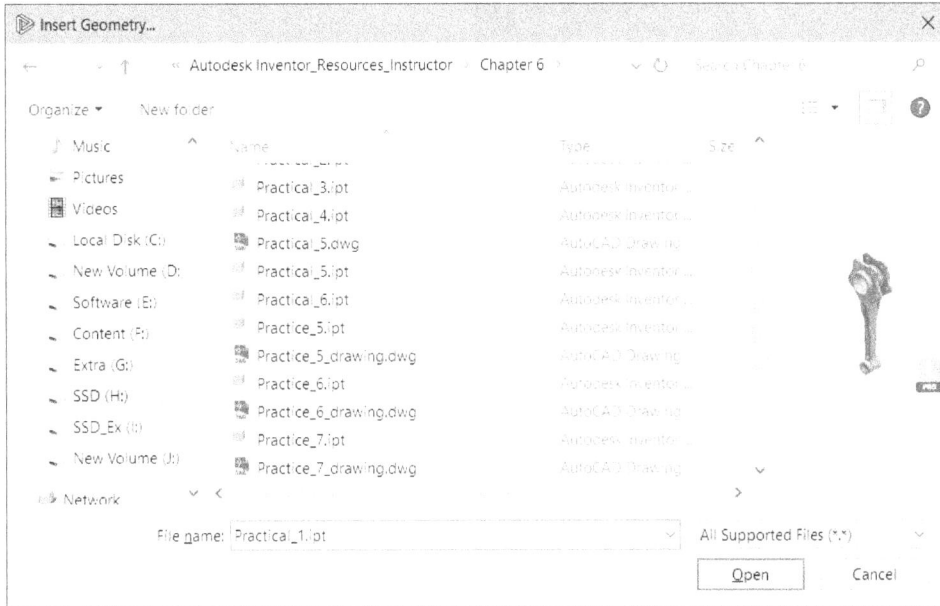

Figure-7. Insert Geometry dialog box

- Select desired extension of model file to be inserted from the **File type** drop-down of the **Insert Geometry** dialog box and then select the model part to be inserted in the project.
- Click on the **Open** button from the dialog box to insert the model. Preview of the model will be displayed with drag handles to reorient the model if required; refer to Figure-8.

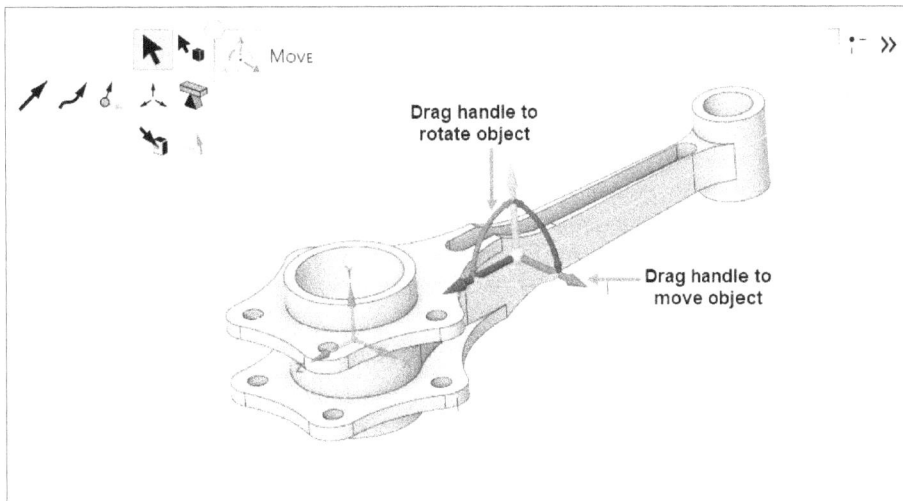

Figure-8. Model inserted

- Use the arrows of drag handle to move the part in respective direction and rotate the curved arrows of handle to rotate the part. Alternatively, select desired selection button from the **HUD** toolbar and perform related orientation action. Press **ESC** to exit the tool and place the model.

Updating Model with new CAD file

The **Update from new CAD file** tool is used to replace current model in graphics area with another model. The procedure to use this tool is given next.

• Click on the **Update from new CAD file** tool from the **File** menu. The **Open** dialog box will be displayed.
• Select desired model file and click on the **Open** button. The model in graphics area will be replaced by model of selected file.

Defining Settings

The **Settings** tool in **File** menu is used to define setting parameters for the application. On clicking this tool, the **Settings** dialog box will be displayed; refer to Figure-9. Various commonly modified parameters of this dialog box are discussed next.

General Settings

• Select the **General** option from left area in the dialog box if not selected by default. The options in the dialog box will be displayed as shown in Figure-9. The options in **General** page are used to define general parameters related to graphics, user interface, appearance, and selection.
• Set desired value in **Rendering quality** drop-down to define the smoothness and sharpness of models in graphics area. Note that increasing this value also increases processing time. Similarly, you can set the value in **Anti-aliasing** drop-down to define sharpness of edges of model.

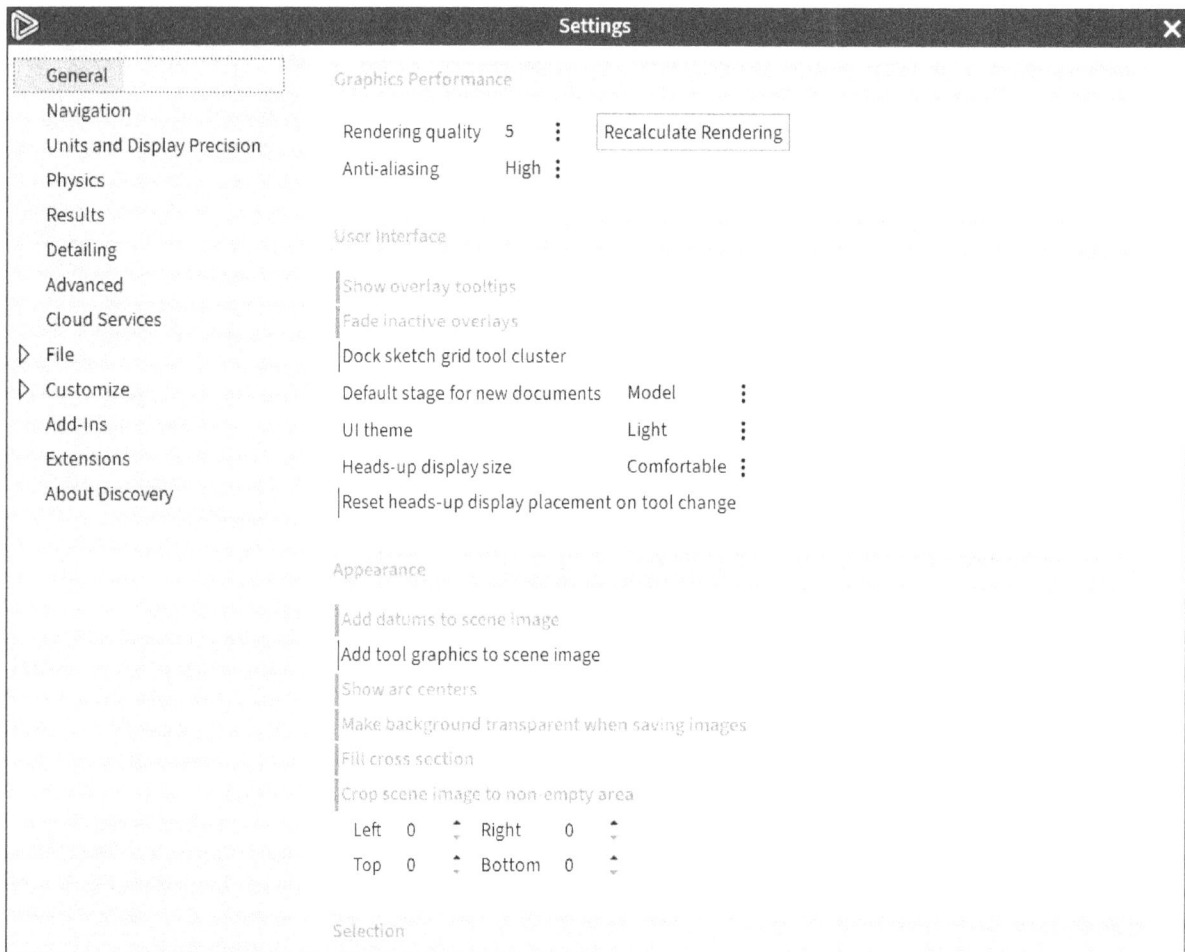

Figure-9. Settings dialog box

- Select the **Show overlay tooltips** toggle button to enable display of tool tips when you hover cursor on them.
- Select desired option from the **Default stage for new documents** drop-down to define in which mode the software will open by default when starting a new document.
- Select desired option from the **UI theme** drop-down to change color of user interface to dark or light.
- Select desired option from the **Heads-up display size** drop-down to define the size of **HUD** toolbar displayed on selecting objects in graphics area.

Navigation Settings

The options in **Navigation** page are used to modify the settings related to various navigation movements like panning, zooming, rotating view, and so on; refer to Figure-10. Various options of this page are discussed next.

- Select desired option from the **Spin** drop-down to define mouse key/gesture for spinning the model. Similarly, select desired options for **Pan** and **Zoom** functions in the **Mouse Mapping** area.

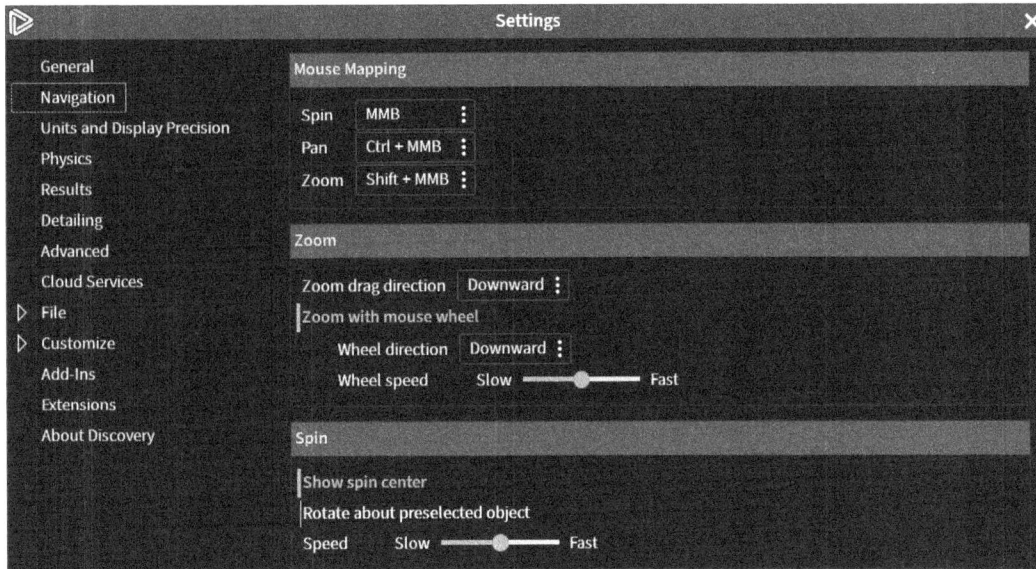

Figure-10. Navigation page in Settings

- Select desired option from the **Zoom drag direction** drop-down to define whether zoom in will be performed by dragging cursor upward or downward (while holding SHIFT+MMB).
- Similarly, select desired option from the **Wheel direction** drop-down to define zoom in direction.
- Set desired value in the **Wheel speed** slider to define how fast the zooming will be performed based on movement of mouse scroll. Similarly, you can set spin speed for model using the **Speed** slider in **Spin** section of page.

Unit System and Precision Settings

Select the **Units and Display Precision** option from left area to set unit system and grid settings for the model; refer to Figure-11. Various options of this page are discussed next.

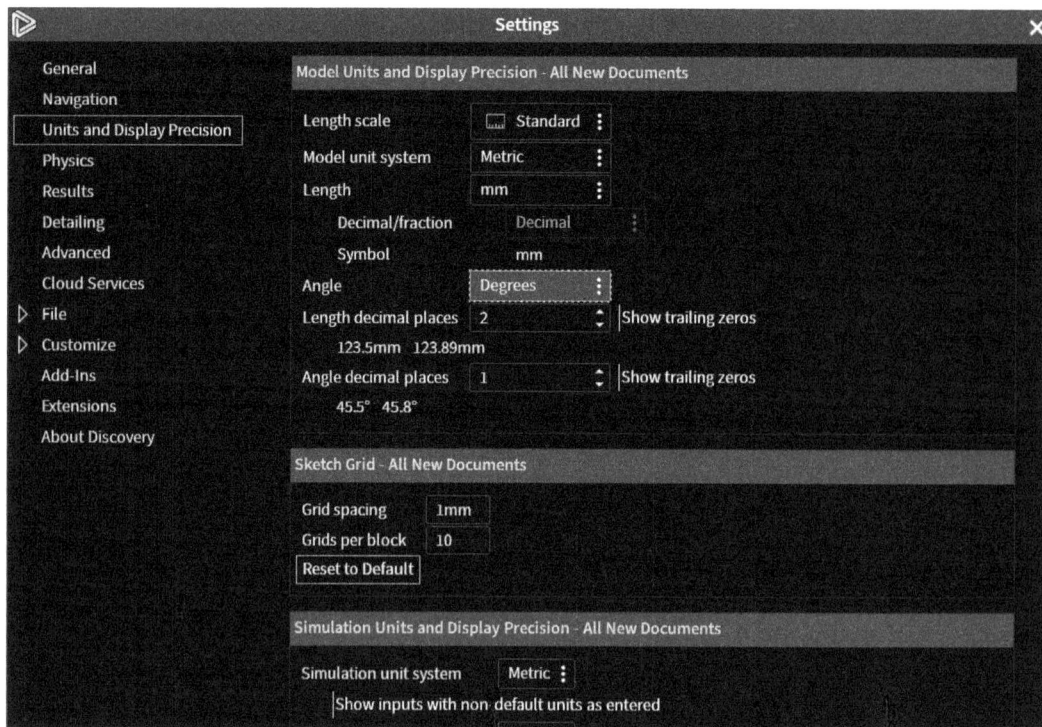

Figure-11. Units and Display Precision page

- Select desired option from the **Model unit system** drop-down to define default unit system for projects in the software. Select **Metric** option to use millimeters, centimeters, or meters unit. Select the **Imperial** option from drop-down to use inches or feet unit.
- Similarly, select unit for angle in the **Angle** drop-down.
- Set desired values in the **Length decimal places** and **Angle decimal places** drop-downs to define number of digits after decimal upto which precision will be displayed when working with numbers in the project.
- Set desired value in the **Grid spacing** edit box to define distance between two consecutive grid lines. Similarly, specify desired value in the **Grids per block** edit box to define the number of grid lines in each block.
- Similarly, you can set units for simulation in the **Simulation Units and Display Precision** section of dialog box.

Physics Settings

Select the **Physics** option from left area to define default analysis parameters like gravity direction, general temperature, default material, and so on; refer to Figure-12. Various options of this page are given next.

- Select the **Include self weight** toggle button if you are performing analysis on objects having enough weight to affect the value of gravity. Generally, objects of astronomy relevance can affect the gravity value.
- Select the **Include buoyancy** toggle button if you want to include buoyancy value of fluid to be included in gravity calculation when working with fluid flow analyses.
- Select desired option from the **Gravity direction** drop-down to define the default direction in which gravity force will work for analysis.
- Specify desired value in the **Working temperature** edit box to define default environment temperature for performing various analysis in the software.

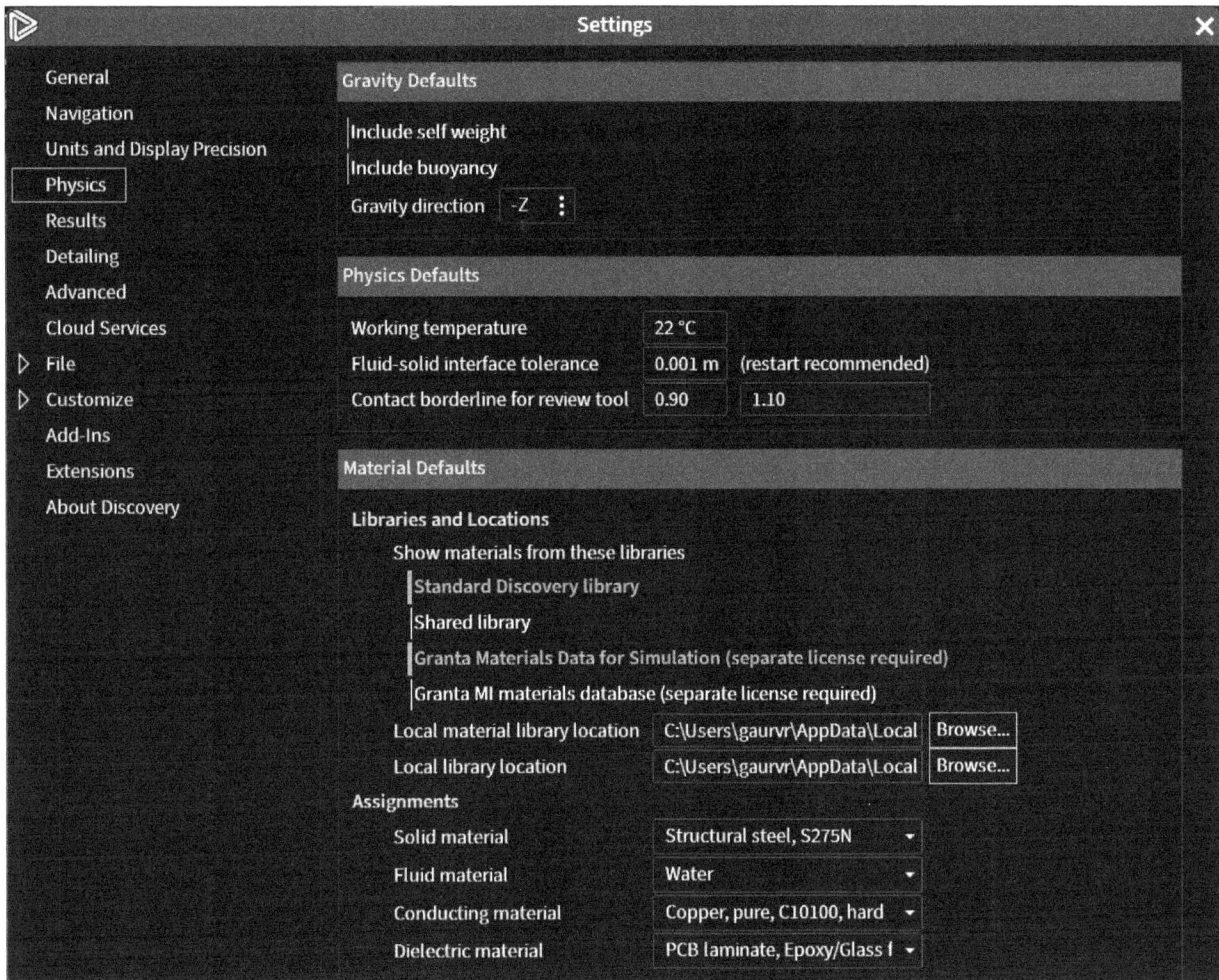

Figure-12. Physics page

- Specify desired value in the **Fluid-solid interface tolerance** edit box to define the amount of error allowed when creating fluid-solid boundary elements.
- Similarly, specify desired parameters in the **Material Defaults** section to define the default materials to be used for assignments like solid material, fluid material, conducting material, and dielectric material.
- Select desired option from the **Include imported bodies by default** drop-down to define whether imported bodies will be included in analysis by default or you want to add them in analysis manually.
- Select the **Automatic** option from the **Solution progression controls** drop-down to automatically define the subsets to be performed when performing analysis to increase accuracy of results and reaching convergence. Select the **Define specific values** option from the drop-down to manually specify number of initial and general subsets to be performed for each analysis.
- Select the **Use LiveGX solver** toggle button to use LiveGX solver for solving the fluid dynamics analysis equations. Select the **LiveGX CPU mode** toggle button to use CPU instead of GPU for solving analysis.
- Select desired option from **Stop on** drop-down to define the criteria for solving analysis. Select the **Engineering convergence** option from the drop-down if you want to engineering parameters of model like pressure, temperature, velocity, etc. to stabilize for marking analysis to be completed. Select the **Numerical Convergence** option from the drop-down if you want to stabilize the difference in volume of fluid throughout the domain.

- Select desired option from the **Modeling method** drop-down to define the method for creating analysis model of fluid dynamics problem. Select the **Laminar** option from the drop-down if there is not turbulence calculation in the CFD model. Select the **Turbulent k-epsilon standard** option from the drop-down if the wall treatment of fully developed and there is very low turbulence. This is the most used industrial option. Select the **Turbulent k-epsilon realizable** option where you want to keep good balance between performance and accuracy while using k-epsilon model. Select the **Turbulent k-omega SST** option from drop-down if you want high accuracy in results at boundary layers. This model is used for free flow external analyses like flow around airfoils and other similar aerodynamics analyses. Select the **Turbulent k-omega standard** option from the drop-down if you want a good compromise between accuracy and performance for turbulent fluid flow with separation.

- Specify desired value in the **Maximum number of iterations** edit box to define maximum limit upto which fluid dynamics equations can be iterated.

- Set desired values in the **Target numerical convergence** and **Target numerical convergence, energy** edit boxes to define accuracy in solution at which analysis will be assumed completed.

- Select desired option from the **Fidelity multiplier (Explore)** drop-down to define quality objects displayed in graphics area. Similarly, set other parameters in this page.

Results Settings

- Select the **Results** option from left area in the dialog box to define settings related to displaying results after solving analysis.

- Select the **Show fluid-flow results during solve (in Refine)** toggle button to check intermediate results when performing fluid flow analyses.

- Select the **Show modal stress results** toggle button to check intermediate analysis results of modal stress analysis.

- Select desired toggle buttons from the **Generate Scene Results** section of the page to define whether you want to compute the results of analysis on saving and on solving.

- Similarly, you can set parameters in other pages of the **Settings** dialog box. You will learn more about other parameters later. Click on the **Close** button (**X**) at top right corner of the dialog box.

SKETCHING

The tools in **SKETCH** panel of **DESIGN** tab in the **Ribbon** are used to create sketch curves. These sketch curves can be used as reference for analysis parameters and you can also use them to create 3D model which represents the specimen for various tests and analyses. Various tools of this panel are discussed next.

Creating Lines

The **Line** tool is used to create a line in the graphics area by specifying two points. The procedure to use this tool is given next.

- Click on the **Line** tool from the **SKETCH** panel in the **DESIGN** tab of the **Ribbon**. You will be asked to specify start point of the line and grid will be displayed in graphics area along with **Line HUD**; refer to Figure-13.

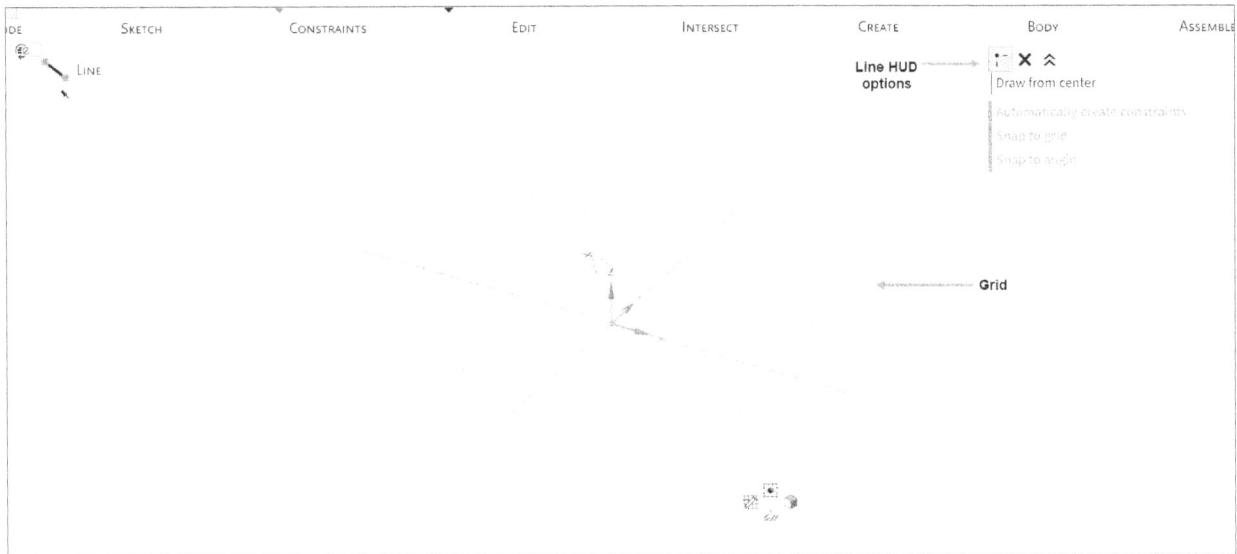

Figure-13. On starting Line tool

- Select desired toggle button from the **HUD** to activate related snaps and modifiers. Select the **Draw from center** toggle button to create line using midpoint. Select the **Automatically create constraints** toggle button to automatically create constraints when creating line. Select the **Snap to grid** toggle button from the HUD to make cursor snap to grid intersection points when drawing objects. Select the **Snap to angle** toggle button to automatically snap to major angles like 30, 45, 60, 75, 90, and so on.
- Click in the graphics area at desired location to specify start point of the line. You will be asked to specify the end point of line.
- Click at desired location to specify the end point of line. The line will be created and you will be asked to specify next point.
- If you want to switch to plan view then click on the **Plan view** button from the toolbar in graphics area; refer to Figure-14.
- If you want to create the sketch on a new plane then click on the **New Plane** button from the toolbar. You will be asked to select reference curve, axis, or plane/face. Hover the cursor on desired reference curve, axis, plane, etc. Preview of the plane will be displayed; refer to Figure-15.

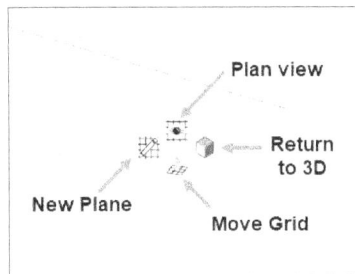

Figure-14. Toolbar in graphics area

- Click on the **Move Grid** button from the toolbar to move grid of plane at desired location.

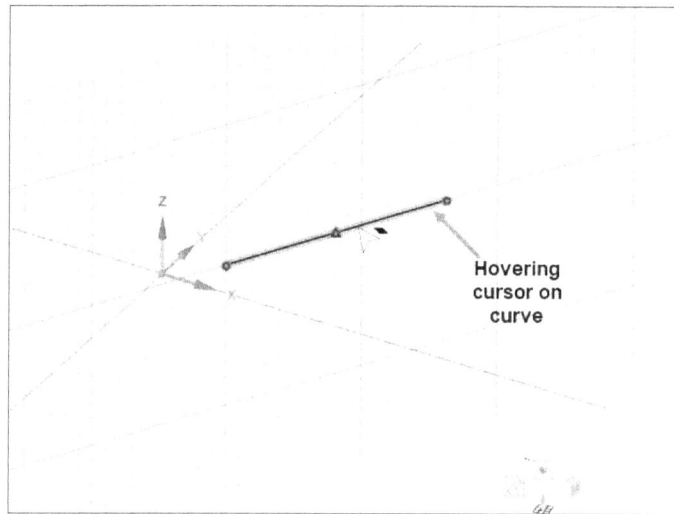
Figure-15. Preview of plane

Creating Tangent Line

The **Tangent Line** tool is used to create a line tangent to selected circle/arc/spline/curve. The procedure to use this tool is given next.

- Click on the **Tangent Line** tool from the **SKETCH** panel in the **DESIGN** tab of the **Ribbon**. You will be asked to select curve.
- Select desired curve/arc/circle/spline from the graphics area. One end point of line will be connected to the curve and you will be asked to specify other end point; refer to Figure-16.

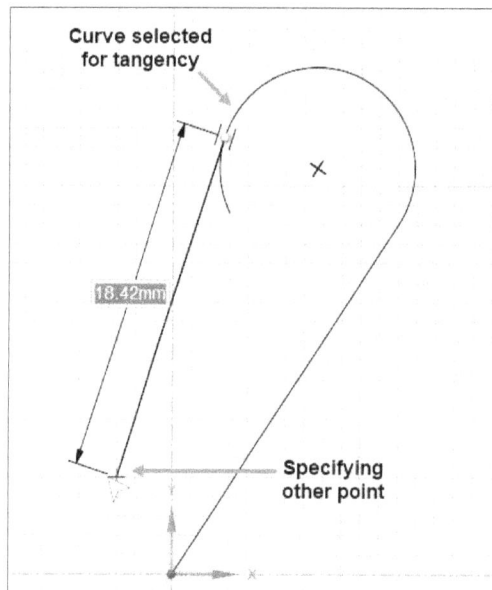
Figure-16. Creating tangent line

- Click at desired location to specify other end point of line.

Creating Construction Line

The **Construction Line** tool is used to create reference line for other sketch entities. The procedure to use this tool is given next.

- Click on the **CONSTRUCTION LINE** tool from the **SKETCH** panel of **DESIGN** tab in the **Ribbon**. You will be asked to specify start point of the construction line.
- Create the construction line as discussed for **LINE** tool.

Creating Rectangle

The **RECTANGLE** tool is used to create rectangle by specifying two corner points or center and corner point. The procedure to use this tool is given next.

- Click on the **RECTANGLE** tool from the **SKETCH** panel in the **DESIGN** tab of the **Ribbon**. You will be asked to specify corner point.
- Select the **Draw from center** toggle button from the **HUD** toolbar in graphics area if you want to create rectangle using center point and corner point. Deselect the **Draw from center** toggle button from **HUD** toolbar if you want to specify diagonal corner points of rectangle.
- Click in the graphics area to specify two points. The rectangle will be created; refer to Figure-17.

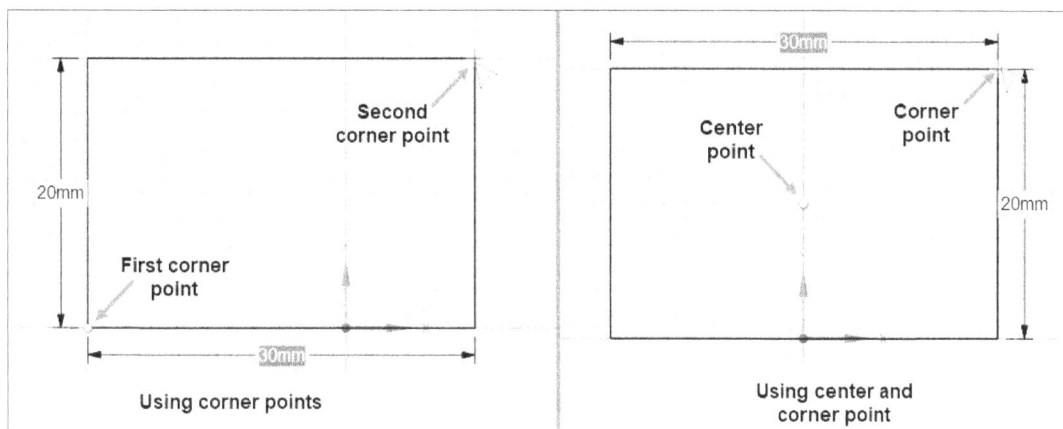

Figure-17. Creating rectangles

- Press **ESC** to exit the tool.

Creating Three Point Rectangle

The **THREE-POINT RECTANGLE** tool is generally used to create slanted rectangles. The procedure to use this tool is given next.

- Click on the **THREE-POINT RECTANGLE** tool from the **SKETCH** panel in the **DESIGN** tab of the **Ribbon**. You will be asked to specify starting point for one size of rectangle.
- Click at desired location to specify the start point and then specify the end point of side of a rectangle. Opposite side of rectangle will be attached to cursor; refer to Figure-18.
- Click at desired location to specify the position of side. The rectangle will be created.

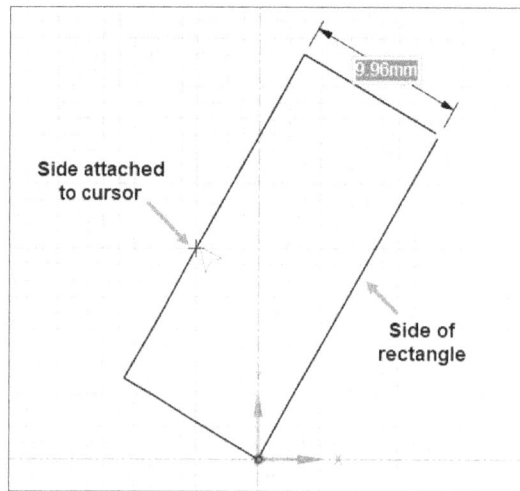

Figure-18. Creating three-point rectangle

Creating Ellipse

The **ELLIPSE** tool is used to create ellipse by specifying center, large axis, and small axis. The procedure to use this tool is given next.

- Click on the **ELLIPSE** tool from the **SKETCH** panel in the **DESIGN** tab of the **Ribbon**. You will be asked to specify location of center point of the ellipse.
- Click at desired location to specify the center point of ellipse. You will be asked to specify length of major axis.
- Move the cursor in desired direction and enter desired length of axis or click at desired location to specify length of major axis of ellipse. You will be asked to specify length of minor axis of ellipse.
- Enter desired value of length or click at desired location to specify the length of minor axis. The ellipse will be created; refer to Figure-19.

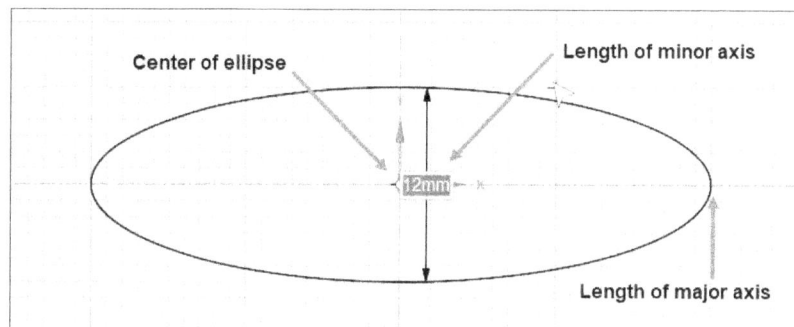

Figure-19. Creating ellipse

- Press **ESC** to exit the tool.

Creating Circle

The **CIRCLE** tool is used to create circle using center point and diameter. The procedure to use this tool is given next.

- Click on the **CIRCLE** tool from the **SKETCH** panel in the **DESIGN** tab of the **Ribbon**. You will be asked to specify center point of the circle.
- Click at desired location to specify the center point. You will be asked to specify diameter of the circle.

• Click at desired location to define diameter of circle or enter the diameter value in input box; refer to Figure-20. The circle will be created.

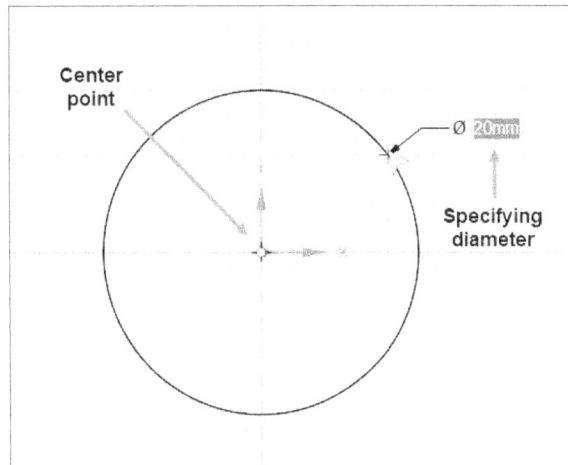

Figure-20. Creating circle

Creating Three Point Circle

The **THREE-POINT CIRCLE** tool is used to create circle by specifying three circumferential points. This tool is generally used to create circle tangent to other sketch entities. The procedure to use this tool is given next.

• Click on the **THREE-POINT CIRCLE** tool from the **SKETCH** panel in the **DESIGN** tab of the **Ribbon**. You will be asked to specify center point of the circle.
• Click at desired location to specify a circumferential point. You will be asked to specify next circumferential point.
• Select desired circumferential point. You will be asked to specify last circumferential point of the circle; refer to Figure-21.

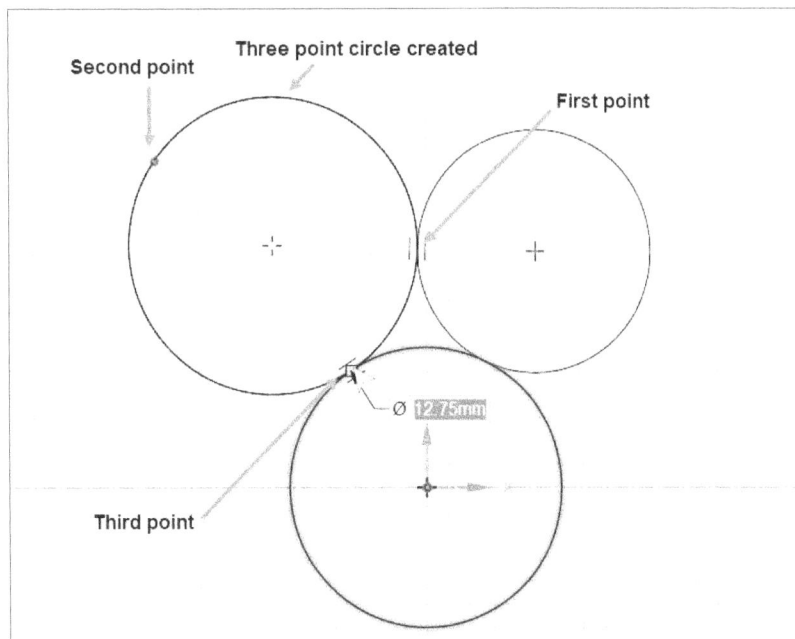

Figure-21. Three point circle created

- Click at desired location to specify the last point. The circle will be created. Note that you can create circle tangent to other entities by selecting points on them.

Creating Polygon

The **POLYGON** tool is used to create polygonal entities like pentagon, hexagon, and so on. The procedure to use this tool is given next.

- Click on the **POLYGON** tool from the **SKETCH** panel in the **DESIGN** tab of the **Ribbon**. You will be asked to specify center point of the circle to be used as reference for defining sides of polygon.
- Select the **Internal radius** toggle button from the HUD toolbar if you want to create the polygon using inscribed circle as reference. Clear this toggle button if you want to use circumscribed circle as reference to create polygon; refer to Figure-22.

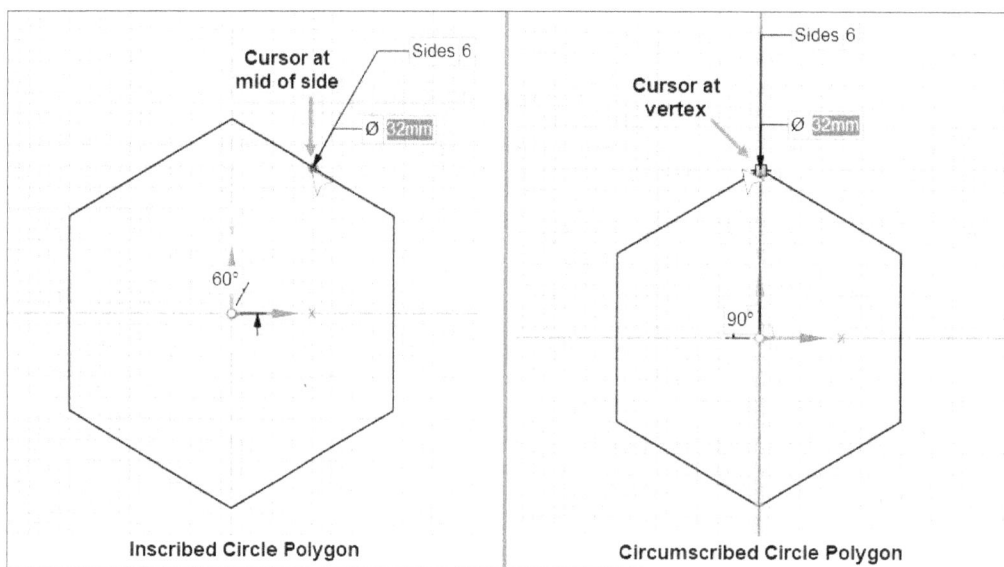

Figure-22. Creating polygon

- Specify desired parameters in the input boxes to define number of sides, diameter of reference circle, and orientation of polygon. You can switch between various input boxes by pressing **TAB** key from keyboard. After specifying parameters, press **ENTER** or click at desired location to create the polygon.

Creating TANGENT ARC

The **TANGENT ARC** tool is used to create an arc tangent to selected entity (line/curve). The procedure to use this tool is given next.

- Click on the **TANGENT ARC** tool from the **SKETCH** panel in the **DESIGN** tab of the **Ribbon**. You will be asked to select entity to which tangent arc is to be created.
- Click at desired location on the entity to define start point of tangent arc. You will be asked to specify end point of the arc.
- Specify desired values in the input boxes or click at desired location to create the tangent arc; refer to Figure-23.

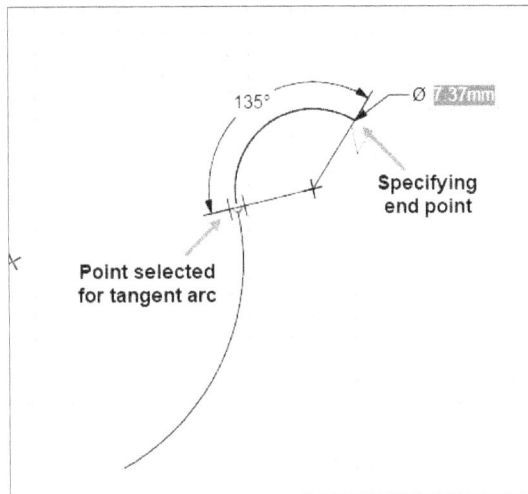

Figure-23. Creating tangent arc

Creating Three-Point Arc

The **THREE-POINT ARC** tool is used to create an arc by specifying three circumferential points. The procedure to use this tool is given next.

- Click on the **THREE-POINT ARC** tool from the **SKETCH** panel in the **DESIGN** tab of the **Ribbon**. You will be asked to specify start point of arc.
- Click at desired location to specify the start point. You will be asked to specify end point of the arc.
- Click at desired location to specify end point and then a circumferential point of arc. The arc will be created.

Creating a Sweep Arc

The **SWEEP ARC** tool is used to create an arc by specifying center point, radial start point, and end point. The procedure to use this tool is given next.

- Click on the **SWEEP ARC** tool from the **SKETCH** panel in the **DESIGN** tab of the **Ribbon**. You will be asked to specify center point of arc.
- Click at desired location to specify center point of the arc. You will be asked to define start point of arc.
- Specify the radius and start angle values in input boxes for the arc or you can click at desired location to specify start point of arc. You will be asked to specify end point of the arc.
- Enter desired angular span of arc in the input box or click at desired location to specify end point of arc. The arc will be created. Press **ESC** to exit the tool.

Creating Spline

The **SPLINE** tool is used to create a spline passing through selected points. The procedure to use this tool is given next.

- Click on the **SPLINE** tool from the **SKETCH** panel in the **DESIGN** tab of the **Ribbon**. You will be asked to specify start point of the spline.
- Click at desired location in the drawing area to specify start point of spline. You will be asked to specify next point of the spline.

• Click at desired locations in the graphics area to specify next consecutive control points of the spline. Press **ESC** to create the spline and exit the tool.

Creating Point

The **POINT** tool is used to create points at specified locations. After activating the tool, click at desired locations in the graphics area to create the points.

Creating Face Curve

The **FACE CURVE** tool is used to create curves on selected solid face. The procedure to use this tool is given next.

• Click on the **FACE CURVE** tool from the **SKETCH** panel in the **DESIGN** tab of the **Ribbon**. You will be asked to select face of the solid.
• Click on face of solid on which you want to create spline curve and define the starting point of spline. Other endpoint of spline will be attached to cursor; refer to Figure-24.

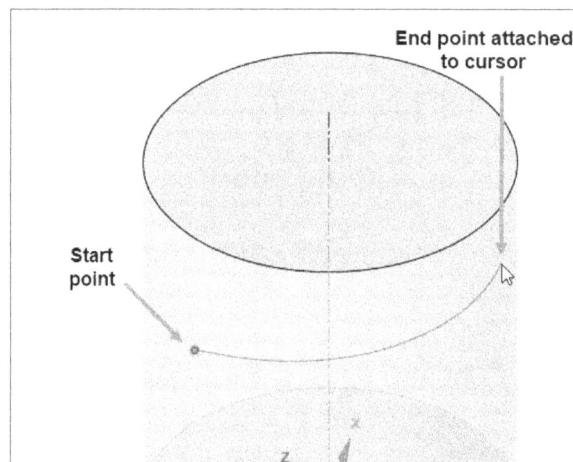

Figure-24. Creating curve on face

• Click at desired locations on selected face to define subsequent points of the curve. After specifying points, press **ESC** to exit the tool and create the curve.

Creating Rounded Corner

The **CREATE ROUNDED CORNER** tool is used to apply fillets or chamfers at sharp corners of the sketch entities. The procedure to use this tool is given next.

• Click on the **Create Rounded Corner** tool from the **SKETCH** panel in the **DESIGN** tab of the **Ribbon**. You will be asked to select corner point.
• By default, fillet is applied at selected corners by using this tool. Select the **Chamfer** toggle button from the **HUD** toolbar to create chamfer instead of fillet.
• Select the **Trim** toggle button from the **HUD** toolbar to remove extra portion of entities at corner after creating round/chamfer; refer to Figure-25.
• Select desired corner point. The round/chamfer feature will be attached to cursor and you will be asked to specify size of feature; refer to Figure-25.

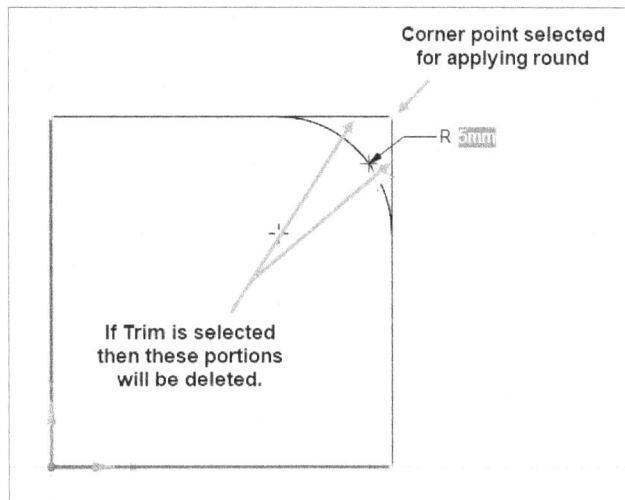
Figure-25. Creating corner round

- Click at desired location to define size of round/chamfer or enter desired value in the input box.

Creating Offset Curves

The **OFFSET CURVE** tool is used to create copy of selected curves at specified distance. The procedure to use this tool is given next.

- Select the curves/entities to be offset and click on the **OFFSET CURVE** tool from the **SKETCH** panel in the **DESIGN** tab of the **Ribbon**. The options in HUD toolbar will be displayed as shown in Figure-26.

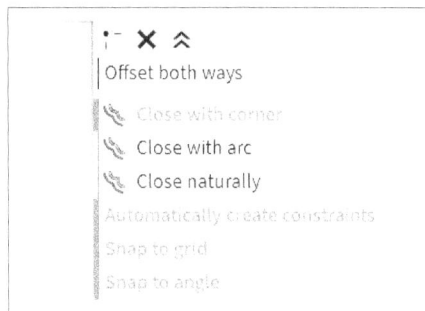
Figure-26. Options in HUD toolbar for Offset

- Select the **Close with corner** toggle button from the toolbar to create closed loop offset with corner points. Select the **Close with arc** toggle button from the toolbar to create offset with rounds applied at corners. Radius of these rounds is equal to offset distance. Select the **Close naturally** toggle button to replicate the corner conditions of original geometry when offsetting.
- After setting desired options in the **HUD** toolbar, click at desired distance from the original geometry to create the offset copy or you can enter the distance value in input box. The offset copy will be created; refer to Figure-27.

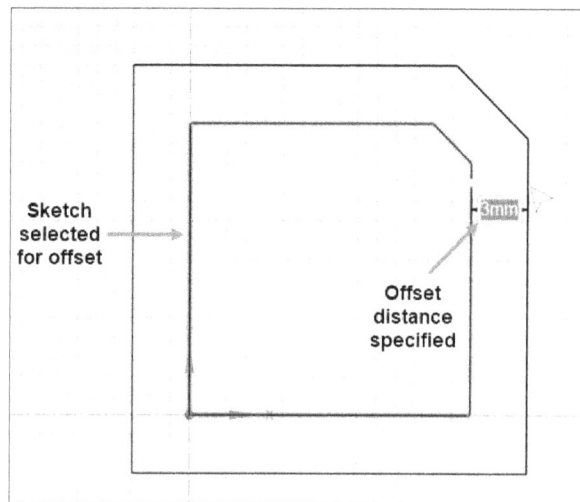

Figure-27. Preview of offset

Projecting Curves on Sketch Plane

The **PROJECT TO SKETCH** tool is used to project selected edges, loops, chains, vertex, and so on. The procedure to use this tool is given next.

• Click on the **PROJECT TO SKETCH** tool from the **SKETCH** panel in the **DESIGN** tab of the **Ribbon**. You will be asked to select entities to be projected.
• Select desired edges/chains/vertices to be projected. The projected entities will be created; refer to Figure-28.

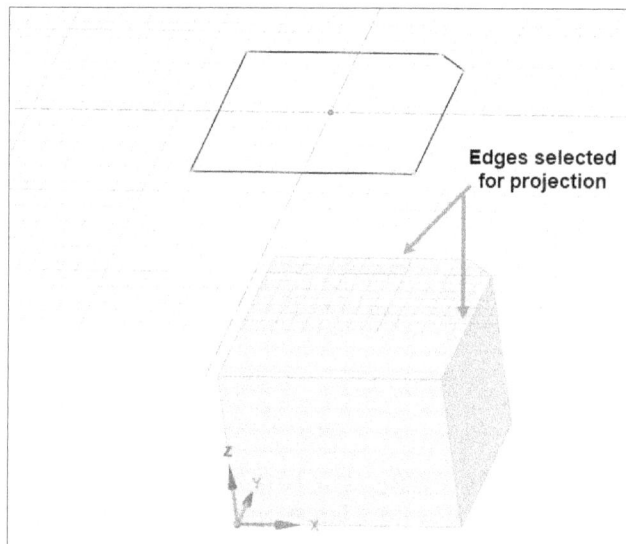

Figure-28. Edges projected on sketch plane

Creating Corners

The **CREATE CORNER** tool is used to trim extra portions of selected lines/curves to form a corner. The procedure to use this tool is given next.

• Click on the **CREATE CORNER** tool from the **SKETCH** panel in the **DESIGN** tab of the **Ribbon**. You will be asked to select the intersecting entities to form a corner.
• Select the two intersecting curves to form corner. The extra portion of selected entities will be trimmed; refer to Figure-29. Note that the selected sides of curves with respect to intersection point will be kept after forming corner.

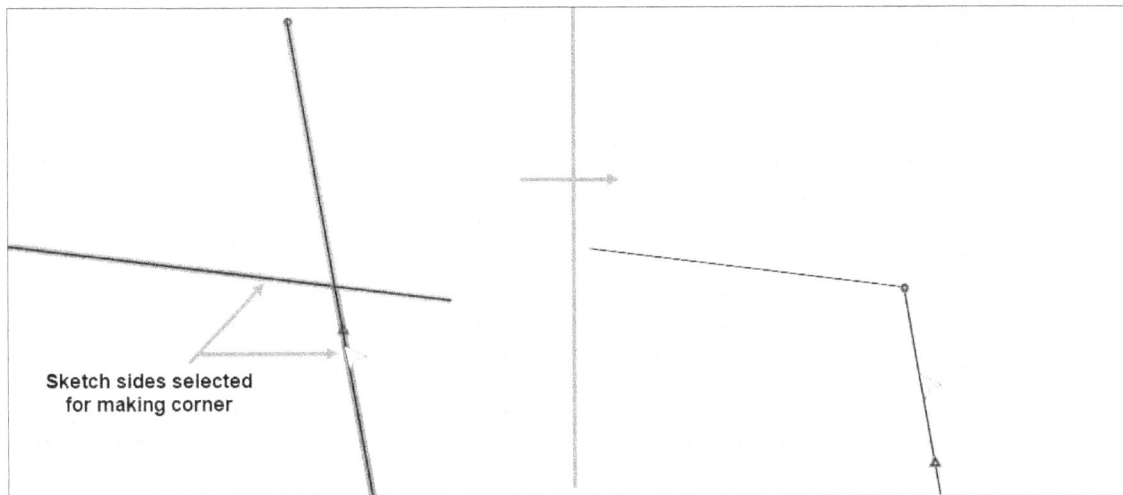

Figure-29. Creating corner

Trimming Segment of Curve

The **TRIM AWAY** tool is used to remove selected segment of curve with respect to nearest intersection/corner point. The procedure to use this tool is given next.

- Click on the **TRIM AWAY** tool from the **SKETCH** panel in the **DESIGN** tab of the **Ribbon**. You will be asked to select entities to be trimmed.
- Hover the cursor on portion of sketch curve to be trimmed. The portion to be removed will be highlighted in red color; refer to Figure-30.

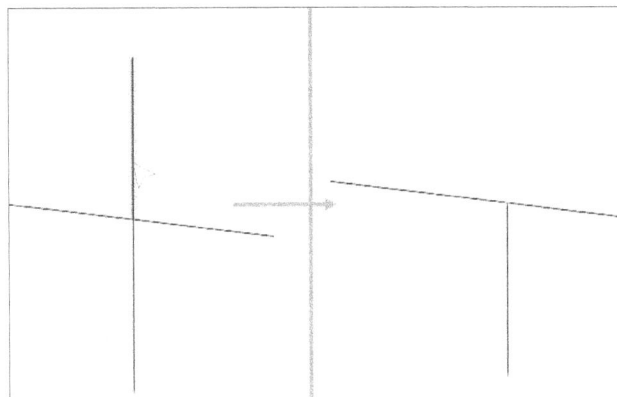

Figure-30. Trimming curve

- Click on the portion to trim it.

Splitting Curves

The **SPLIT CURVE** tool is used to split a sketch curve at desired location. The procedure to use this tool is given next.

- Click on the **SPLIT CURVE** tool from the **SKETCH** panel in the **DESIGN** tab of the **Ribbon**. You will be asked to select the entity to be split.
- Select desired sketch entity that is to be split. You will be asked to specify the split point.
- Click at desired locations to create the splits; refer to Figure-31. Press **ESC** to exit the tool.

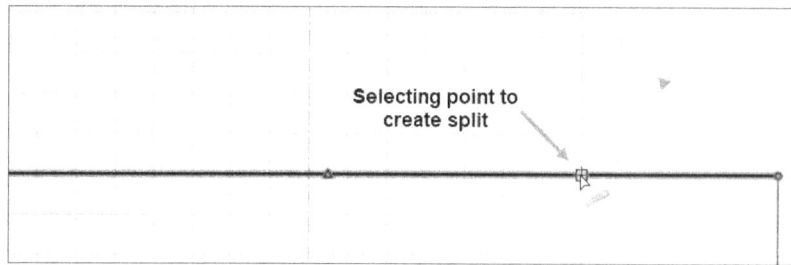

Figure-31. Splitting curve

Applying Constraints

Constraints are used to restrict the orientation and shape of selected sketch entities. The tools to apply various constraints are available in the **Constraints** drop-down of **Ribbon**; refer to Figure-32. These tools are discussed next.

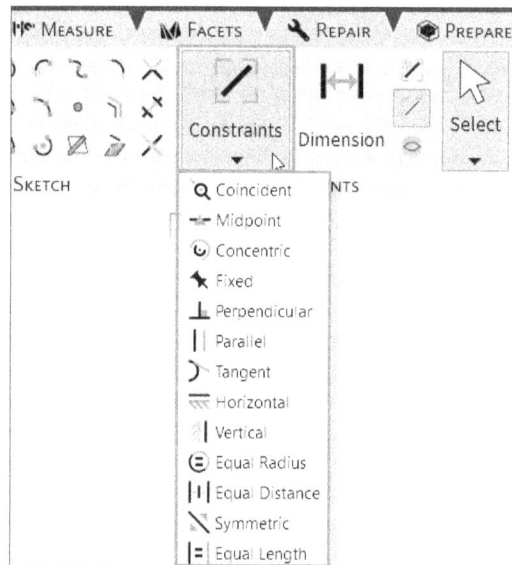

Figure-32. Constraints drop-down

Creating Coincident Constraint

The **Coincident** tool is used to connect selected entities at common selected point. The procedure to use this tool is given next.

- Click on the **Coincident** tool from the **Constraints** drop-down in the **CONSTRAINTS** panel of **DESIGN** tab in the **Ribbon**. You will be asked to select points on the sketch entities to be made coincident.
- Select the two points to be made coincident. The constraint will be applied; refer to Figure-33.

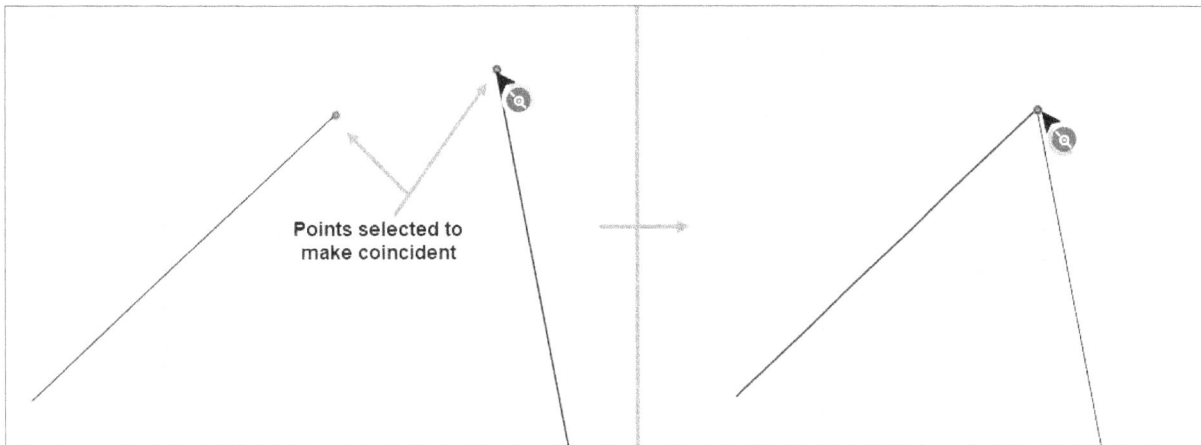

Figure-33. Making lines coincident

Apply Midpoint Constraint

The **Midpoint** tool is used to place selected vertex at midpoint of another entity. The procedure to use this tool is given next.

*   Click on the **Midpoint** tool from the **Constraints** drop-down in the **CONSTRAINTS** panel in the **DESIGN** tab of the **Ribbon**. You will be asked to select entities to apply midpoint constraint.
*   Select the vertex of first entity that is to be placed on another entity. You will be asked to select another entity.
*   Select the other entity whose midpoint is to be used. The midpoint constraint will be applied; refer to Figure-34.

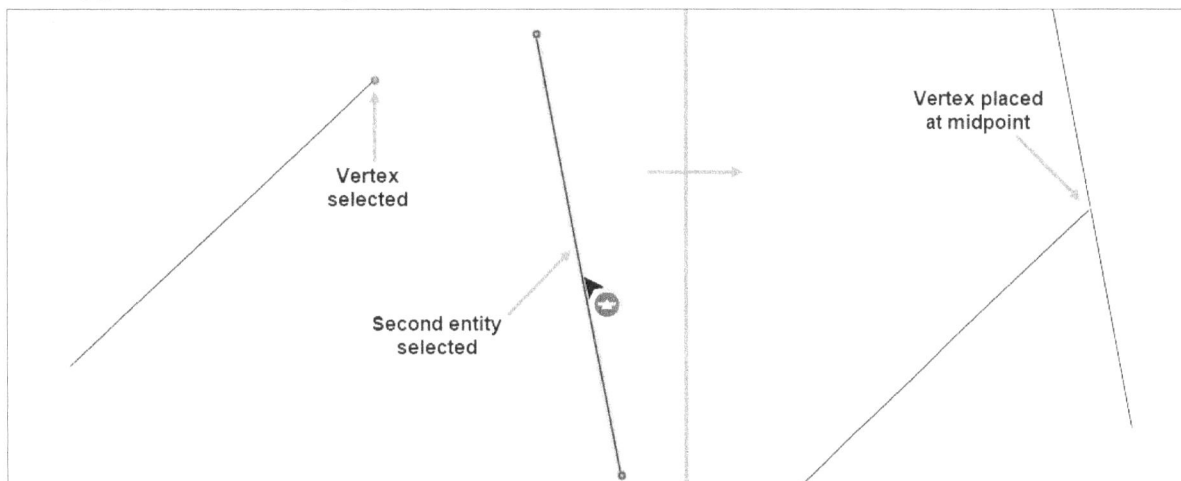

Figure-34. Applying midpoint constraint

Applying Concentric Constraint

The **Concentric** tool is used to make two circular sketch entities share the same center point. The procedure to use this tool is given next.

*   Click on the **Concentric** tool from the **Constraints** drop-down in the **CONSTRAINTS** panel of the **DESIGN** tab in the **Ribbon**. You will be asked to select the entities to make them concentric.
*   Select the two circular entities to be made concentric; refer to Figure-35.

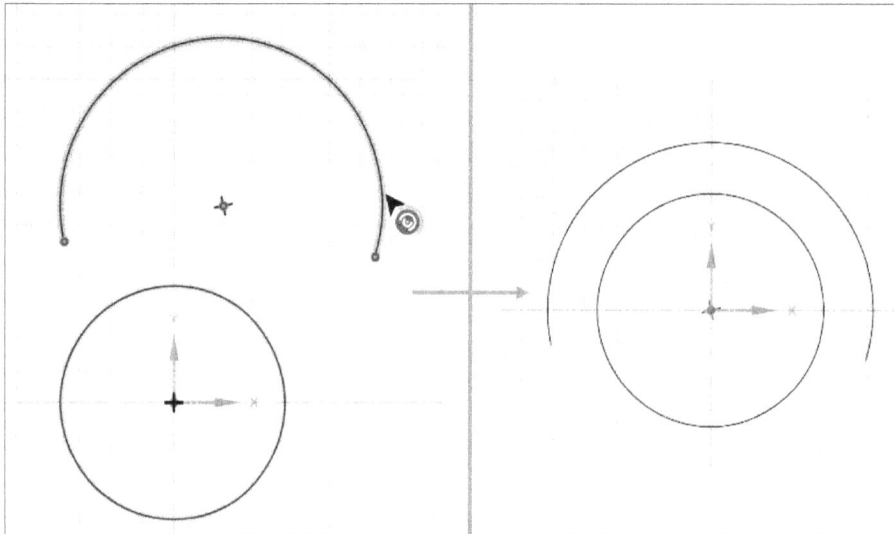

Figure-35. Applying concentric constraint

Applying Fixed Constraint

The **Fixed** tool is used to fix selected sketch entity at its current location. To use this tool, select the sketch entity to be fixed at its location and then click on the **Fixed** tool from the **Constraints** drop-down in the **CONSTRAINTS** panel of **DESIGN** tab in the **Ribbon**.

Applying Perpendicular Constraint

The **Perpendicular** constraint is used to make two selected sketch curves perpendicular at selected locations. The procedure to use this tool is given next.

*   Click on the **Perpendicular** tool from the **Constraints** drop-down in the **CONSTRAINTS** panel of the **Ribbon**. You will be asked to select sketch entity.
*   Select the first entity which is allowed to move if both entities are not constrained earlier. You will be asked to select the second entity.
*   Select the second entity to apply the perpendicular constraint; refer to Figure-36. Press **ESC** to exit the tool.

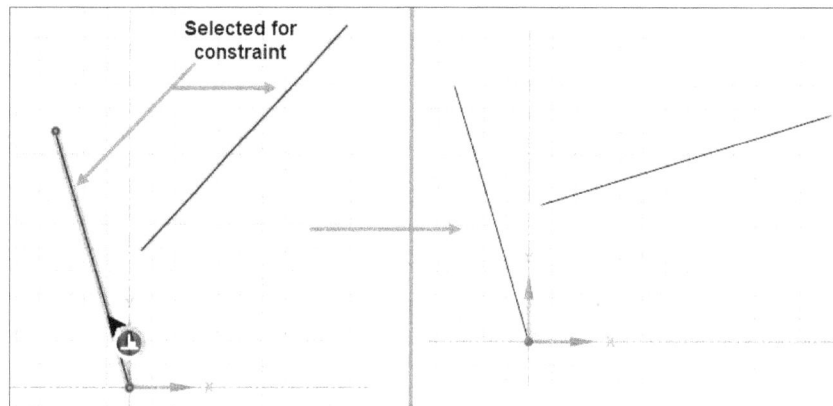

Figure-36. Applying perpendicular constraint

Applying Parallel Constraint

The **Parallel** tool is used to make two selected entities parallel to each other. The procedure to use this tool is given next.

- Click on the **Parallel** tool from the **Constraints** drop-down in the **CONSTRAINTS** panel of **DESIGN** tab in the **Ribbon**. You will be asked to select entities to be made parallel.
- Select two lines/curves to be made parallel. The constraint will be applied; refer to Figure-37.

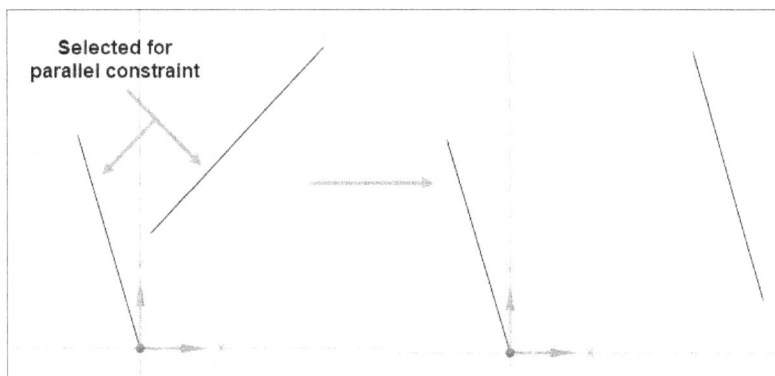

Figure-37. Applying parallel constraint

Applying Tangent Constraint

The **Tangent** constraint is used to make selected entity tangent to other curve. The procedure to use this tool is given next.

- Click on the **Tangent** tool from the **Constraints** drop-down in the **CONSTRAINTS** panel of **DESIGN** tab in the **Ribbon**. You will be asked to select sketch entity.
- Select the two entities to be made tangent. The constraint will be applied; refer to Figure-38. Note that at least one entity should be a non-linear curve.

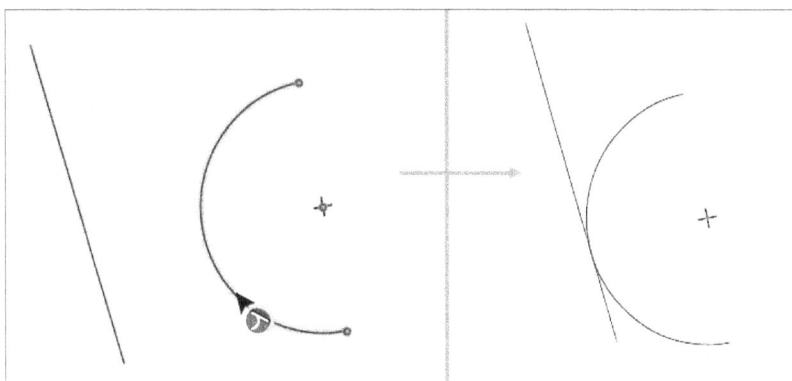

Figure-38. Applying tangent constraint

Applying Horizontal and Vertical Constraints

The Horizontal constraint is used to make selected lines aligned with horizontal axis. To apply this constraint, click on the **Horizontal** tool from the **Constraints** drop-down in the **CONSTRAINTS** panel of **DESIGN** tab in the **Ribbon** and select sketch lines to be made horizontal.

Similarly, you can use the **Vertical** tool in **Constraints** drop-down to apply vertical constraint.

Applying Equal Radius Constraint

The **Equal Radius** constraint is used to make radius of two selected circular entities equal. The procedure to apply this constraint is given next.

- Click on the **Equal Radius** tool from the **Constraints** panel in the **CONSTRAINTS** tab in the **Ribbon**. You will be asked to select the entities whose radii are to be equalled.
- Select two circular entities whose radii are to be matched. The constraint will be applied; refer to Figure-39.

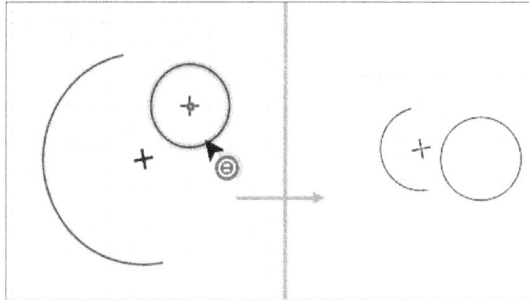

Figure-39. Applying equal radius constraint

Applying Equal Distance Constraint

The **Equal Distance** constraint is used to make distance between two pairs of sketch entities equal. The procedure to apply this constraint is given next.

- Click on **Equal Distance** tool from the **Constraints** drop-down in the **CONSTRAINTS** panel in the **DESIGN** tab of the **Ribbon**. You will be asked to select sketch entities.
- Select the two sketch entities (points, lines, etc.) of first pair. You will be asked to select the entities of second pair.
- Select the second pair entities; refer to Figure-40. The constraint will be applied.

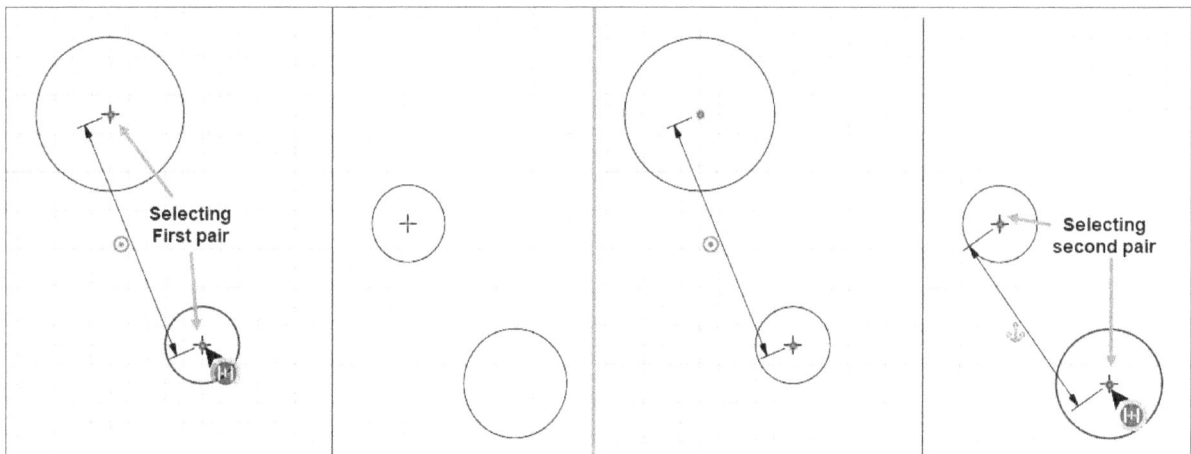

Figure-40. Selecting entities for equal distance constraint

Applying Symmetric Constraint

The **Symmetric** constraint is used to make two selected entities mirror copy about an axis/line. The procedure to apply the constraint is given next.

- Click on the **Symmetric** tool from the **Constraints** drop-down in the **CONSTRAINTS** panel of the **DESIGN** tab in the **Ribbon**. You will be asked to select symmetry axis.
- Select the line to be used as mirror line. You will be asked to select entities to be made symmetric.
- Select the two entities on opposite sides of mirror line. The symmetric constraint will be applied; refer to Figure-41.

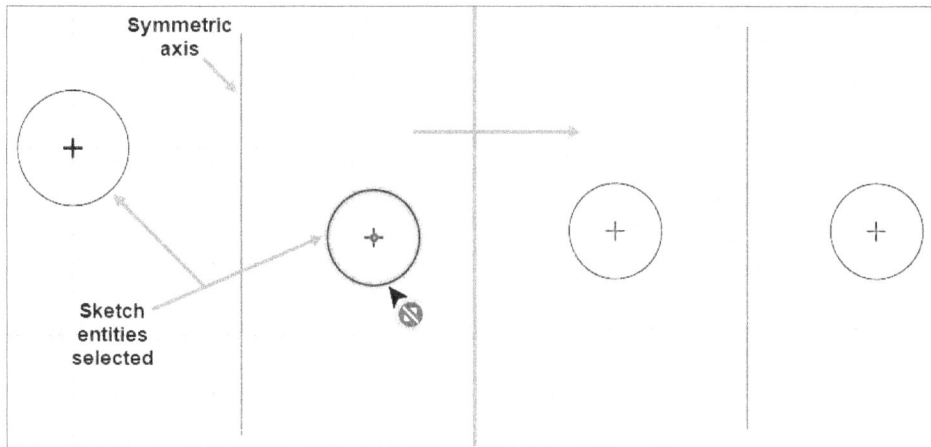

Figure-41. Applying symmetric constraint

Applying Equal Length Constraint

The **Equal Length** constraint is applied to make length of two selected lines equal. The procedure to use this tool is given next.

• Click on the **Equal Length** tool from the **Constraints** drop-down in the **CONSTRAINTS** panel of the **DESIGN** tab in the **Ribbon**. You will be asked to select lines to be made equal.
• Select the two lines of unequal length. The constraint will be applied; refer to Figure-42.

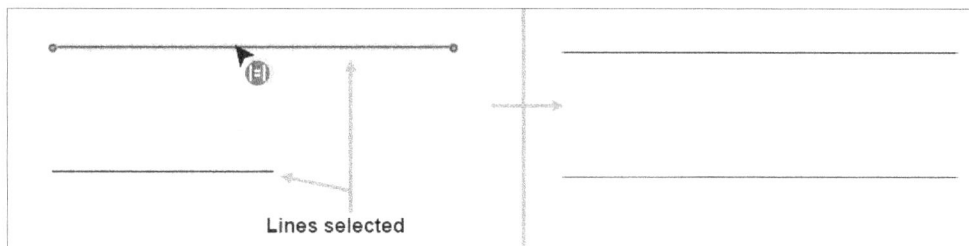

Figure-42. Applying equal length constraint

Applying Dimensions

The **Dimension** tool is used to apply various different dimensions like linear dimensions, circular dimensions, and so on. The procedure to apply dimensions are given next.

• Click on the **Dimension** tool from the **CONSTRAINTS** panel in the **DESIGN** tab of the **Ribbon**. You will be asked to select entity to be dimensioned.
• Select an arc to apply radius dimension, select a circle to apply diameter dimension, select a line to dimension length of line, or select two points to apply dimension between the points. The dimension will be attached to cursor.
• Click at desired location to place the dimension; refer to Figure-43. If you want to modify the size of entity then double-click on the dimension value and enter desired value.

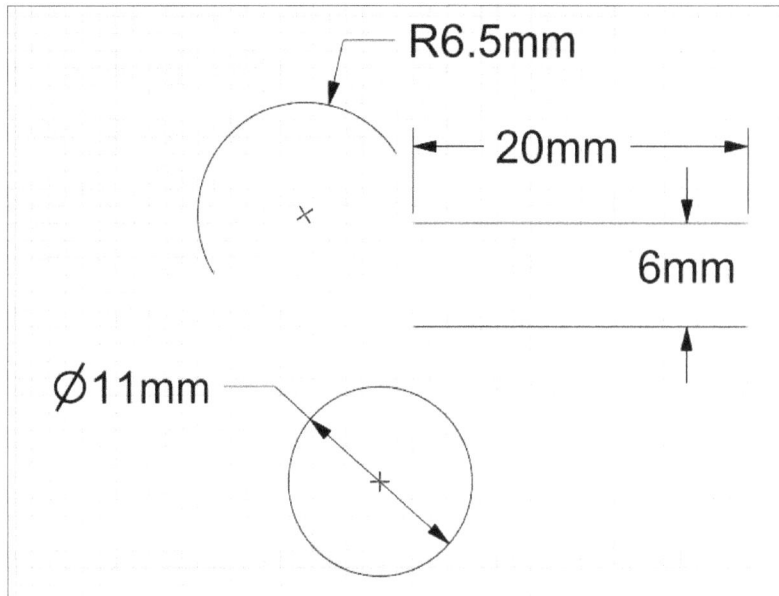

Figure-43. Applying dimensions

Applying Constrains Automatically

The **AUTOCONSTRAIN** tool is used to apply potential constrains to all the entities in current sketch. Click on this tool from the **CONSTRAINTS** panel in the **DESIGN** tab of the **Ribbon** to apply possible constraints in the sketch.

Showing/Hiding Constraint Color

The **Show Constraint Color** tool in **CONSTRAINTS** panel of **Ribbon** is used to display fully constrained sketch entities in bold blue color and underconstrained sketch entities in light blue color. If the sketch is over-constrained then it will be displayed in red color. Refer to Figure-44.

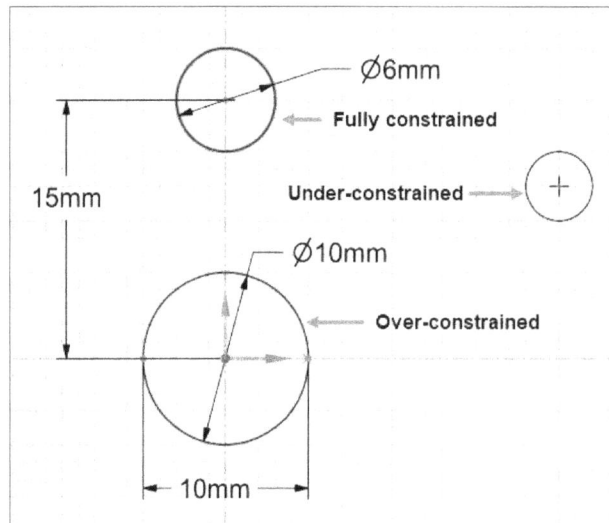

Figure-44. Showing constraint status

If you want to hide the constraint colors then click on the **Hide Constraint Color** tool which is displayed in place of **Show Constraint Color** tool.

Showing/Hiding Constraints

Click on the **SHOW CONSTRAINTS** tool in **CONSTRAINTS** panel of **DESIGN** tab in the **Ribbon** to display constraints applied in the sketch; refer to Figure-45. If you want to hide the constraints then click on the button again.

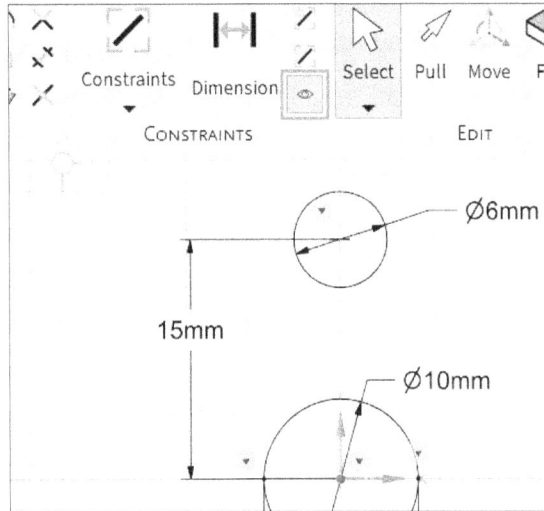

Figure-45. Showing constraints

Selection Tools

The tools in **Select** drop-down are used to perform different types of selections; refer to Figure-46. Various tools in this drop-down are discussed next.

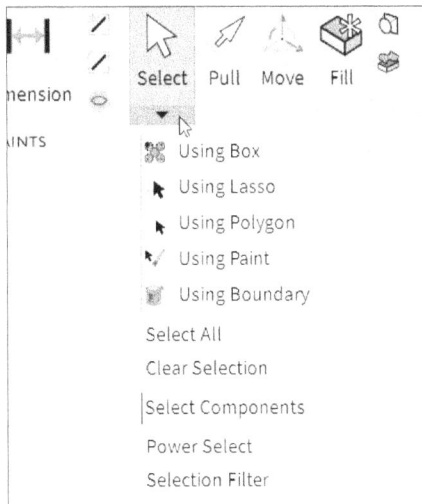

Figure-46. Select drop-down

Using Box Selection

The **Using Box** tool in **Selection** drop-down is used to select objects by drawing box in the graphics area. There are two ways to make a box for box selection: Moving cursor left to right while drawing box and moving cursor right to left while drawing box. If you move cursor from left to right while making box for selection then all the objects that completely fall inside the box will get selected and if you move cursor from right to left then objects that touch the boundary of box will also get selected; refer to Figure-47.

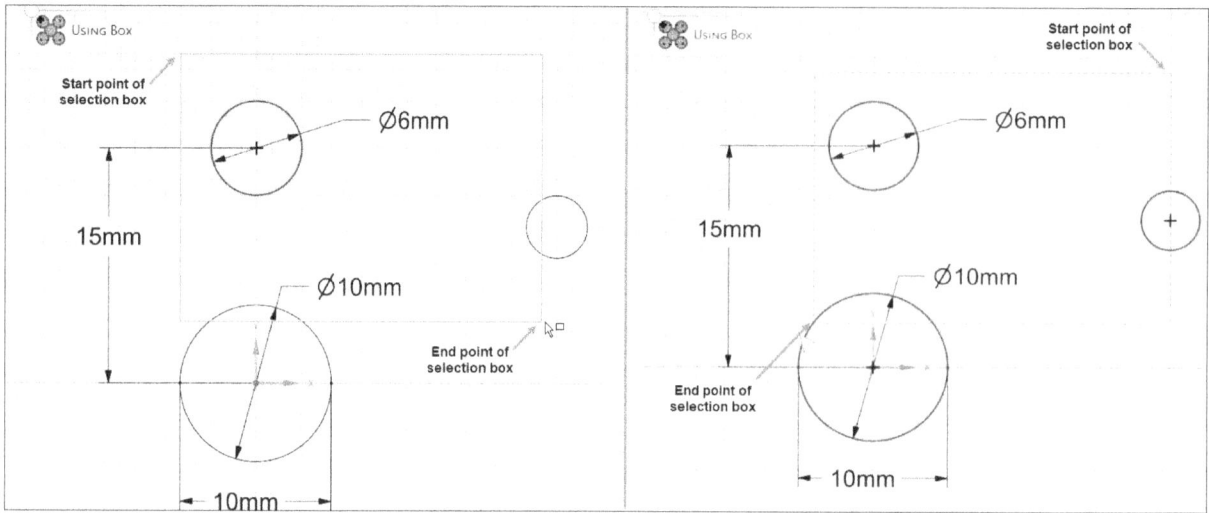

Figure–47. Box selection

Using Lasso Selection

Click on the **Using Lasso** tool from the **Select** drop-down if you want to use a freeform selection boundary. After activating tool, click and drag the cursor to draw the boundary. All the objects inside this boundary will get selected; refer to Figure-48.

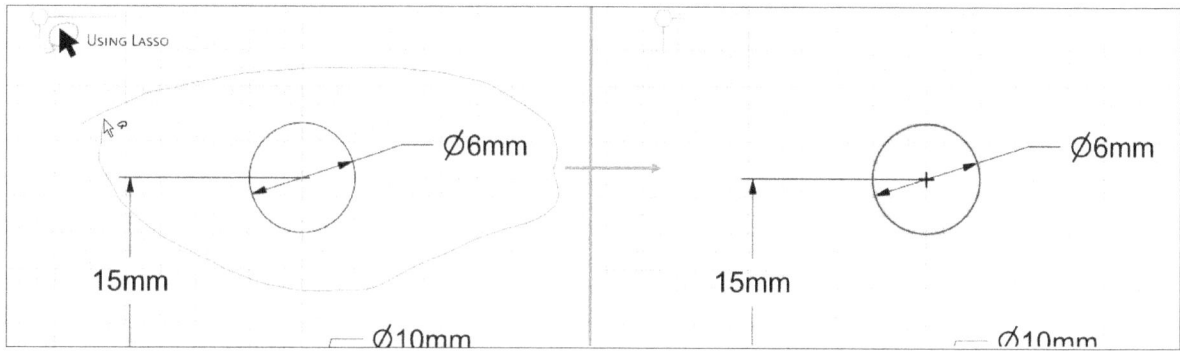

Figure–48. Lasso selection

Using Polygon Selection

The **Using Polygon** tool is used to create a polygonal boundary to be used for selecting objects. After activating this tool, draw polygon by specifying corner points. The objects inside the polygon will be selected; refer to Figure-49.

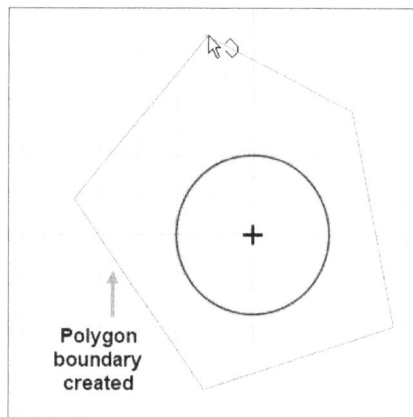

Figure–49. Polygon boundary created for selection

Using Paint Selection

The **Using Paint** tool in **Select** drop-down of **EDIT** panel in the **DESIGN** tab of **Ribbon** is used to select objects by dragging cursor over them.

Using Boundary Selection

The **Using Boundary** tool is used to select objects bound inside selected faces. This selection is useful when selecting 3D objects bound by selected faces. You will learn more about this tool later when working with 3D objects.

Selecting All Objects

The **Select All** tool is used to select all the objects in the currently active component.

Deselecting All Objects

The **Clear Selection** tool in **Select** drop-down of **EDIT** panel in the **DESIGN** tab of **Ribbon** is used to deselect all the objects currently in selection.

Selecting Components

The **Select Components** tool is used to select components from the graphics area. By default, this selection mode is active.

Using Power Select

The **Power Select** tool of **Select** drop-down is used to select all the objects similar to selected object. The procedure to use this tool is given next.

- Click on the **Power Select** tool from the **Select** drop-down in the **EDIT** panel of **DESIGN** tab in the **Ribbon**. The **Power Selection** dialog box will be displayed; refer to Figure-50.

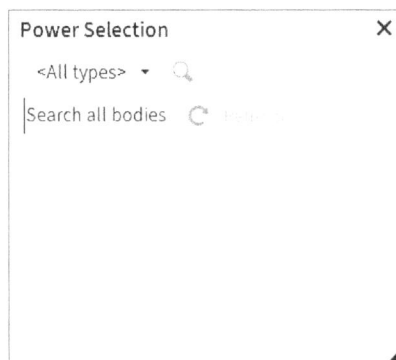

Figure-50. Power Selection dialog box

- Select an object from graphics area to select similar objects. The options to select similar objects based on different criteria will be displayed in the dialog box. Select desired criteria(s) from the dialog box to perform selection; refer to Figure-51.

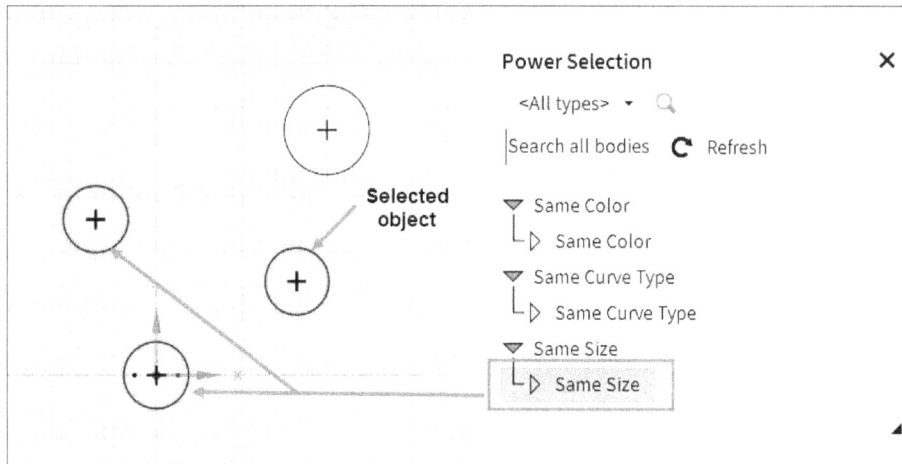

Figure-51. Selecting similar objects

• Close the **Power Selection** dialog box to exit the tool.

Selection Filters

The **Selection Filter** tool is used to set filters for selecting objects. The flyout will be displayed; refer to Figure-52. By default, the **Smart** toggle button is active in **Selection Filter** section of flyout. Click again on the toggle button to activate other filters of the flyout. The **All** node will get selected. Expand the **All** node and de-select the filters for objects that you do not want to be selected.

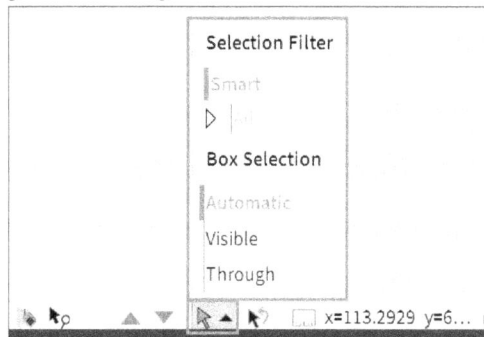

Figure-52. Selection Filter flyout

PERFORMING PULL OPERATIONS

The **Pull** tool of **EDIT** panel is used to perform various operations like extrude, revolve, offset, sweep, draft, and so on. The procedure to use this tool is given next.

• Click on the **Pull** tool from the **EDIT** panel in the **DESIGN** tab of the **Ribbon**. The options in the HUD toolbar will be displayed as shown in Figure-53.

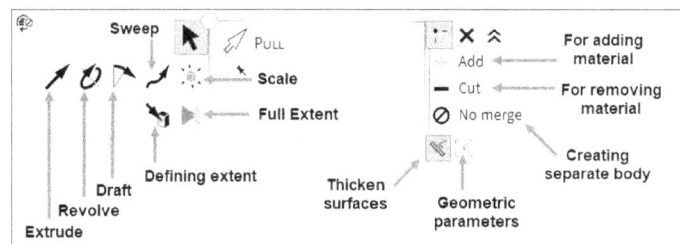

Figure-53. PULL HUD toolbar

Various options of the **HUD** toolbar are discussed next.

Performing Extrude Pull Operation

The **Extrude** operation is performed to form solids by sweeping selected closed loop sketch in vertical direction. The procedure to perform extrude operation is given next.

- After activating the **Pull** tool, click on the **Extrude** button from the **HUD** toolbar. You will be asked to select face.
- Select the closed loop sketch (face), drag it upto desired height for creating extrude feature, and enter desired height of extrude feature in the input box; refer to Figure-54.

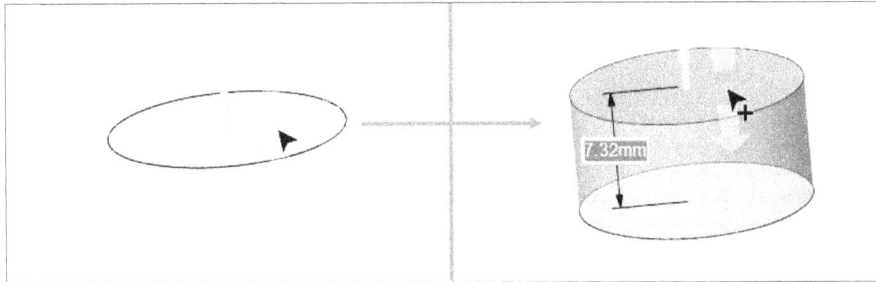

Figure-54. Extrude using Pull

- Press **ESC** to exit the extrude mode.

Performing Revolve Pull Operation

The Revolve operation is performed to form a solid by revolving selected closed loop sketch about an axis. The procedure to perform this operation is given next.

- After activating **Pull** tool from the **EDIT** panel in **Ribbon**, select the **Revolve** button from **HUD** toolbar. You will be asked to select an axis about which the selected sketch will be revolved.
- Select desired axis from the graphics area. You will be asked to select face(s) to be revolved.
- Select the face and drag it. The revolve feature will be created; refer to Figure-55. If you want to create full 360 degree revolve feature then click on **Full Extent** button after selecting the face of feature to be revolved; refer to Figure-56.

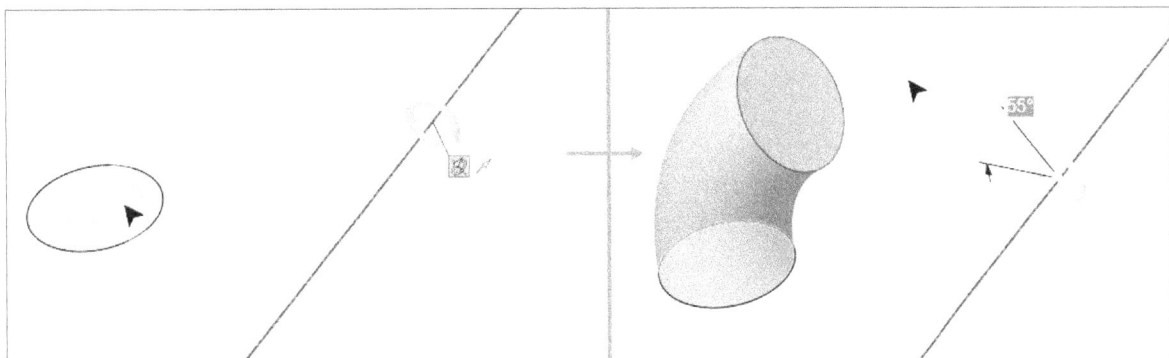

Figure-55. Creating revolve feature

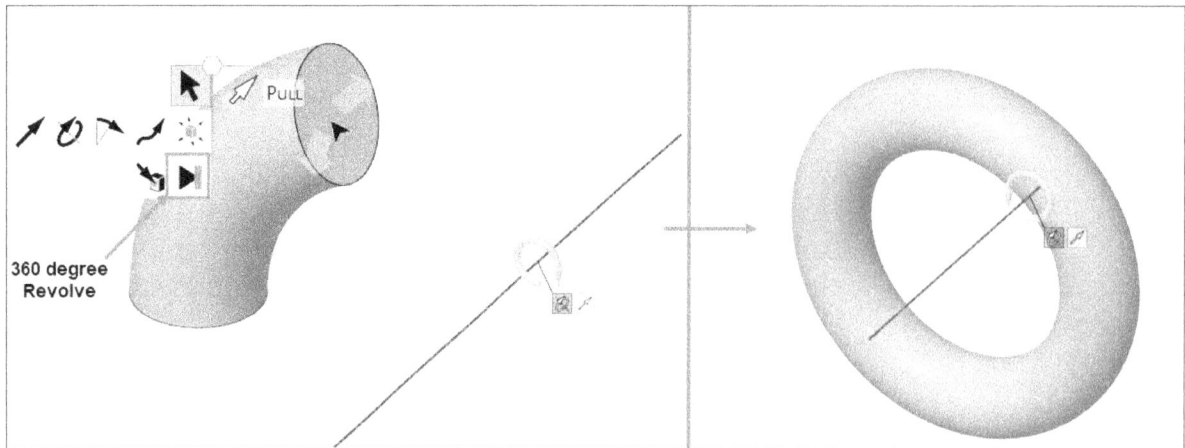

Figure-56. Full revolve feature

Applying Draft Angle to Faces

The Draft operation is performed to apply taper angle to faces of the selected body. The procedure to perform this operation is given next.

*   Click on the **Draft** button from the **HUD** toolbar and select the face to be used as reference for applying draft. You will be asked to select faces on which draft angle will be applied.
*   Select the wall faces to be apply draft angle; refer to Figure-57. You will be asked to specify the draft angle.

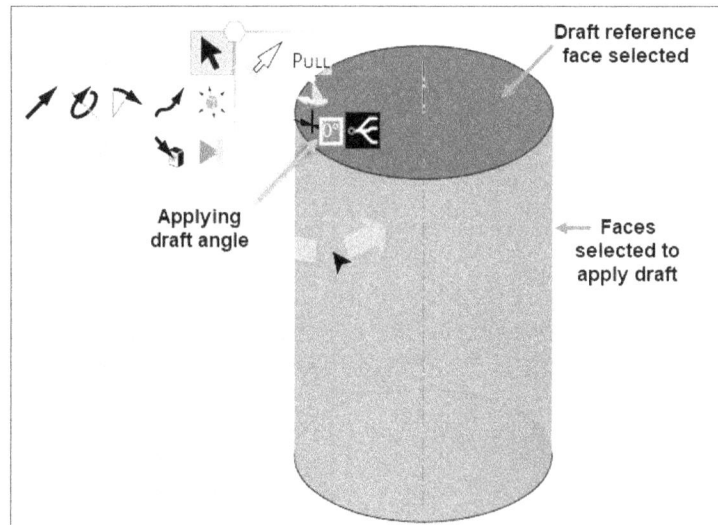

Figure-57. Applying draft

*   Enter desired angle value in the input box or drag the handle to apply draft. The draft angle will be applied; refer to Figure-58. Press **ESC** to exit the draft mode.

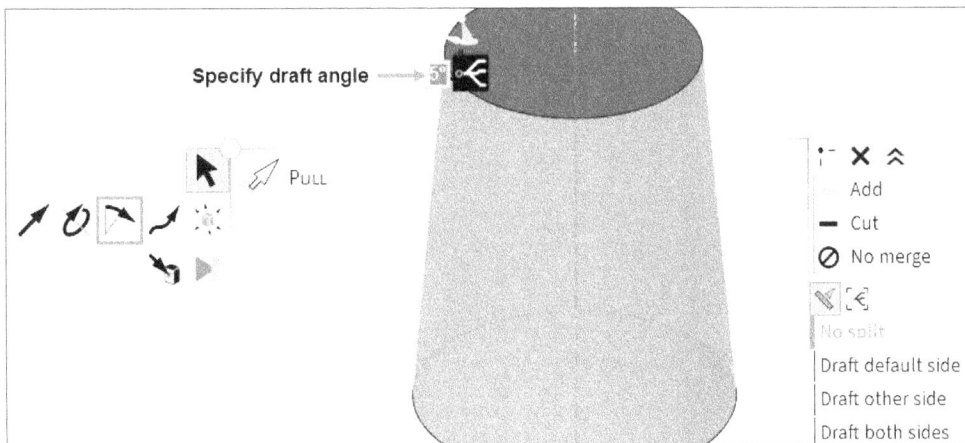

Figure-58. After applying draft

Performing Sweep Pull Operation

The Sweep operation is performed to create solid object by sweeping closed loop sketch along selected curve. The procedure to perform this operation is given next.

- Select the closed loop sketch section and click on the **Sweep** button from the **HUD** toolbar. You will be asked to select trajectory curve for sweep feature.
- Select the open curve and then drag the earlier selected closed loop sketch along the curve to create the feature; refer to Figure-59. Press **ESC** to exit.

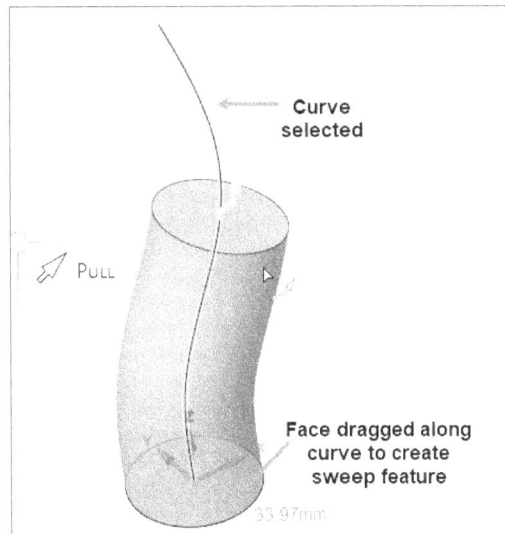

Figure-59. Creating sweep feature

Performing Scale Operation

The Scale operation is performed to increase/decrease the size of selected objects. The procedure to perform this operation is given next.

- Click on the **Scale** button from the **HUD** toolbar and select the point to be used as reference for scaling operation. You will be asked to select the body to be scaled up/down.
- Select the face of body to be scaled. The handle to scale selected body will be displayed; refer to Figure-60. Select and drag the handle to increase/decrease the size of body.

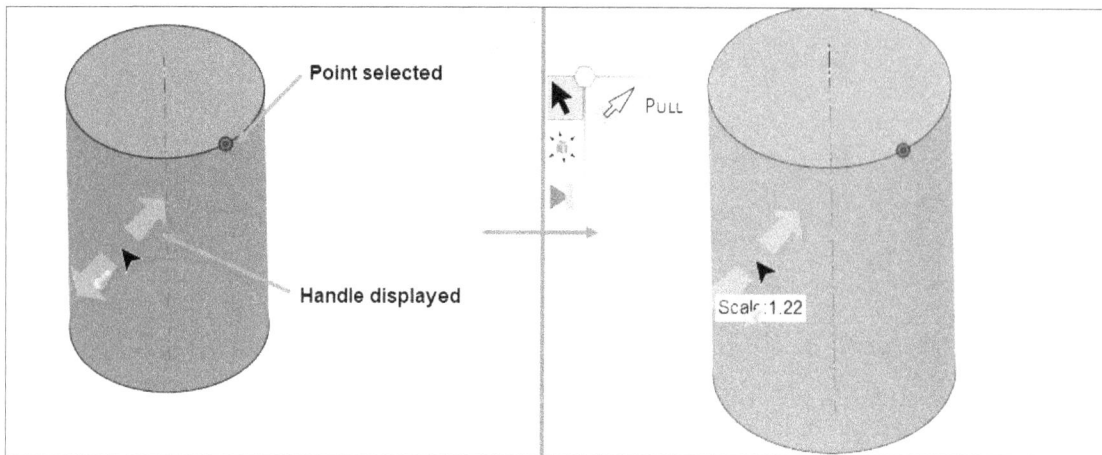

Figure-60. Scaling body

• Note that when you perform scale operation then **Resize holes** and **Preserve holes** options are displayed in the right **HUD** toolbar; refer to Figure-61. Select the **Resize holes** option from the toolbar if you want to increase the size of hole as well when increasing size of body. Select the **Preserve holes** option from the toolbar to keep the hole size fixed.

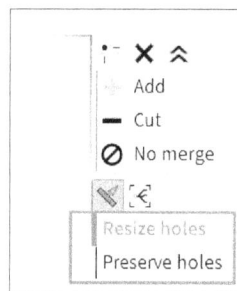

Figure-61. Options related to scaling holes

• Press **ESC** to exit the operation mode.

Applying Rounds

• On selecting an edge from the model, options to apply rounds and chamfers will be displayed; refer to Figure-62.

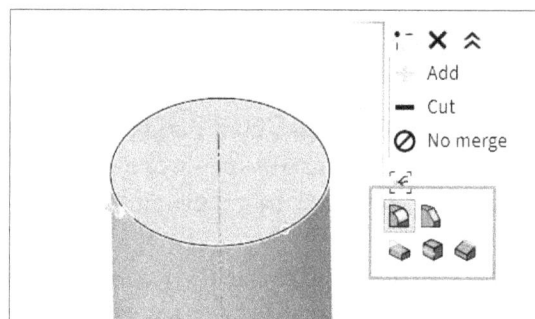

Figure-62. Options for edge

• Select the **Round** button 🗋 from the right toolbar to apply fillet (round) and drag the edge to apply round. Preview of round will be displayed with input box for specifying radius and other related parameters; refer to Figure-63.

- Specify desired value of radius in the input box; refer to Figure-63. If you want to create round with varying radius then select the **Variable Radius Round** button from the right **HUD** toolbar and drag the handles to change radius value; refer to Figure-64.

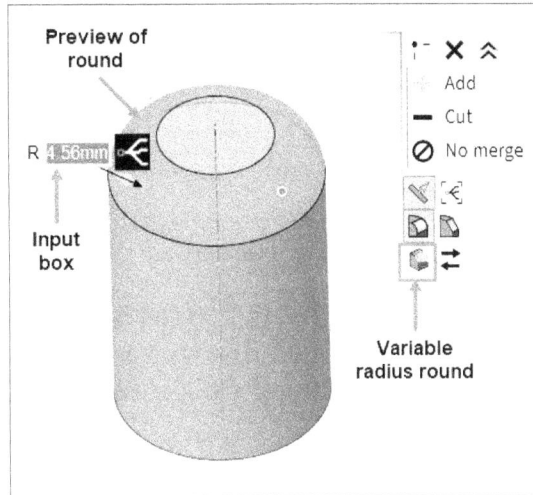

Figure-63. Preview of round

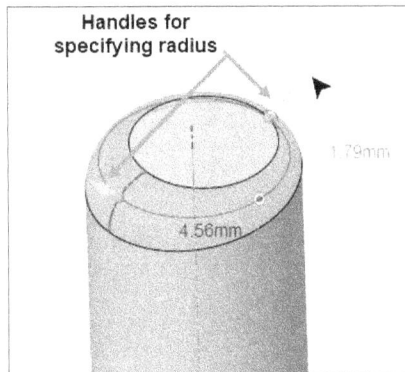

Figure-64. Variable radius round

Applying Chamfers

- Select the **Chamfer** button from the toolbar to apply chamfer and drag it outside. Preview of chamfer will be displayed; refer to Figure-65.

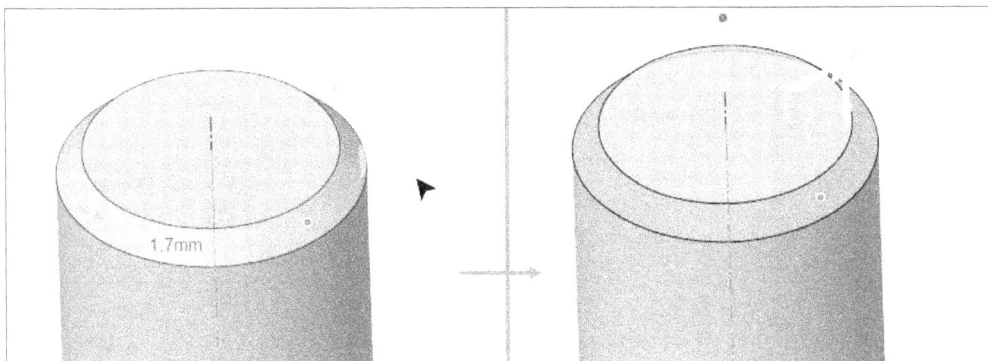

Figure-65. Preview of chamfer

- Drag the handles to modify size of chamfer in different directions or click on one of the handle. Input boxes to specify size of chamfer will be displayed; refer to Figure-66. Specify desired values in the input boxes to define chamfer size.

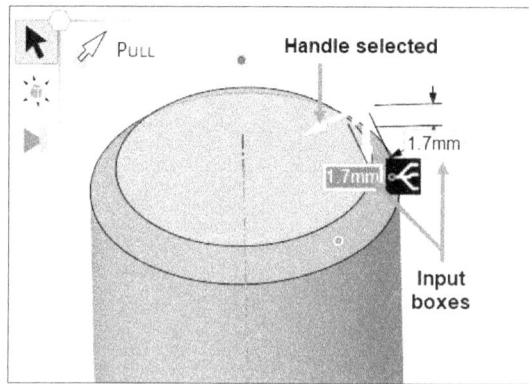

Figure-66. Input boxes for specifying size of chamfer

Extruding Edges

- After activating the **Pull** tool, select the edge that you want to extrude. The options in the right **HUD** toolbar will be displayed as shown in Figure-67.

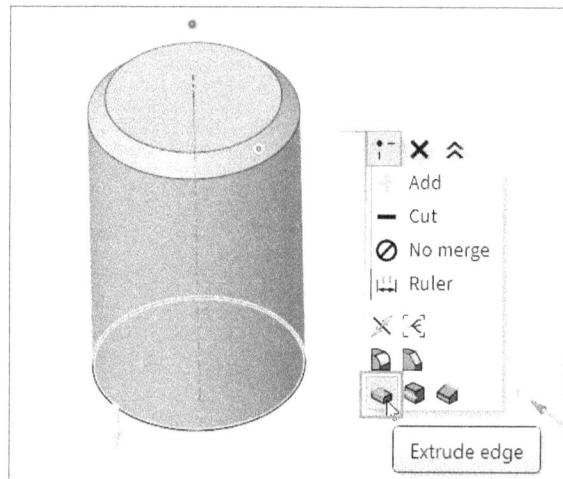

Figure-67. Extrude edge option

- Select the **Extrude edge** button from the toolbar. Handles to extrude edge will be displayed on the edge.
- Drag the handles to create the feature; refer to Figure-68.

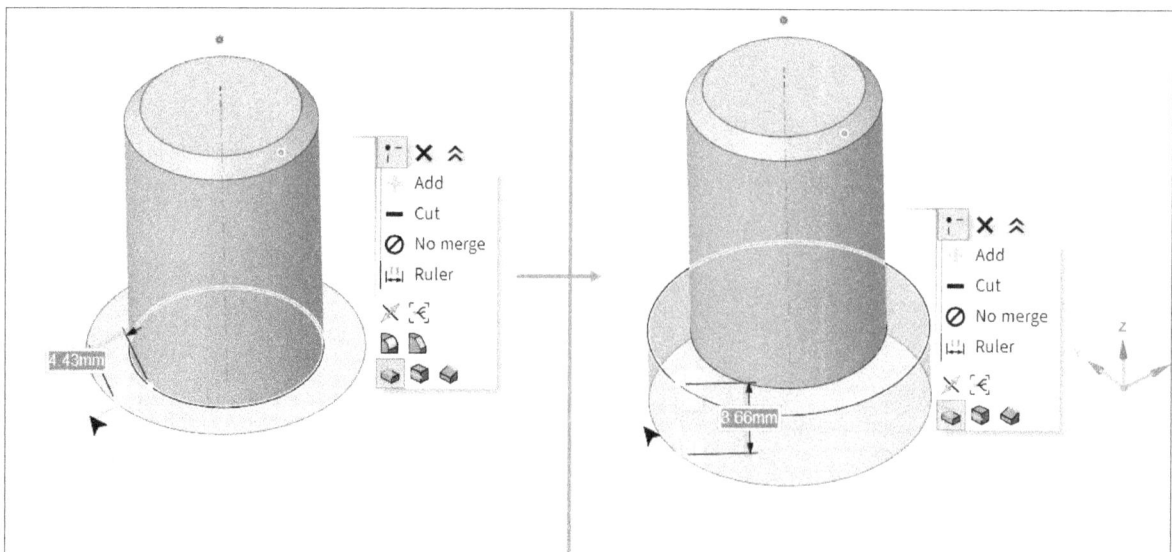

Figure-68. Extruding edge

- Similarly, you can use the **Copy edge** and **Pivot edge** buttons from the toolbar to perform related operations.

MOVING OBJECTS

The **Move** tool is used to move selected objects using different references. The procedure to use this tool is given next.

- Click on the **Move** tool from the **EDIT** panel in the **DESIGN** tab of the **Ribbon**. The **HUD** toolbar for moving objects will be displayed; refer to Figure-69.

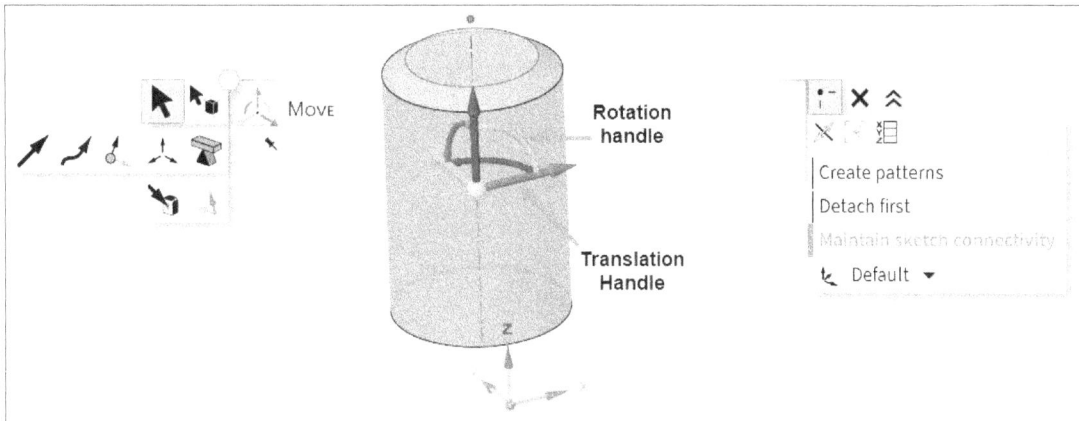

Figure-69. HUD toolbar for moving objects

- Select the rotation or translation handle on the model and drag to modify position/orientation of the model.
- If you want to follow a curve when moving object then click on the Move along curve button from the left **HUD** toolbar and select the curve to be used as reference. The handle to move object along selected path will be displayed; refer to Figure-70. Drag the handle to change the position.

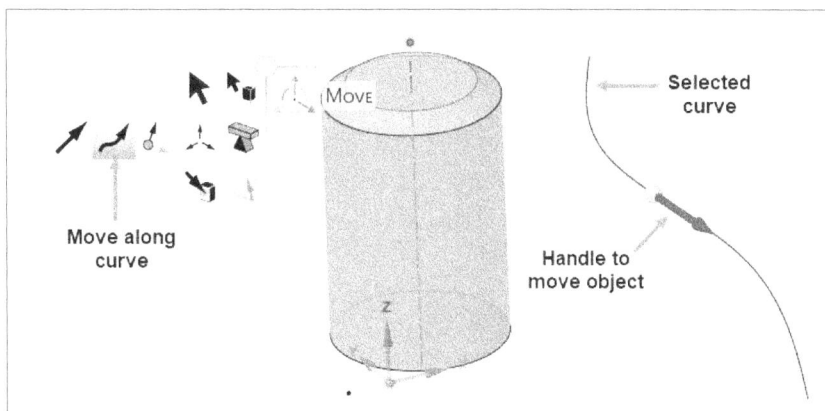

Figure-70. Moving object along curve

Creating Pattern

- You can also create pattern of selected object while performing move operation. To do so, select the **Create patterns** toggle button from the right HUD toolbar after selecting the body (make sure all faces of body are selected).
- Drag the body using desired handle. Copy of selected body will be displayed with input boxes; refer to Figure-71.

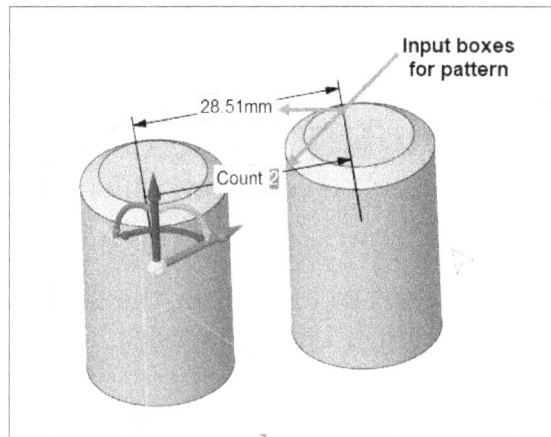

Figure-71. Creating pattern

- Set desired values in the input boxes to define number of instances and total span for pattern or gap between two consecutive instances; refer to Figure-72.

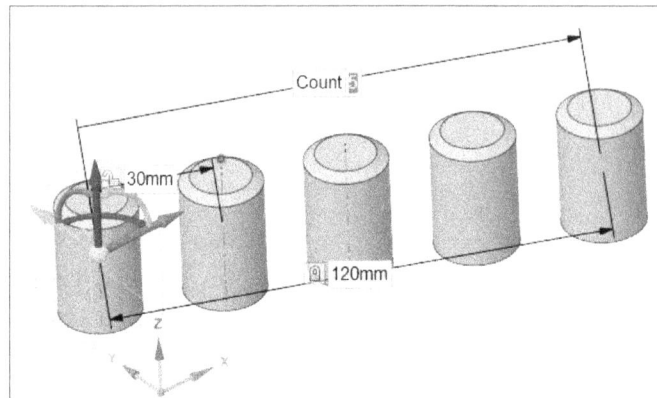

Figure-72. Specifying pattern values

- Press **ESC** twice to exit the tool.

VIEW ARC TOOLS

The tools in **View Arc** are used to orient and manage views of objects in the graphics area; refer to Figure-73. When you work on 3D models, there will be times when you need to change the view from front to top or to isometric and so on. In such cases, the tools of **View Arc** come to the rescue. Various tools of the **View Arc** are discussed next.

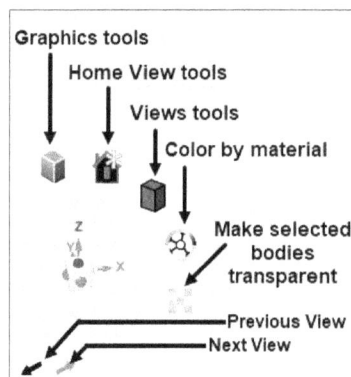

Figure-73. Tools in View Arc

Graphics Tools

The options in the **Graphics** flyout of **View Arc** are used to modify display style of objects in the graphics area; refer to Figure-74.

- Select the **Shaded** option from the flyout to display 3D objects like solids and surfaces as shaded (filled volumes).
- Select the **Enhanced Shaded** option from the flyout to display objects shaded along with shadows and lighting effects.
- Select the **Wireframe** option from the flyout to display 3D objects in bare shell form with only boundary edges. Refer to Figure-75 for different graphic styles.

Figure-74. Graphics flyout

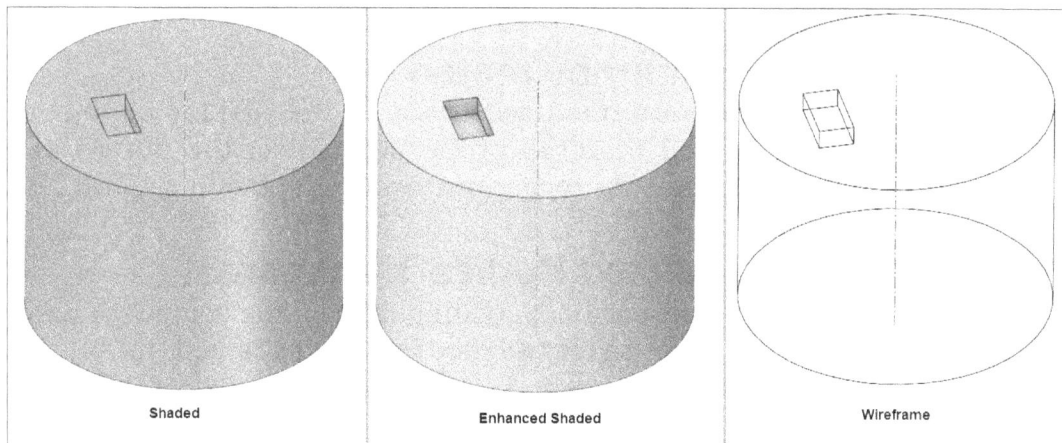

Figure-75. Graphic styles

Home View Tools

The tools in **Home** flyout are used to set and reorient view to specified home orientations; refer to Figure-76.

- After setting model in desired orientation, click on the **Set As Home View** option from the flyout to define the current orientation as Home view.
- If the model is in different orientation then you can select the **Home** option from the flyout to return to home view.
- Click on the **Reset Home View** option from the flyout to return home settings to default.
- Click on the **Snap View** option from the flyout if you want to orient a face parallel to screen. After selecting this option, click on the face to be oriented parallel.

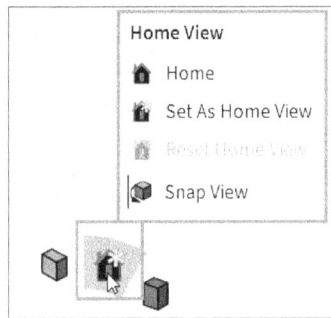

Figure-76. Home View flyout

Setting Views

The options in **Views** flyout are used to switch to different orientations like front view, left view, right view, and so on; refer to Figure-77. Select desired button from the flyout to reorient the model in respective orientation.

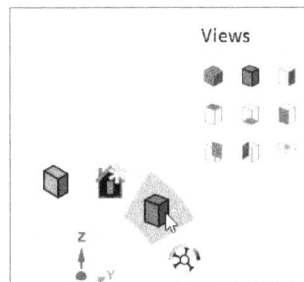

Figure-77. Views flyout

Color by material

The **Color by material** toggle button is used to toggle the display of objects in the colors assigned according to their material properties. Select the button to display material assigned colors and clear the toggle button to display in default colors.

Making Selected Bodies Transparent

If you want to make a solid/surface body transparent then select the body from graphics area and select the **Make selected bodies transparent** button from the **View Arc**. The body will become transparent; refer to Figure-78. Select the transparent body and click on the button again to return body to its previous state.

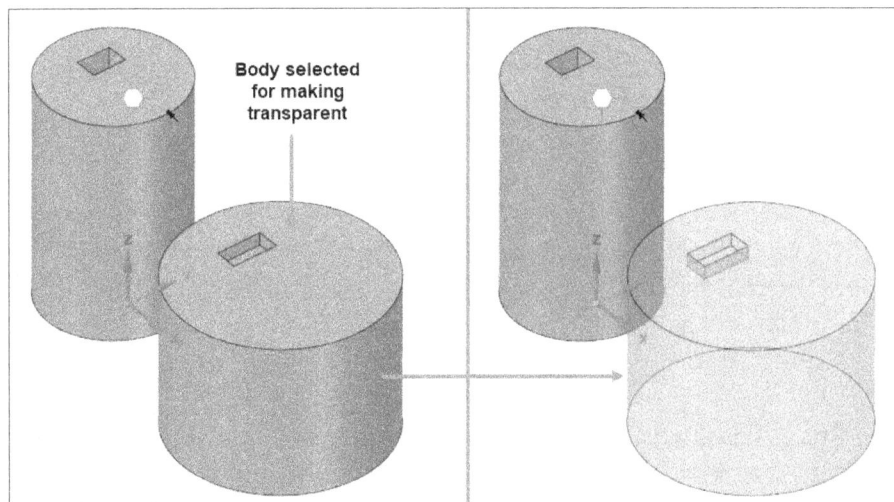

Figure-78. Making body transparent

SELF-ASSESSMENT

Q1. What is Ansys Discovery primarily used for?
A. Only geometry modeling
B. Only physics simulation
C. Rapid designing and simulation
D. 3D printing

Q2. Which feature of Ansys Discovery allows collaboration with Ansys Workbench?
A. Physics Tree
B. Ribbon
C. File Menu
D. Standalone application

Q3. What is the default selection in the First page of Ansys Discovery?
A. Model stage
B. Welcome option
C. Light theme
D. View controls

Q4. Which section in the First page allows setting shortcut keys for orientation operations?
A. Select your theme
B. Getting Started
C. View controls
D. Learn

Q5. What is the purpose of the Stage Navigator in Ansys Discovery?
A. To switch between various workspaces
B. To change view style
C. To manage elements of a model
D. To define material properties

Q6. Where can you find options to check basic operation information about Ansys Discovery?
A. Getting Started
B. What's New
C. Learn
D. File menu

Q7. What does the Ribbon in Ansys Discovery contain?
A. A list of recent files
B. A palette of tools grouped in panels
C. A collection of saved projects
D. A section for defining unit systems

Q8. In Ansys Discovery, which section is used to manage model elements like sketches and solids?
A. Design Tree
B. Physics Tree
C. View Arc
D. Status Bar

Q9. What is the purpose of the Physics Tree in Ansys Discovery?
A. To apply colors to a model
B. To manage elements of the model
C. To manage parameters and properties of analysis
D. To create new simulations

Q10. Which option allows you to start a new file in Ansys Discovery?
A. CTRL+N
B. CTRL+O
C. Save tool
D. Insert Geometry

Q11. How can you save a copy of the current file with a different name?
A. Save tool
B. Save As tool
C. Open tool
D. Create Report

Q12. What is the purpose of the Create Report tool in the File menu?
A. To import models from other CAD software
B. To generate an HTML report of analysis results
C. To modify user interface settings
D. To create a new file

Q13. Which tool is used to import models from CAD software like AutoCAD and CATIA?
A. Create Report
B. Save As
C. Insert Geometry
D. Update from new CAD file

Q14. What is the function of the Update from new CAD file tool?
A. To replace the current model with another model
B. To modify simulation parameters
C. To save a copy of the file
D. To perform mesh analysis

Q15. In the General Settings of Ansys Discovery, what does the Rendering quality option define?
A. The color scheme of the user interface
B. The smoothness and sharpness of models
C. The shortcut keys for view controls
D. The type of material assigned to a model

Q16. In Navigation settings, what does the Zoom drag direction option control?
A. The default zoom level of the model
B. The speed of zooming
C. The direction of zooming with SHIFT+MMB
D. The rotation of the model

Q17. How can you define the unit system for projects in Ansys Discovery?
A. By selecting the Model unit system option in Settings
B. By using the Ribbon menu
C. By modifying the Stage Navigator
D. By clicking on the Insert Geometry tool

Q18. In Physics Settings, what does the Include self weight toggle button do?
A. It adds gravity effects for heavy objects
B. It enables object buoyancy calculations
C. It changes the user interface theme
D. It modifies the number of iterations in simulations

Q19. What is the purpose of the Working temperature setting in Physics Settings?
A. To define the ambient temperature for simulations
B. To modify the rendering quality
C. To change the default material properties
D. To define the grid spacing

Q20. Which fidelity multiplier option should be selected for the best balance between performance and accuracy in fluid dynamics simulations?
A. Turbulent k-epsilon realizable
B. Laminar
C. Turbulent k-omega SST
D. Turbulent k-epsilon standard

Q21. Which panel in the DESIGN tab contains tools for creating sketch curves?
A. MODIFY panel
B. SKETCH panel
C. CREATE panel
D. ASSEMBLE panel

Q22. What is the primary function of the Line tool?
A. To create a construction line
B. To create a line by specifying two points
C. To create a rectangle using two diagonal points
D. To create a circle using three points

Q23. Which toggle button should be selected to create a line using its midpoint?
A. Snap to grid
B. Snap to angle
C. Draw from center
D. Automatically create constraints

Q24. The Tangent Line tool is used to create a line that is tangent to which of the following entities?
A. Rectangle
B. Circle/Arc/Spline/Curve
C. Polygon
D. Point

Q25. What is the function of the Construction Line tool?
A. To create a permanent sketch entity
B. To create a reference line for other sketch entities
C. To create an offset copy of a curve
D. To split a curve into two segments

Q26. Which tool is used to create a rectangle by specifying two corner points or a center and corner point?
A. POLYGON
B. CIRCLE
C. RECTANGLE
D. TANGENT ARC

Q27. What is the purpose of the Three-Point Rectangle tool?
A. To create an equilateral triangle
B. To create a rectangle that is slanted
C. To create a rectangle with rounded corners
D. To create a polygon with three sides

Q28. The Ellipse tool requires which three points to define an ellipse?
A. Center, major axis, minor axis
B. Start, midpoint, endpoint
C. Three circumferential points
D. Two diameters and center

Q29. What is the primary method for defining a circle using the Circle tool?
A. Using three circumferential points
B. Using the center point and diameter
C. Using two diagonal points
D. Using the midpoint and an edge

Q30. How is a Three-Point Circle created?
A. By specifying center and two radii
B. By specifying three circumferential points
C. By specifying the center and diameter
D. By specifying two diagonal points

Q31. The POLYGON tool allows the user to create a polygon using which reference?
A. Only an inscribed circle
B. Only a circumscribed circle
C. Either an inscribed or a circumscribed circle
D. A rectangle as a reference

Q32. The Tangent Arc tool is used to create an arc that is tangent to:
A. A line/curve
B. A rectangle
C. A polygon
D. A point

Q33. How many points are required to create a Three-Point Arc?
A. One
B. Two
C. Three
D. Four

Q34. What are the required inputs for creating a Sweep Arc?
A. Center point, start point, end point
B. Major axis, minor axis, center
C. Three circumferential points
D. Start, midpoint, endpoint

Q35. Which tool is used to create a spline passing through selected points?
A. POLYGON
B. SPLINE
C. TANGENT ARC
D. OFFSET CURVE

Q36. The FACE CURVE tool is used to create curves on:
A. A sketch plane
B. A selected solid face
C. A construction line
D. A point

Q37. What is the function of the CREATE ROUNDED CORNER tool?
A. To create an inscribed polygon
B. To apply fillets or chamfers at sharp corners
C. To create a three-point rectangle
D. To project curves onto a sketch plane

Q38. Which tool is used to create a copy of selected curves at a specified distance?
A. OFFSET CURVE
B. SPLINE
C. TANGENT ARC
D. CREATE CORNER

Q39. The PROJECT TO SKETCH tool is used to:
A. Project selected edges, loops, chains, and vertices onto a sketch plane
B. Create a three-point arc
C. Create a construction line
D. Offset a curve

Q40. The CREATE CORNER tool is used to:
A. Trim extra portions of lines/curves to form a corner
B. Offset a selected curve
C. Create a fillet at the intersection of two lines
D. Define a construction line

Q41. The TRIM AWAY tool removes:
A. The entire curve
B. Selected segment of a curve with respect to the nearest intersection
C. Only construction lines
D. The entire sketch

Q42. The SPLIT CURVE tool is used to:
A. Remove an entire curve
B. Split a curve at a specified location
C. Create a rounded corner
D. Project curves onto a sketch plane

Q43. Which tool is used to apply various types of dimensions in a sketch?
A. Pull Tool
B. Dimension Tool
C. Extrude Tool
D. Selection Tool

Q44. What happens when you select an arc using the Dimension tool?
A. It applies a diameter dimension
B. It applies a length dimension
C. It applies a radius dimension
D. It applies a chamfer dimension

Q45. How can you modify the size of an entity after applying a dimension?
A. By right-clicking on it
B. By double-clicking on the dimension value and entering a new value
C. By selecting it and pressing Delete
D. By selecting the Move tool

Q46. What is the purpose of the AUTOCONSTRAIN tool?
A. To apply possible constraints to all entities in the current sketch
B. To automatically delete unwanted constraints
C. To modify the dimensions of entities
D. To hide constraints

Q47. What color represents a fully constrained sketch entity?
A. Light blue
B. Red
C. Bold blue
D. Green

Q48. Which selection method allows selecting objects by drawing a freeform boundary?
A. Box Selection
B. Lasso Selection
C. Polygon Selection
D. Boundary Selection

Q49. In Box Selection, what happens if you move the cursor from left to right?
A. It selects objects touching the boundary
B. It selects only objects fully inside the box
C. It selects all objects in the graphics area
D. It deselects all objects

Q50. What is the purpose of the Power Select tool?
A. To select all objects in the scene
B. To select similar objects based on criteria
C. To apply constraints automatically
D. To delete selected objects

Q51. Which tool is used to select all objects in a currently active component?
A. Power Select
B. Select All
C. Select Components
D. Clear Selection

Q52. What does the Pull tool allow you to perform?
A. Only extrude operations
B. Various operations like extrude, revolve, offset, and sweep
C. Only scale and chamfer operations
D. Only selection-based operations

Q53. Which operation is performed to create a solid by revolving a closed loop sketch about an axis?
A. Extrude
B. Scale
C. Sweep
D. Revolve

Q54. What is the function of the Draft operation?
A. To extrude edges
B. To apply a taper angle to faces of a body
C. To revolve a face
D. To apply a fillet

Q55. Which operation is used to create a solid object by sweeping a closed loop sketch along a curve?
A. Extrude
B. Sweep
C. Scale
D. Draft

Q56. What is the function of the Scale operation?
A. To apply constraints
B. To increase or decrease the size of an object
C. To create a fillet
D. To apply a chamfer

Q57. Which toolbar option should be selected to keep the hole size fixed while scaling an object?
A. Resize holes
B. Preserve holes
C. Scale factor
D. Draft angle

Q58. How do you apply a round (fillet) to an edge?
A. Select an edge and click the Round button
B. Click the Chamfer button
C. Use the Box Selection tool
D. Use the AUTOCONSTRAIN tool

Q59. What is the purpose of the Variable Radius Round tool?
A. To apply a constant radius fillet
B. To apply a varying radius fillet by dragging handles
C. To delete constraints from a sketch
D. To perform an extrude operation

Q60. What is the primary purpose of the Chamfer tool?
A. To create a fillet
B. To add a tapered edge
C. To select multiple objects
D. To revolve a face

Q61. Which operation is performed when you drag an edge using the Extrude Edge tool?
A. It moves the edge
B. It deletes the edge
C. It creates a new extruded feature
D. It applies a chamfer

Q62. Which tools can be used along with the Pull tool to modify edges?
A. Copy Edge and Pivot Edge
B. Power Select and Selection Filter
C. Lasso Selection and Polygon Selection
D. Show Constraints and Hide Constraints

Q63. What is the primary function of the Move tool?
A. To delete objects
B. To rotate the entire model
C. To move selected objects using different references
D. To scale objects

Q64. How can you move an object along a curve?
A. Select the Move along curve button and specify the reference curve
B. Rotate the object manually
C. Use the Snap View tool
D. Change the view orientation

Q65. What happens when you enable the Create patterns toggle while moving an object?
A. The object is deleted
B. A copy of the selected object is created in a pattern
C. The object scales automatically
D. The object gets locked in place

Q66. How do you exit the Move tool?
A. Press ESC twice
B. Click on the Snap View tool
C. Right-click and select Close
D. Restart the software

Q67. What is the purpose of the View Arc tools?
A. To edit objects
B. To orient and manage views of objects in the graphics area
C. To create new 3D models
D. To apply material properties

Q68. Which option in the Graphics flyout displays objects with shadows and lighting effects?
A. Shaded
B. Wireframe
C. Enhanced Shaded
D. Transparent

Q69. What does the Wireframe option in the Graphics flyout do?
A. Displays 3D objects in bare shell form with only boundary edges
B. Fills the model with a texture
C. Removes all edges from the view
D. Converts the model into a 2D sketch

Q70. What does the Set As Home View tool do?
A. Saves the current orientation as the default home view
B. Moves the model to a new location
C. Deletes the home view
D. Locks the model in place

Q71. How can you return a model to its saved home view?
A. By selecting the Home option from the Home flyout
B. Using the Move tool
C. Rotating the model manually
D. Resetting the software

Q72. What does the Snap View tool do?
A. Aligns a selected face parallel to the screen
B. Moves objects to a new position
C. Switches to a wireframe view
D. Deletes all constraints

Q73. Which tool allows switching between different orientations such as front view and right view?
A. Views flyout
B. Home View tool
C. Wireframe tool
D. Snap View tool

Q74. What does the Color by material toggle button do?
A. Toggles display of objects in colors assigned by material properties
B. Removes material properties from objects
C. Converts objects into wireframe mode
D. Assigns a new color randomly

Q75. How can you make a solid body transparent?
A. Select the body and click on the Make selected bodies transparent button
B. Delete the body
C. Scale down the body
D. Use the Wireframe tool

Q76. How do you restore a transparent body to its previous state?
A. Select the transparent body and click on the Make selected bodies transparent button again
B. Reset the home view
C. Change the material properties
D. Use the Views flyout

Chapter 2

Ansys Discovery-Design II

Topics Covered

The major topics covered in this chapter are:

- *Filling Holes and Voids*
- *Creating Blend Feature*
- *Replacing Face*
- *Performing Combine Operation*
- *Splitting Body*
- *Splitting Face/Surface*
- *Creating Projection*
- *Creating Datum Features*
- *Creating Patterns*
- *Creating Shell Feature*
- *Performing Body Modification Operations*
- *Assembly Constraints*
- *Display Tools*
- *Measurement Tools*

INTRODUCTION

In previous chapter, you have learned to performing sketching and editing operations. You also learned to perform various view related options. In this chapter, you will learn to create datum features like plane, axis, point, and so on. You will also learn to create patterns, copy features, bodies, and assemblies.

FILLING HOLES AND VOIDS

The **Fill** tool is used to fill holes and voids in bodies surrounded by walls. The procedure to use this tool is given next.

*   Click on the **Fill** tool from the **EDIT** panel in the **DESIGN** tab of the **Ribbon**. The **HUD** toolbars will be displayed.
*   Select the faces to be filled and click on the **OK** button from the **HUD** toolbar. Selected voids/holes will be filled; refer to Figure-1.

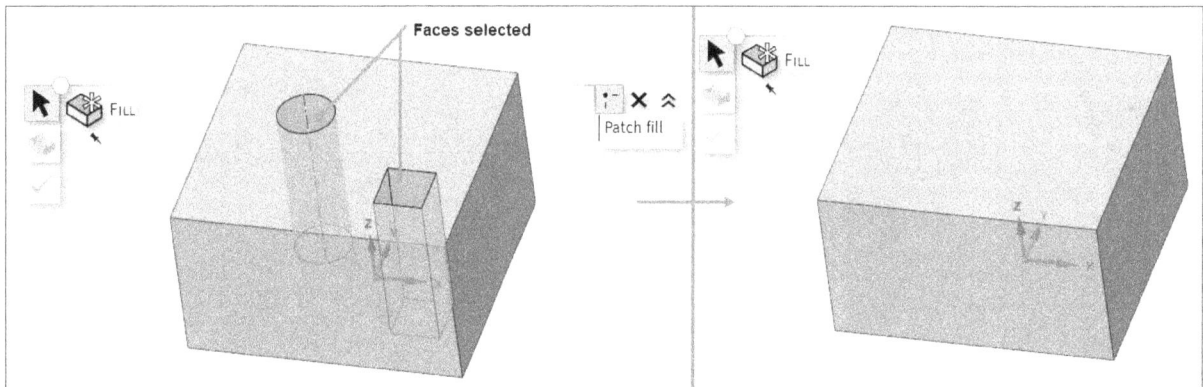

Figure-1. Filling holes and voids

CREATING BLEND FEATURE

The **BLEND** tool is used to create a solid feature by joining two or more closed loop curves. You can use guide-curves and centerline curve to further control the shape of blend feature. You can also use this tool to blend two or more curves to form single joint curve. The procedure to create a blend feature is given next.

*   Click on the **BLEND** tool from the **EDIT** panel in the **DESIGN** tab of the **Ribbon**. The **HUD** toolbars will be displayed with options related to blend feature creation.
*   Select two or more vertices to join them using a single curve; refer to Figure-2.

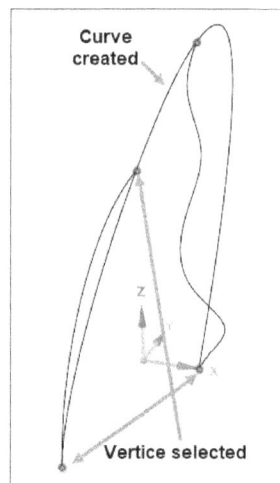

Figure-2. Creating curve by blending

- Select two or more curves to create a blending surface; refer to Figure-3.

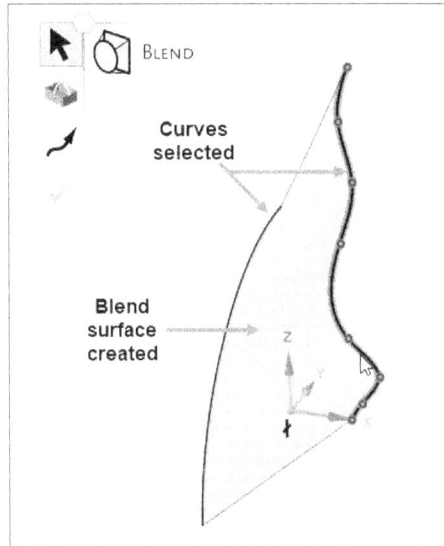

Figure-3. Creating Blending surface

- Select the closed loop sketches in order to create the feature. Preview of blend feature will be displayed; refer to Figure-4.
- If you want to use a guide curve then select the **Guide curve** button from the left **HUD** toolbar and select the guide curve to change shape of blend feature; refer to Figure-5.

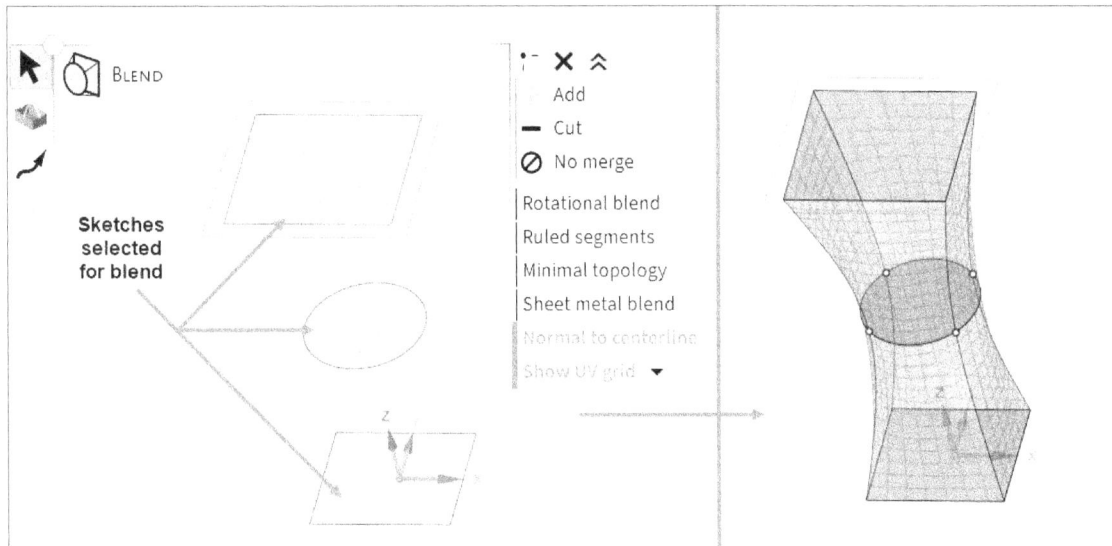

Figure-4. Creating blend feature

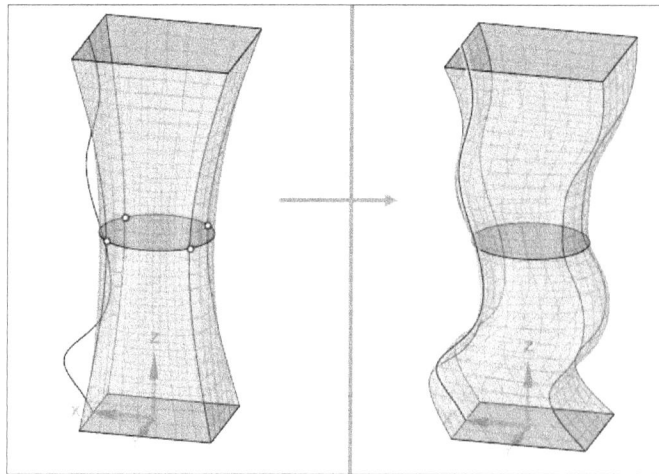

Figure-5. Using guide curves in blend feature

- Similarly, you can use **Centerline** button ⟋ from the left **HUD** toolbar to use selected curve as center reference for blend feature; refer to Figure-6.

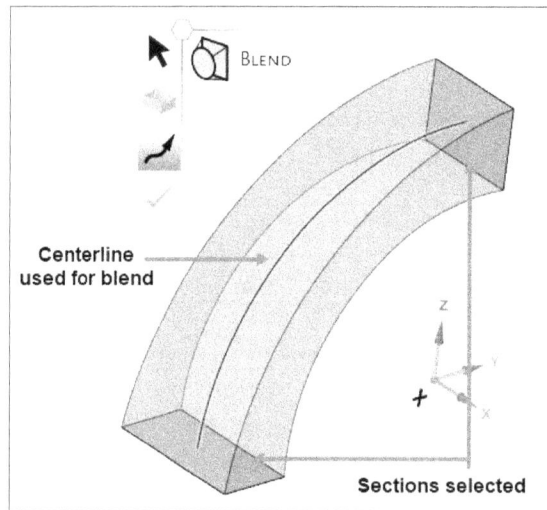

Figure-6. Blend using centerline

- Select the **Rotational blend** toggle button from the right **HUD** toolbar if you want to blend the sections in cylindrical shape about a common axis; refer to Figure-7.

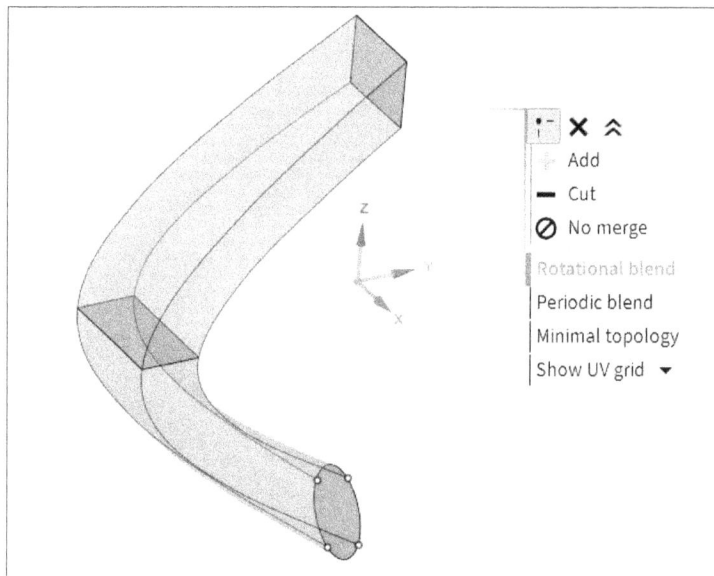

Figure-7. Rotational blend preview

- Select the **Periodic blend** toggle button from the right **HUD** toolbar to create a closed loop blend (ring-like feature) using selected sections so that first and last sections are also blended; refer to Figure-8.

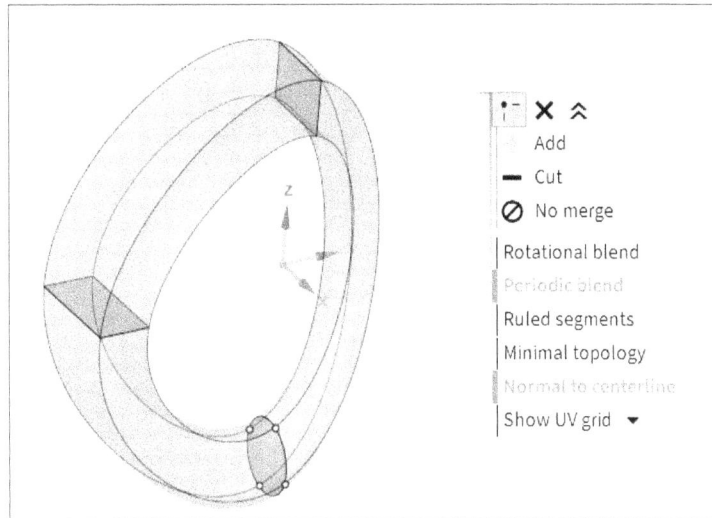

Figure-8. Periodic blend

- Select the **Ruled segments** toggle button from the right **HUD** toolbar if you want to create blend with straight line edges; refer to Figure-9.

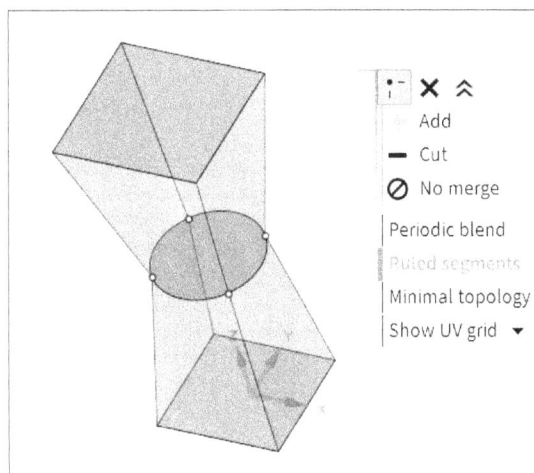

Figure-9. Ruled segments blend

- Select the **Minimal topology** toggle button if you do not want to create face and edge data when creating blend. This will generate smooth blend with only corner vertices.
- Set the other parameters as discussed earlier and click on the **OK** button from the toolbar. The blend will be created.

REPLACING FACE

The **REPLACE** tool is used to replace selected face with another surface in projected direction of the selected face. The procedure to use this tool is given next.

- Click on the **REPLACE** tool from the **EDIT** panel in the **DESIGN** tab of the **Ribbon**. You will be asked to select face to be replaced and **HUD** toolbar will be displayed as shown in Figure-10.

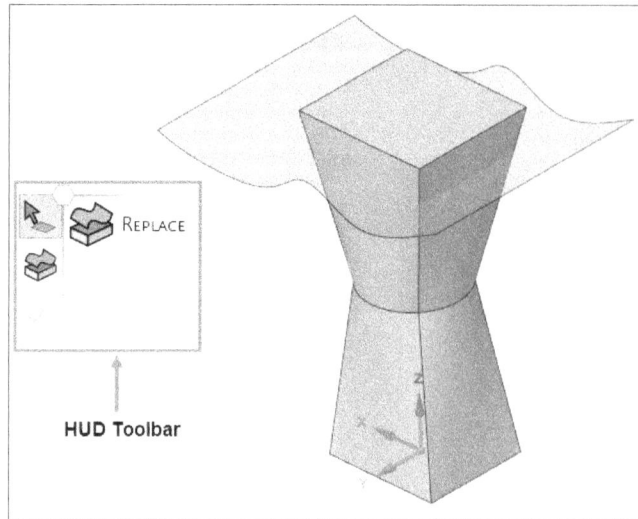

Figure-10. REPLACE HUD toolbar

*   Select the face to be replaced and then select the surface by which selected face will be replaced. The preview of replaced face will be displayed; refer to Figure-11.

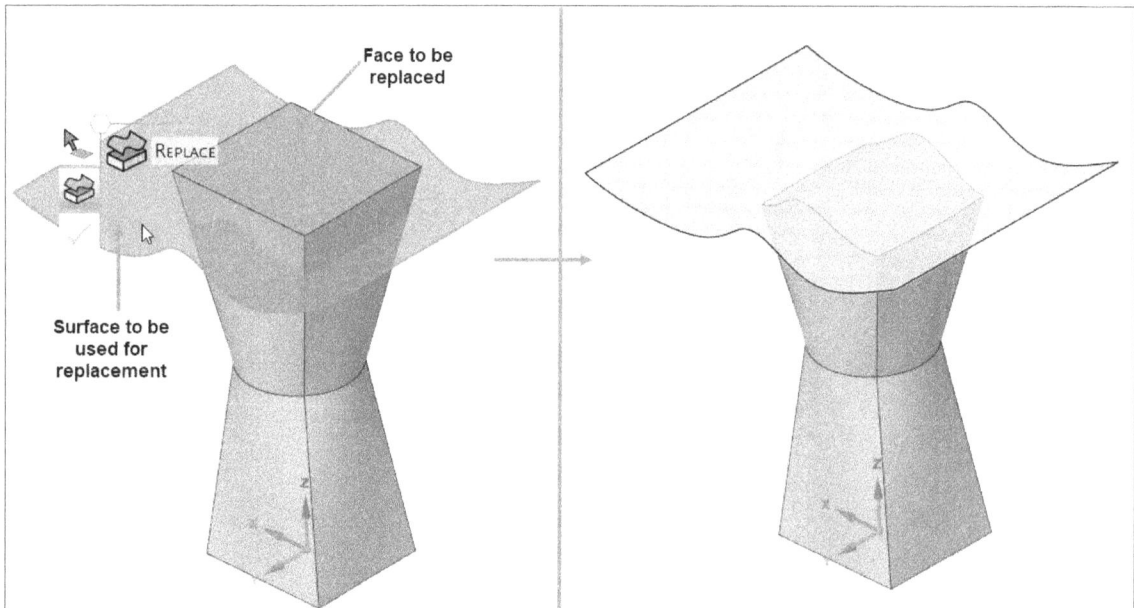

Figure-11. Replacing face

*   Press **ESC** to exit the tool.

PERFORMING COMBINE OPERATION

The **Combine** tool is used to perform various boolean operations like merge, split, subtract, and so on. Note that there must be two or more bodies to perform combine operation. The procedure to use this tool is given next.

*   Click on the **Combine** tool from the **INTERSECT** panel in the **DESIGN** tab of the **Ribbon**. The **HUD** toolbars will be displayed as shown in Figure-12.

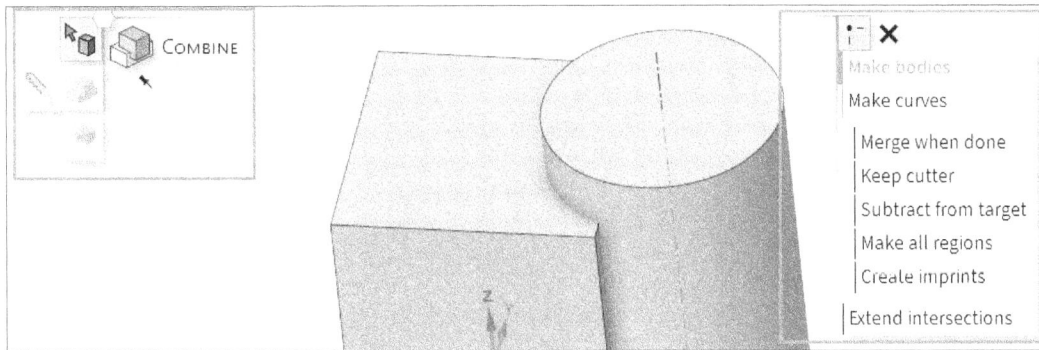

Figure-12. HUD toolbars for Combine

- Select the **Make bodies** toggle button from the right **HUD** toolbar if you want to create bodies after performing combine operation.
- Select the **Merge when done** toggle button from the toolbar if you want to merge two selected bodies.
- Select the **Keep cutter** toggle button if you do not want to remove second selected body. Note that after activating this tool, first selected body becomes target body and second selected body becomes cutter body.
- Select the **Subtract from target** toggle button if you want to remove selected cutter body from the target body.
- Select the **Make all regions** toggle button if you want to keep all the regions of intersecting bodies after performing combine operation.
- Select the **Create imprints** toggle button if you want to generate edge curves at the intersections of target and cutter bodies.
- Select the **Extend intersections** toggle button if you want to extend the surfaces/ faces to possible intersection point for performing combine operation.
- Select the **Make curves** toggle button if you want to generate only curves at intersections of two surface bodies.
- After setting desired parameters, select the target body and cutter body. Preview of operation will be displayed; refer to Figure-13.

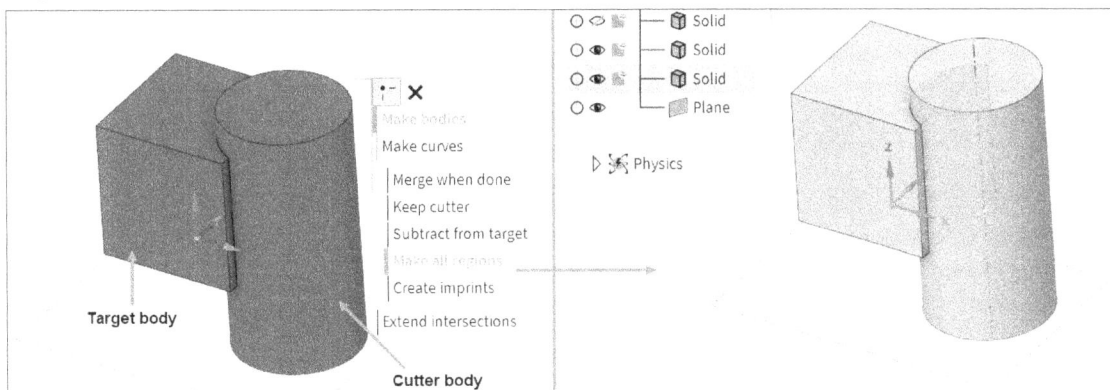

Figure-13. After performing combine

SPLITTING BODY

The **Split Body** tool is used to split selected body using a face/surface/edge loop. The procedure to use this tool is given next.

- Click on the **Split Body** tool from the **INTERSECT** panel in the **DESIGN** tab of the **Ribbon**. The **SPLIT BODY HUD** toolbars will be displayed; refer to Figure-14 and you will be asked to select the target body to be split.

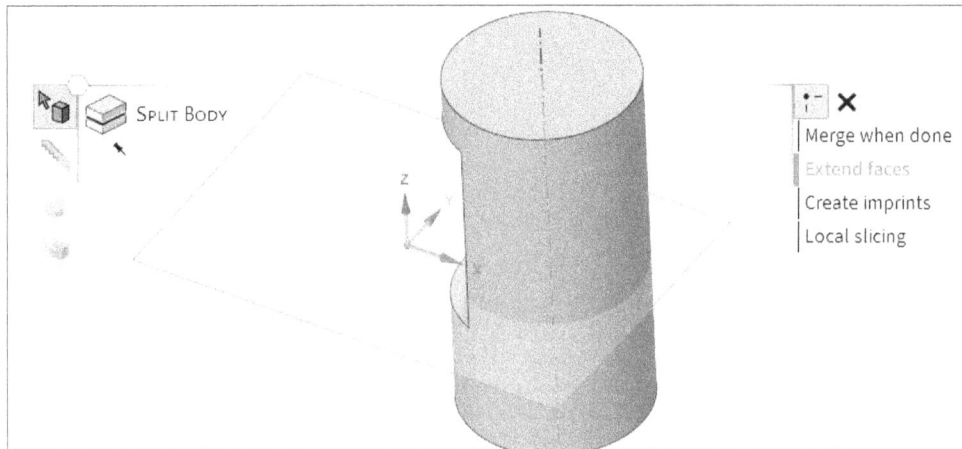

Figure-14. SPLIT BODY HUD toolbars

- Select the **Merge when done** option from the right toolbar to create a merged body after splitting. (Well this option does not make sense here so even I do not know why Ansys has added the option here. When we are splitting, there is no question of merging it with others. So in simple words, selecting this option does nothing.)
- Select the **Extend faces** toggle button from the right toolbar if you want to extend selected splitting tool (face/surface/plane) for performing split operation; refer to Figure-15.

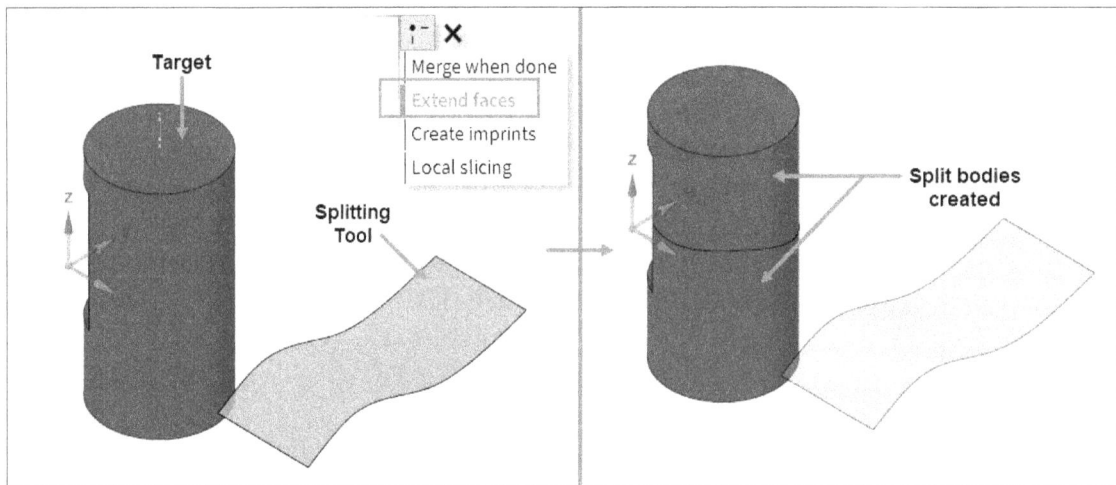

Figure-15. Split bodies created

- Select the **Create imprints** toggle button from the right toolbar if you want to generate edge curves at split location.
- Select the **Local slicing** toggle button from the right toolbar if you want to use objects native to selected target body like reference plane or surface of target body.
- After selecting desired toggle button(s) from the toolbar, select the target body and splitting tool. The split feature will be created.

SPLITTING FACE/SURFACE

The **Split** tool is used to split selected face/surface at selected point. The procedure to use this tool is given next.

- Click on the **Split** tool from the **INTERSECT** panel in the **DESIGN** tab of the **Ribbon**. The **SPLIT HUD** toolbar will be displayed and you will be asked to select surface/face to be split.

- Click on desired surface/face. You will be asked to select point for defining split face.
- Click at desired location to specify the splitting point. The split surface/face will be created; refer to Figure-16.

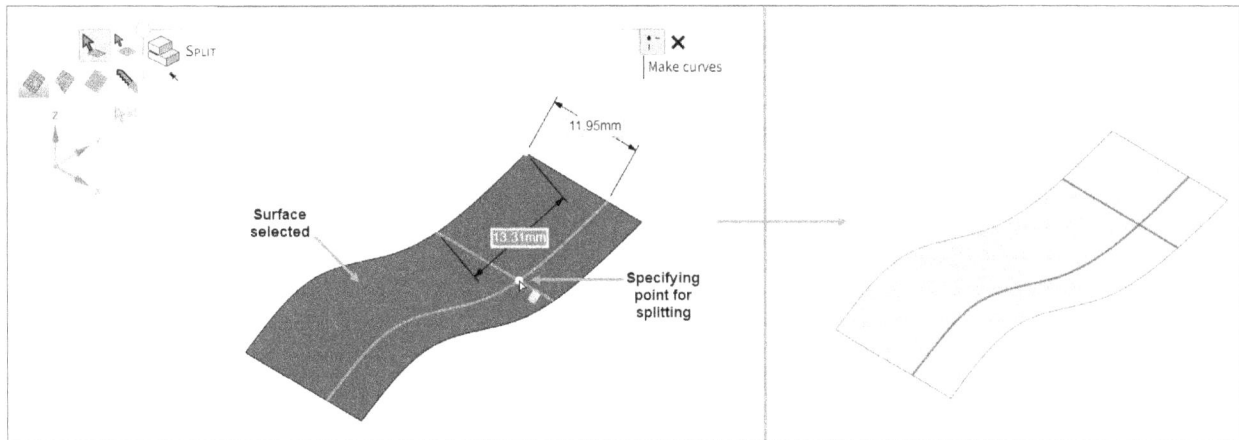

Figure-16. Specifying split point

CREATING PROJECTION

The **Project** tool is used to project (make impression) selected face/surface/curve onto another face/surface in direction normal to selected face/surface/curve. The procedure to use this tool is given next.

- Click on the **Project** tool from the **INTERSECT** panel in the **DESIGN** tab of the **Ribbon**. The **PROJECT HUD** toolbar will be displayed and you will be asked to select faces/ curves to be projected.
- Select the faces/curves (for multiple selection, hold the **CTRL** key while selecting) from the graphics area. Preview of projection will be displayed; refer to Figure-17.

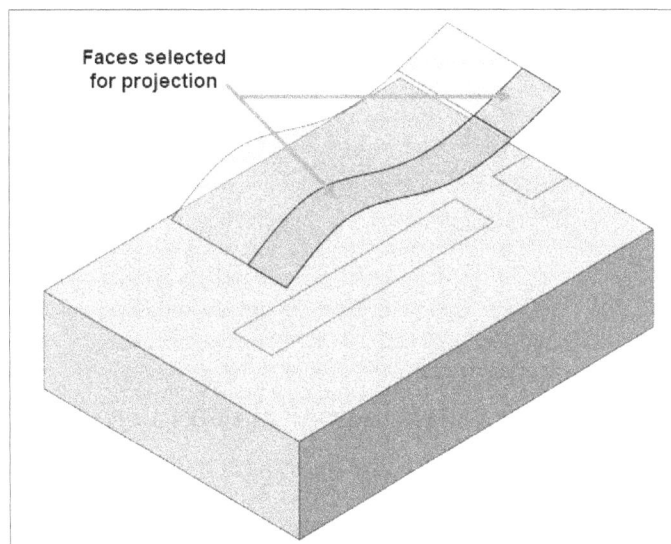

Figure-17. Faces selected for projection

- Select the **Project through solids** toggle option from the right **HUD** toolbar to create projection on all the faces falling in normal direction to selected faces; refer to Figure-18.

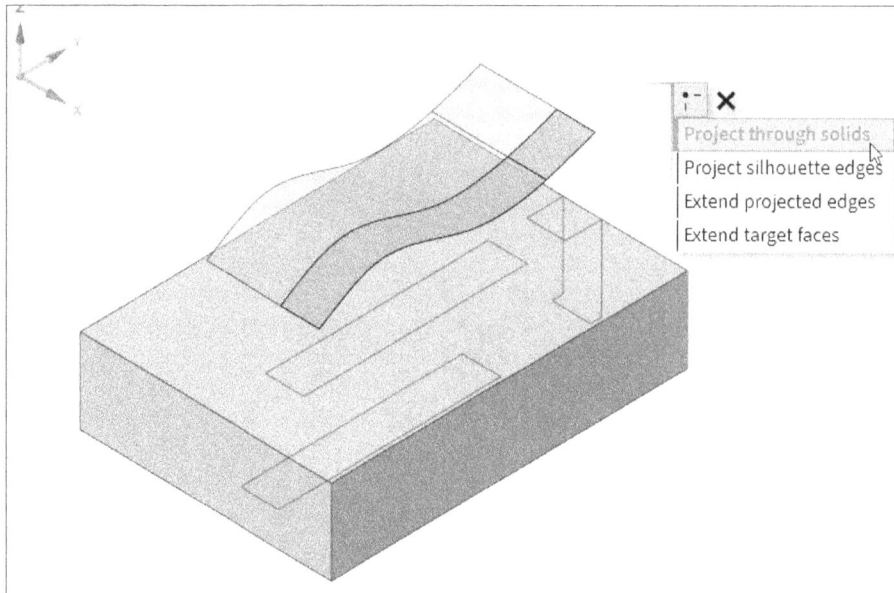

Figure-18. Projection created using Project through solids option

- Similarly, you can use the **Project silhouette edges** toggle button to project boundary edges of selected faces/object like sphere which have no defined edges, the **Extend projected edges** toggle button to extend the length of projection to cover target face; refer to Figure-19 and select the **Extend target faces** toggle button to automatically extend target faces when selected faces for projection are larger than the target faces.

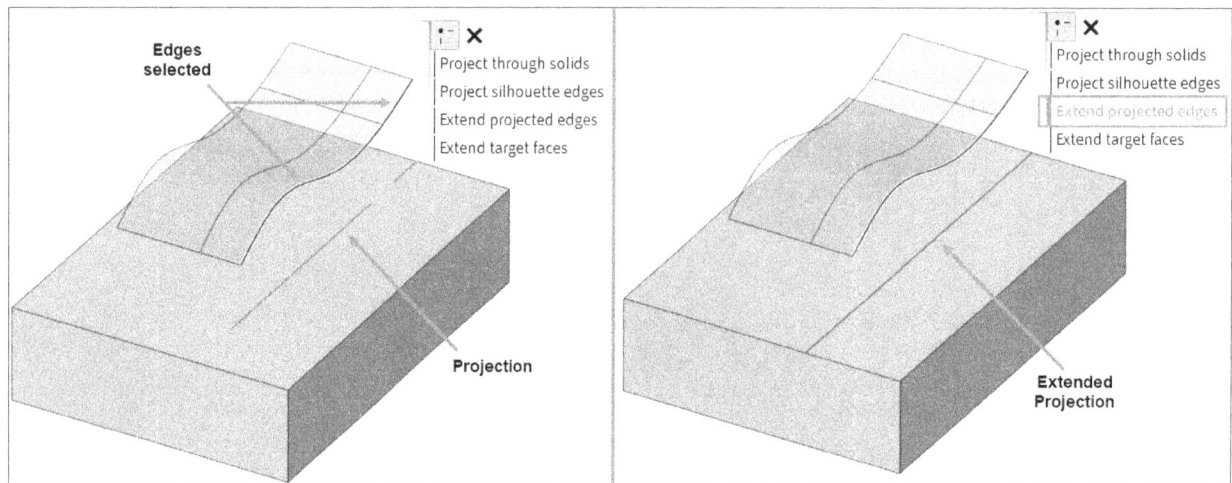

Figure-19. Using Extended projected edges toggle button

- Click on the **OK** button from the left toolbar of **HUD** to create the feature and press **ESC** to exit the tool.

CREATING DATUM FEATURES

The datum features are used as base features and references for placement and orientation of other objects. There are four tools available in the **CREATE** panel of **DESIGN** tab in the **Ribbon** to create datum features: **PLANE**, **AXIS**, **POINT**, and **ORIGIN**. These tools are discussed next.

Creating Datum Plane

The **PLANE** tool is used to create datum plane in the graphics area. The procedure to use this tool is given next.

- Click on the **PLANE** tool from the **CREATE** panel in the **DESIGN** tab of the **Ribbon**. You will be asked to select a reference for creating plane and the **PLANE HUD** toolbar will be displayed.
- Select desired reference feature (face, plane, axis, line, point, or vertex) to create a passing through it.
- If you want to select multiple reference then select the **Progressive selection** button ◼ from the left **HUD** toolbar.
- Depending on selection of entities, you can create different types of planes. Select two edges/axes to create planes passing through selected edges or plane perpendicular to imaginary plane connecting two edges/axes and at the mid of selected edges/axes; refer to Figure-20. Select desired plane preview in dashed lines to create respective plane.

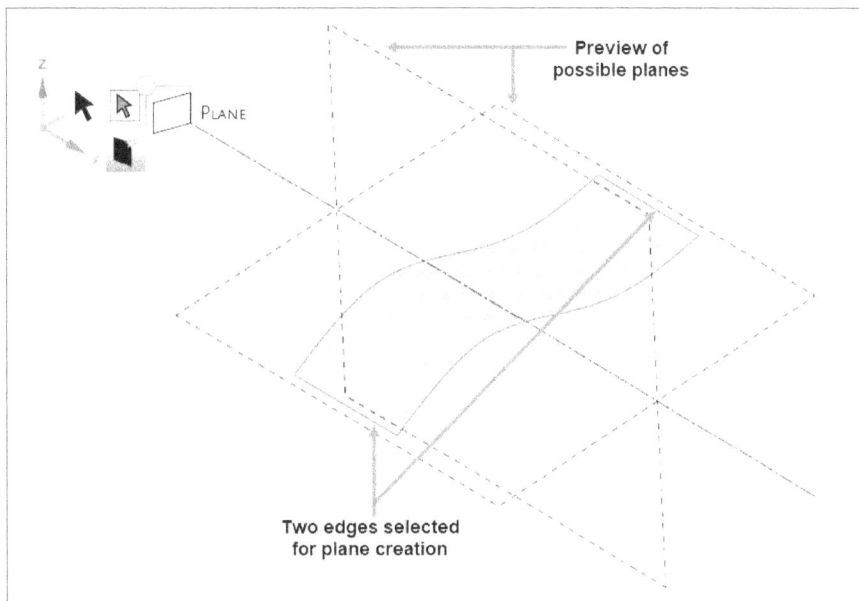

Figure-20. Preview of plane using two edges

- Select three vertices or points to create a plane passing through selected points; refer to Figure-21. Select desired preview to create plane.

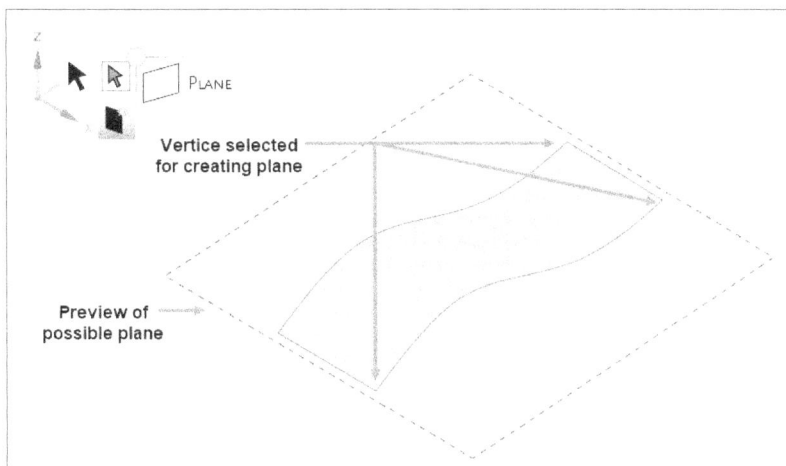

Figure-21. Plane preview using three points

- Select an axis/edge and a vertex/point to create plane passing through or perpendicular to selected edge/axes and passing through selected vertex/point; refer to Figure-22.

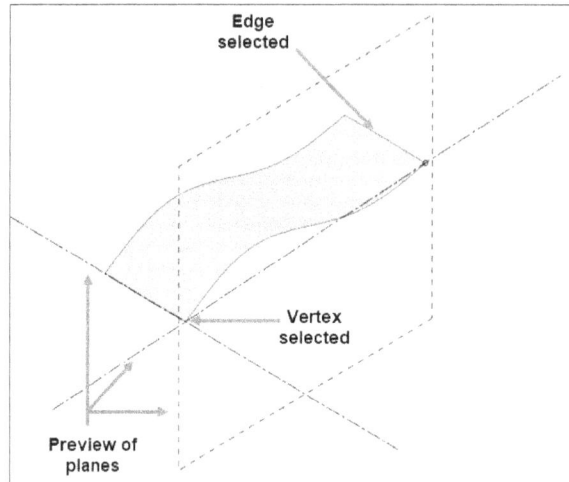

Figure-22. Preview of plane using edge and vertex

- Make sure that **Selection** button is active in the **HUD** toolbar and select curved face of model to create plane tangent to selected face; refer to Figure-23.

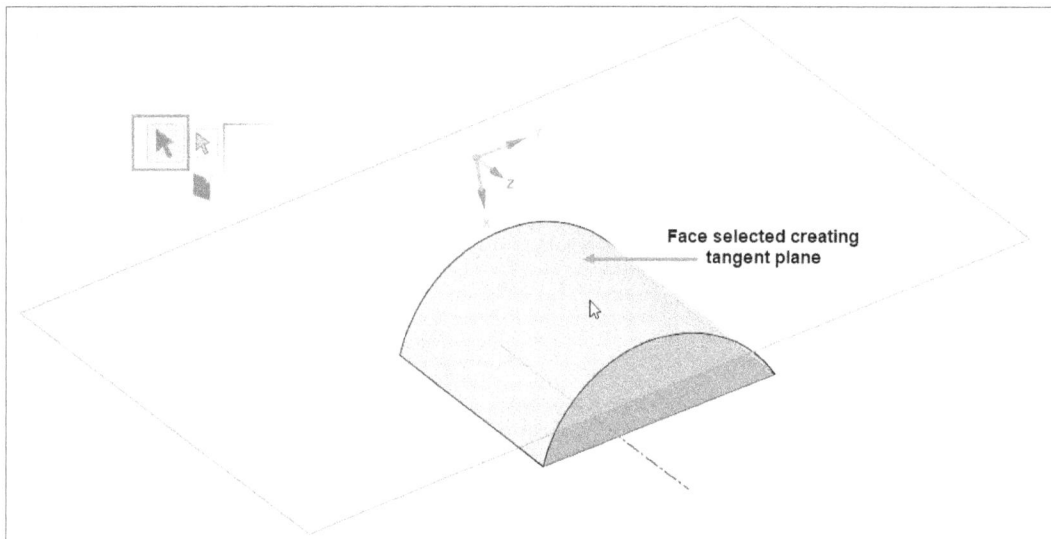

Figure-23. Preview of tangent plane

- Select the **Through Selection** button to create a plane parallel to screen and passing through selected object.
- Press **ESC** to exit the tool.

Creating Datum Axis

The **Axis** tool is used to create a datum axis based on selected references. The procedure to use this tool is given next.

- Click on the **Axis** tool from the **CREATE** panel in the **DESIGN** tab of the **Ribbon**. The **Axis HUD** toolbar will be displayed; refer to Figure-24 and you will be asked to select reference for creating axis.

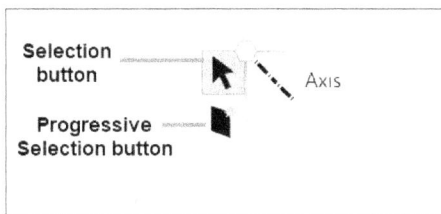

Figure-24. Axis HUD toolbar

- Select a cylindrical face/edge to create axis passing through its center line; refer to Figure-25.

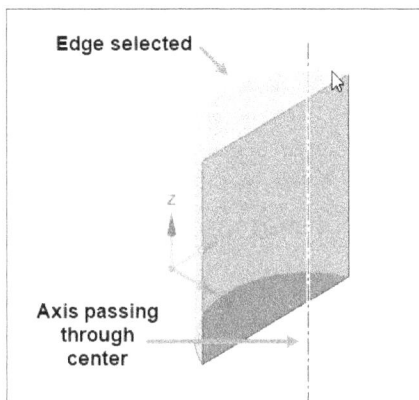

Figure-25. Axis passing through centerline

- Select an edge or axis to create axis passing through it.

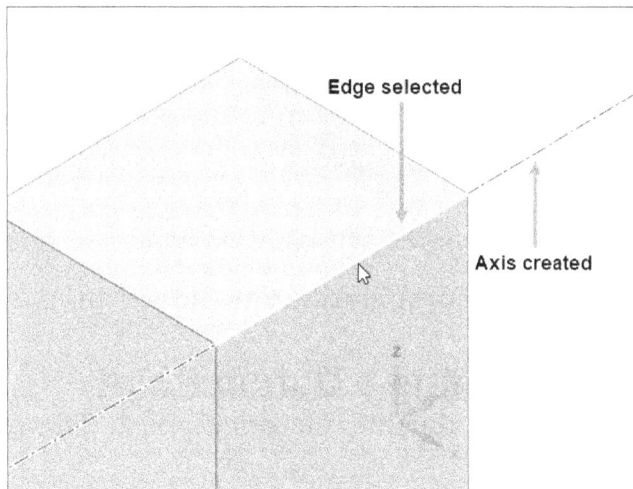

Figure-26. Axis passing through edge

- Select the **Progressive Selection** button from the **HUD** toolbar to select multiple reference objects for creating axis. Select two points from the model to create an axis passing through selected points. Preview of axis will be displayed; refer to Figure-27. Select desired axis preview to create the axis.

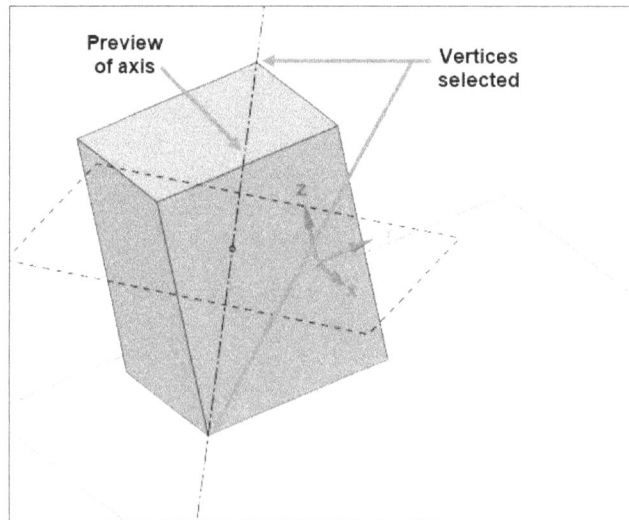

Figure-27. Preview of axis passing through points

- Select two intersecting faces to create an axis at the intersection of those faces; refer to Figure-28.

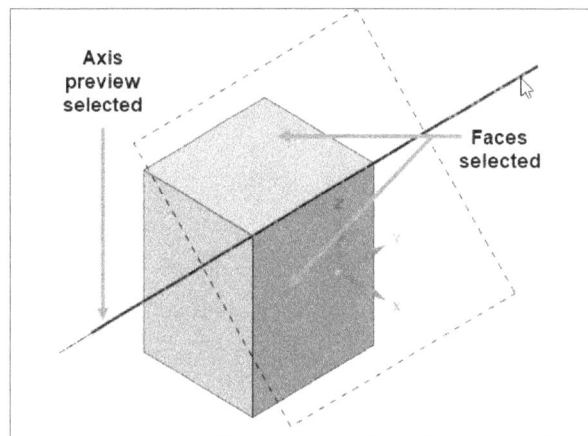

Figure-28. Creating axis at intersection of two faces

- Similarly, you can use other combinations to create an axis.

Creating a Datum Point

The **POINT** tool is used to create datum reference point. The procedure to use this tool is given next.

- Click on the **POINT** tool from the **CREATE** panel in the **DESIGN** tab of the **Ribbon**. The **Point HUD** toolbar will be displayed.
- Select desired vertex or location on the model to place datum point.

Placing Origin

The **ORIGIN** tool is used to place a user coordinate system at selected reference. The coordinate system is used to define directions of axes and zero coordinate point in the graphics area when creating/placing feature. The procedure to use this tool is given next.

- Click on the **ORIGIN** tool from the **CREATE** panel in the **DESIGN** tab of the **Ribbon**. You will be asked to select location for placing coordinate system.
- Click at desired face to create coordinate system with Z axis direction normal to selected face. Click on an axis/line/edge to align Z axis of coordinate system with selected axis/line/edge when placing origin; refer to Figure-29.

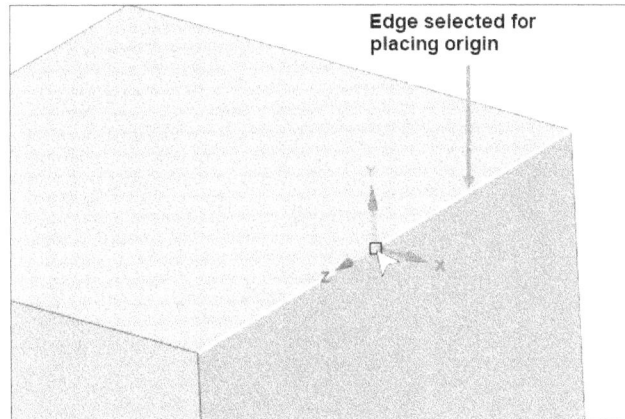

Edge selected for placing origin

Figure-29. Edge selected for origin creation

CREATING LINEAR PATTERN

The **LINEAR PATTERN** tool is used to create multiple copies of selected object along X and Y axes. The procedure to use this tool is given next.

- Select the object(s) you want to pattern and click on the **LINEAR PATTERN** tool from the **CREATE** panel in the **DESIGN** tab of the **Ribbon**. The **LINEAR PATTERN HUD** toolbar will be displayed; refer to Figure-30.

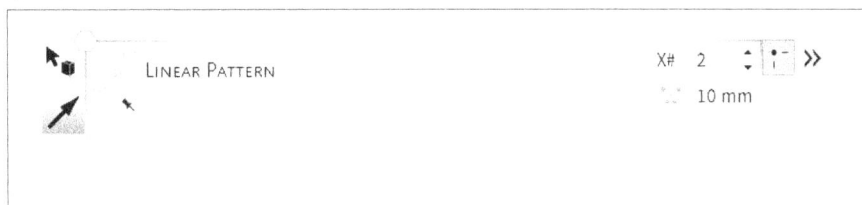

Figure-30. LINEAR PATTERN HUD toolbar

- Click on the **>>** button in the right **HUD** toolbar to expand the toolbar. **One-dimensional** and **Two-dimensional** options will be displayed in the flyout.
- Select the **Two dimensional** option from the flyout if you want to create instances of selected object in Y direction as well apart from X direction; refer to Figure-31.

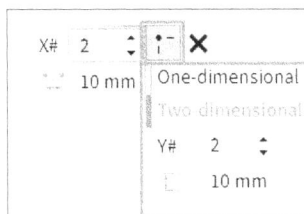

Figure-31. Two-dimensional pattern option

- Set desired values in the **X#** and **Y#** spinners to define the number of instances to be created along X and Y axes, respectively.
- Similarly, specify distance between two instances along X and Y axes in respective distance edit boxes in the **HUD** toolbar.

- After specifying instance parameters, select a linear edge/curve/axis to define direction for X axis; refer to Figure-32. Preview of pattern will be displayed on selecting direction reference. Select desired handle from X and Y direction handles displayed on the body to define + directions for the axes.

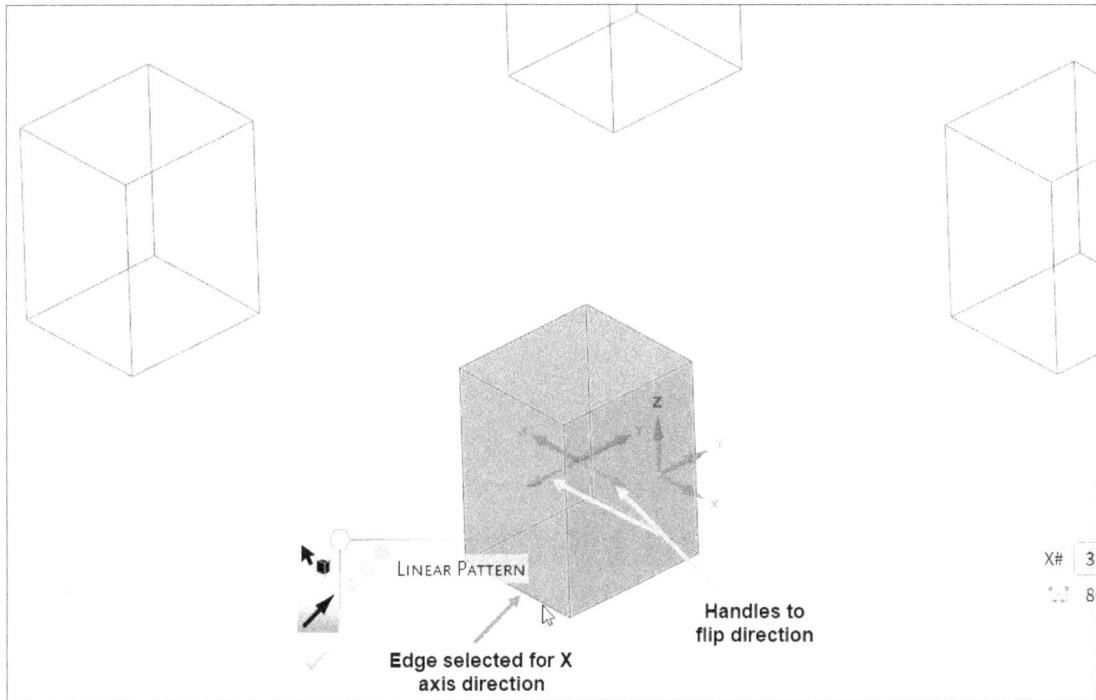

Figure-32. Defining direction for pattern

- After setting desired parameters, click on the **OK** button from the **HUD** toolbar to create the pattern. Press **ESC** to exit the tool.

CREATING CIRCULAR PATTERN

The **CIRCULAR PATTERN** tool is used to create instances of selected object around selected center axis/point/object. The procedure to use this tool is given next.

- Select the object to be patterned and click on the **CIRCULAR PATTERN** tool from the **CREATE** panel in the **DESIGN** tab of the **Ribbon**. The **CIRCULAR PATTERN HUD** toolbars will be displayed.
- Specify desired value in the **Circular #** spinner to define number of instances in circular order. Specify desired value in the **Angle** edit box to define angular span within which pattern will be created.
- Expand the **HUD** toolbar and select the **Two-dimensional** option from the expanded toolbar if you want to create instances in radial direction. The **Linear #** spinner and **Distance** edit boxes will be displayed for defining number of instances and distance between two instances, respectively.
- Set desired parameters as discussed earlier and select the axis. Preview of circular pattern will be displayed; refer to Figure-33.
- Click on the **OK** button from the **HUD** toolbar. The pattern will be created.

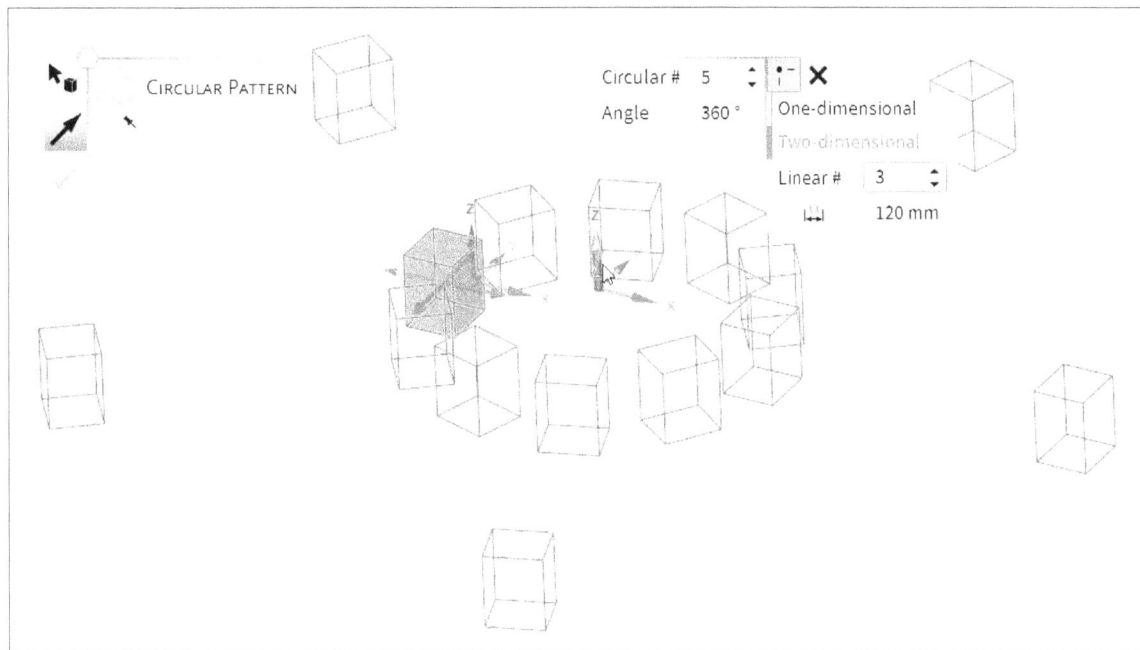

Figure-33. Preview of circular pattern

CREATING FILL PATTERN

The **FILL PATTERN** tool is used to create multiple instances of selected body within specified boundary. The procedure to use this tool is given next.

- Select the object to be patterned and click on the **FILL PATTERN** tool from the **CREATE** panel in the **DESIGN** tab of the **Ribbon**. The **FILL PATTERN HUD** toolbar will be displayed.
- Select the feature that you want to pattern and then select a linear edge/axis/curve to define direction of X axis for the pattern. Preview of pattern will be displayed; refer to Figure-34.

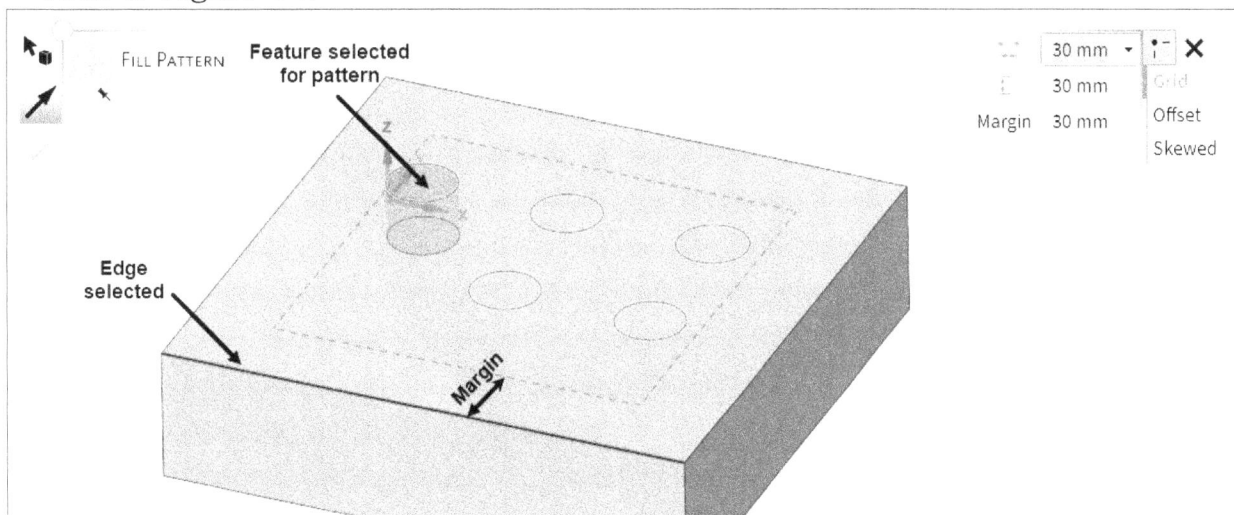

Figure-34. Preview of fill pattern

- Select the **Grid** option from expanded **HUD** toolbar to create pattern in the form of 2D grid along X and Y axes with instances created at intersections of grid lines. Select the **Offset** option if you want to create instances of selected object at specified distance along X and Y axes. You also need to specify the distance from the edge of base feature for creating imaginary pattern fill region. Set desired parameters in the edit boxes to create the pattern. Similarly, select the **Skewed** option from the **HUD** toolbar to create instances in grid pattern with specified offset values; refer to Figure-35.

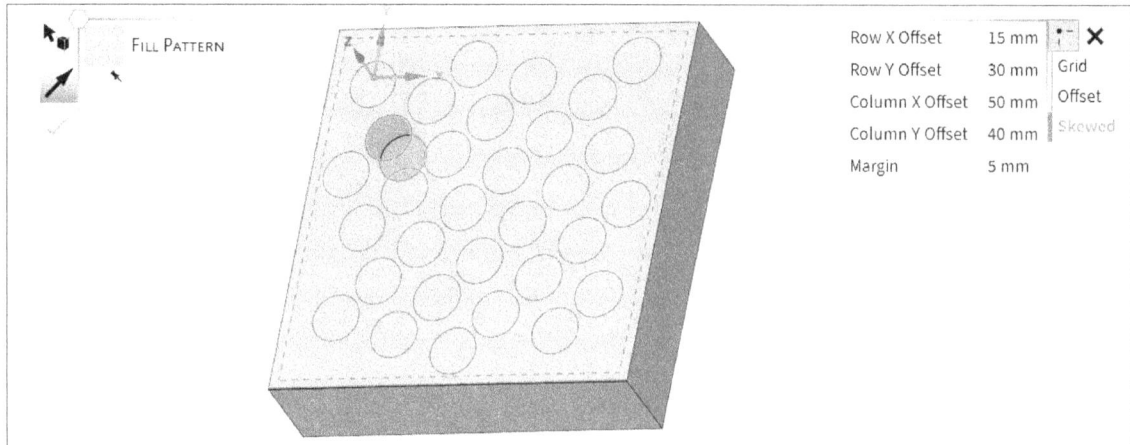

Figure-35. Skewed pattern

- After setting desired parameters, click on the **OK** button to create the feature.

CREATING SHELL FEATURE

The **Shell** tool is used to scoop out material from solid features and create thin shell of specified thickness. The procedure to use this tool is given next.

- Click on the **Shell** tool from the **CREATE** panel in the **DESIGN** tab of the **Ribbon**. The **SHELL HUD** toolbar will be displayed; refer to Figure-36.

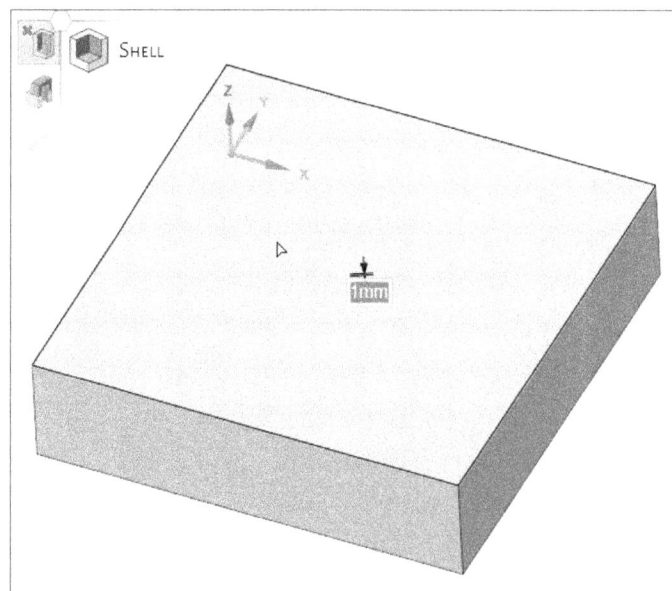

Figure-36. Shell HUD Toolbar

- Specify desired thickness value for shell feature in the input box displayed on hovering cursor at a face and press **ENTER**.
- Click on the face to be removed after applying shell feature. The shell feature preview will be displayed; refer to Figure-37.

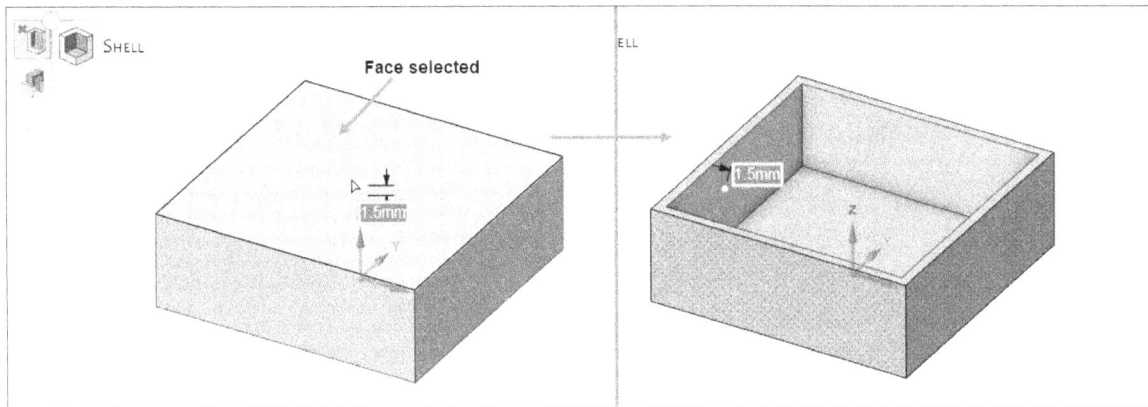

Figure-37. Creating shell feature

- Click on the **OK** button from the **HUD** toolbar to create the feature and press **ESC** to exit the tool.

APPLYING OFFSET RELATION

The **Offset** tool is used to make two selected faces bound by offset relation so that when one face is displaced/modified, the other face also gets displaced/modified by equal value. The procedure to use this tool is given next.

- Click on the **Offset** tool from the **CREATE** panel in the **DESIGN** tab of the **Ribbon**. The **OFFSET HUD** toolbar will be displayed.
- Select two faces to be bound by offset relation. The relation will be applied and selected faces will be displayed with dots; refer to Figure-38.

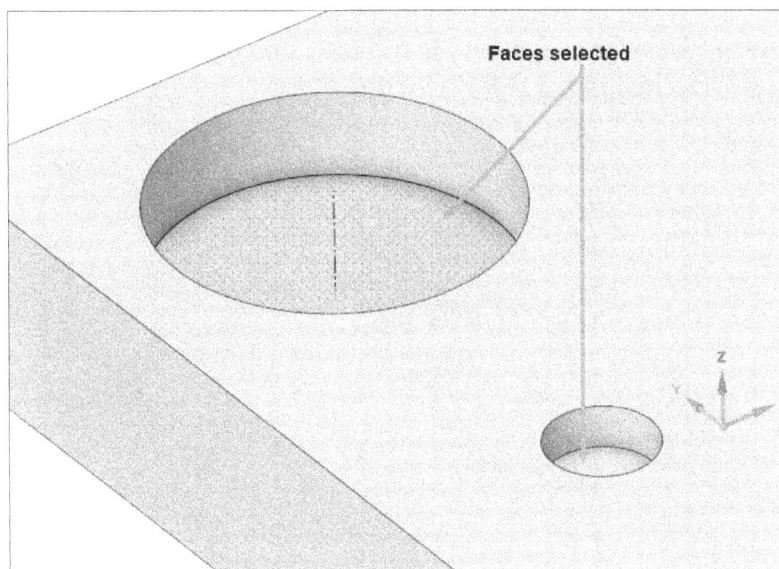

Figure-38. Faces applied with offset relation

- If you change depth of one face then depth of other face will automatically change.

CREATING MIRROR COPY

The **Mirror** tool is used to create mirror copy of selected objects with respect to selected mirror plane. The procedure to use this tool is given next.

- Click on the **Mirror** tool from the **CREATE** panel in the **DESIGN** tab of the **Ribbon**. The **Mirror HUD** toolbar will be displayed and you will be asked to select a plane to be used as reference for mirroring.
- Select desired face/plane to be used as mirror plane. You will be asked to select the object to be mirrored.
- Select the **Solid(s)/Surface(s)** selection button [icon] to create mirror copy of selected solid/surface bodies, select the **Face(s)** selection button [icon] to mirror copy selected faces and select the **Component(s)** selection button [icon] to create mirror copy of selected features like holes, boss features, etc. After selecting desired selection button, hover the cursor on the object to be mirror copied. Preview of feature will be displayed; refer to Figure-39. Click on the object to create mirror copy.

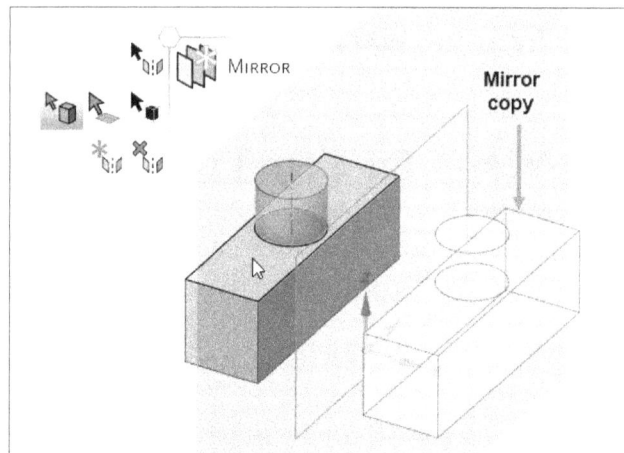

Figure-39. Preview of Mirror copy

- Select the **Create Mirror Relationship** button from the right **HUD** toolbar to apply mirror relationship between two faces. You will be asked to select two faces. Select the first face and second face; refer to Figure-40. The relationship will be applied.

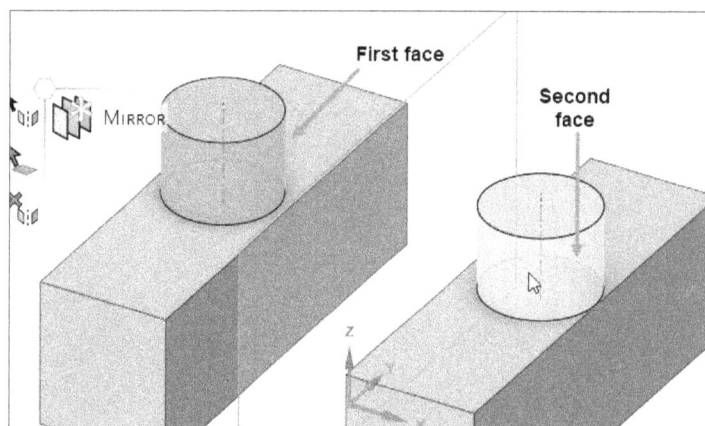

Figure-40. Selecting faces to apply mirror relationship

- If you want to remove mirror relationship then select the **Remove mirror relationship** button from the **HUD** toolbar and select the faces.
- Press **ESC** to exit the tool.

CREATING BODIES BY EQUATIONS

The **Equation** tool is used to create custom surface by specified parameters. The procedure to use this tool is given next.

- Click on the **Equation** tool from the **BODY** panel in the **DESIGN** tab of the **Ribbon**. The **Equation HUD** toolbar will be displayed; refer to Figure-41.

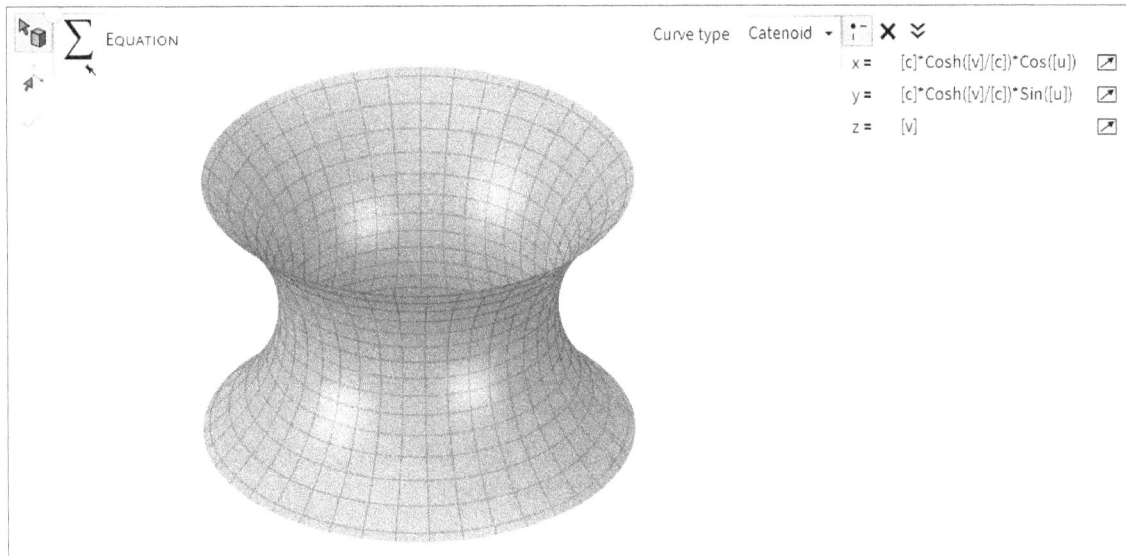

Figure-41. Equation HUD toolbars

- Select desired option from the **Curve type** drop-down to define the type of curve to be used for generating the surface. The equations will be automatically generated based on selected option; refer to Figure-42. Set desired parameters to modify the shape and size of created surface.

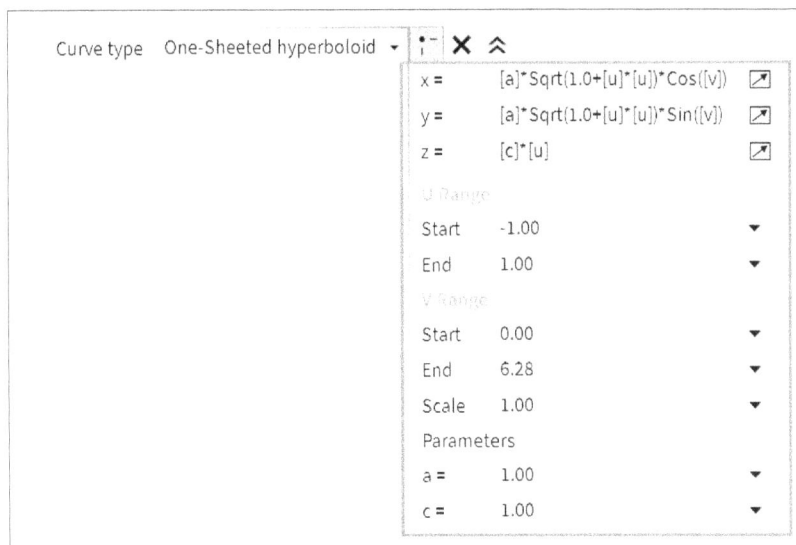

Figure-42. Equations and related parameters

- Select the **Custom** option from the **Curve type** drop-down to create parametric user defined equations for generating curves and surfaces. For example, a circular disc has equation:

$$x(u, v) = u * cos(v)$$
$$y(u, v) = u * sin(v)$$
$$z(u, v) = 0$$
where,
u: 0 to 5
v: 0 to 2*pi

Refer to Figure-43 for creating this surface in Ansys.

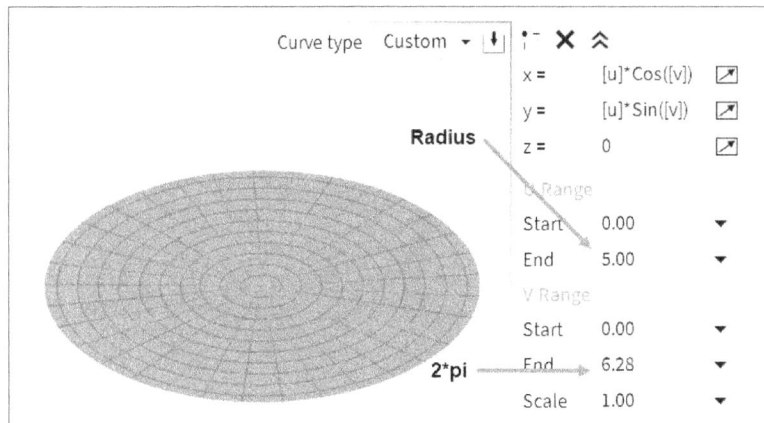

Figure-43. Creating disc using equation

- After setting desired parameters, click on the **OK** button from the **HUD** toolbar to create the feature and press **ESC** to exit the tool.

CREATING CYLINDRICAL BODY

The **Cylinder** tool is used to create cylindrical solid body. The procedure to use this tool is given next.

- Click on the **Cylinder** tool from the **BODY** panel in the **DESIGN** tab of the **Ribbon**. You will be asked to specify start point of vertical axis of cylinder.
- Click at desired location in the graphics area to specify start point of the axis. You will be asked to specify the end point of axis.
- Click at desired location to define height of the cylinder. You will be asked to specify diameter of the cylinder; refer to Figure-44.

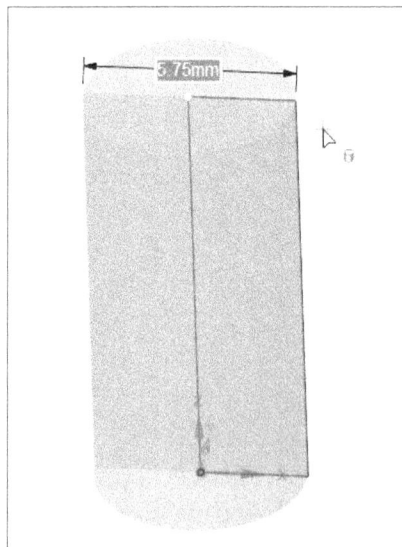

Figure-44. Creating cylindrical body

- Click at desired location to define diameter or enter desired value. The cylinder will be created.

Similarly, you can use the **Sphere** tool in **BODY** panel of **DESIGN** tab in the **Ribbon** to create the sphere feature.

ASSEMBLY CONSTRAINTS

The tools in the **ASSEMBLE** panel of **DESIGN** tab in the **Ribbon** are used to apply various constraints to bind two or more bodies in the form of a functional assembly. For example, car is an assembly of various components like tires, brakes, frame, engine, fuel system, and so on. The tools in **Assemble** panel are discussed next.

Applying Tangent Constraint

The **Tangent** tool is used to make two selected faces tangent. The procedure to use this tool is given next.

* Click on the **Tangent** tool from the **ASSEMBLE** panel in the **DESIGN** tab of the **Ribbon**. The **TANGENT HUD** toolbar will be displayed and you will be asked to select the geometries.
* Select the two faces to which tangent constraint is to be applied; refer to Figure-45.

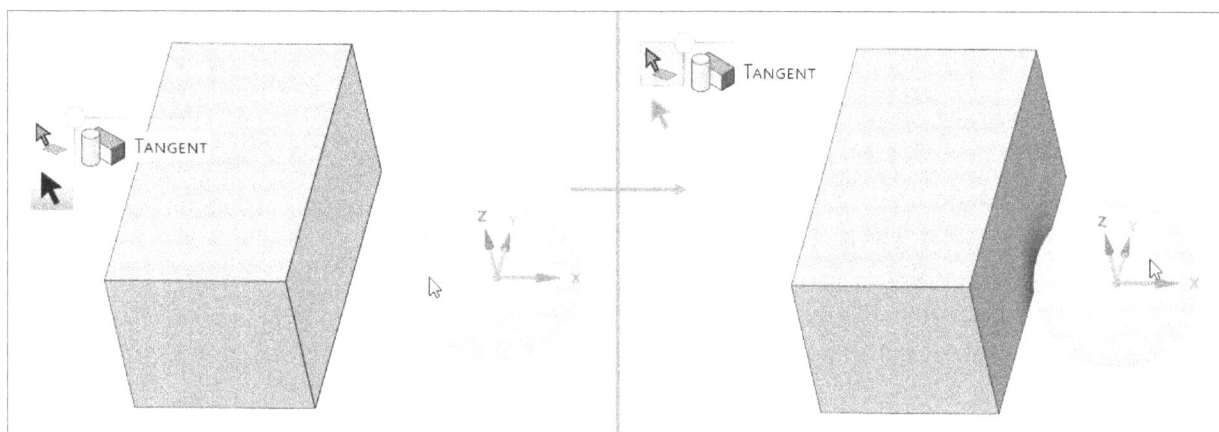

Figure-45. Applying tangent constraint

* Select next pairs to apply tangent constraint and press **ESC** to exit the tool.

Applying Align Constraint

The **Align** tool is used to make selected entities aligned in same plane/axis. The procedure to use this tool is given next.

* Click on the **Align** tool from the **ASSEMBLE** panel in the **DESIGN** tab of the **Ribbon**. The **ALIGN HUD** toolbar will be displayed and you will be asked to select the faces.
* Select two faces to be aligned. The constraint will be applied; refer to Figure-46.

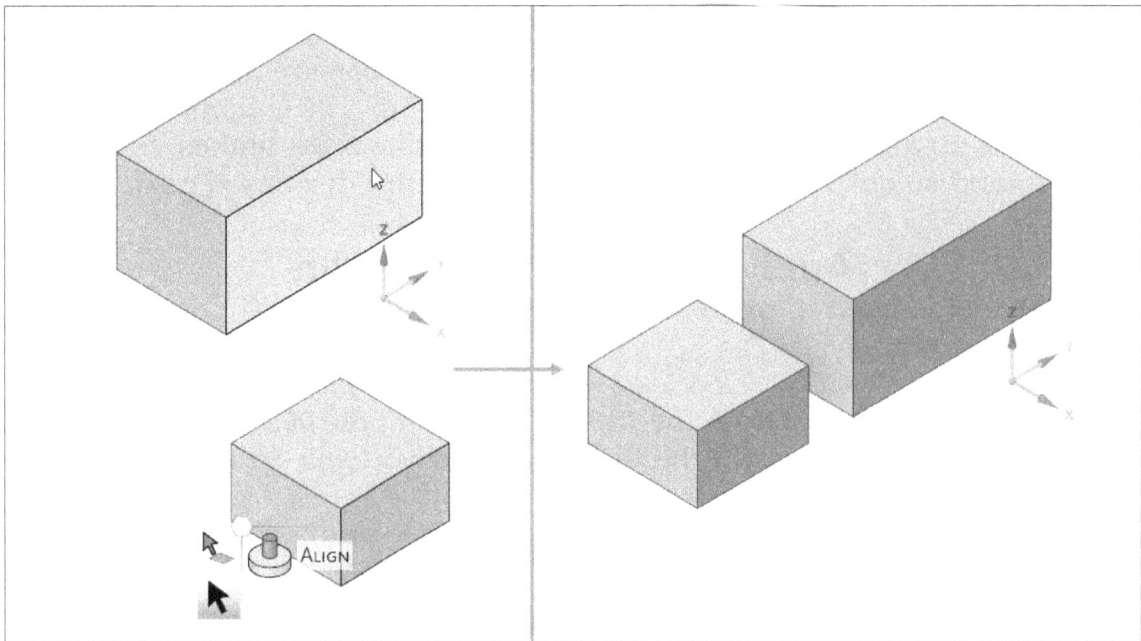

Figure-46. Aligning faces

Applying Orient Constraint

The **Orient** tool is used to make two selected faces oriented in same direction/orientation. The procedure to use this tool is given next.

- Click on the **Orient** tool from the **ASSEMBLE** panel in the **DESIGN** tab of the **Ribbon**. The **ORIENT HUD** toolbar will be displayed and you will be asked to select faces to be oriented.
- Select two faces to be oriented. The constraint will be applied; refer to Figure-47.

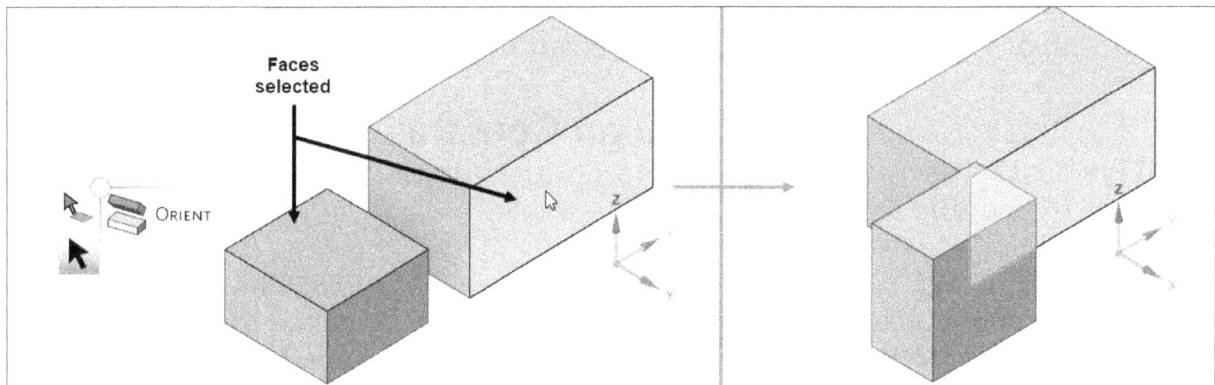

Figure-47. Applying orient constraint

Applying Rigid Constraint

The **Rigid** tool is used to keep orientation of selected faces of two different components fixed with respect to each other. The procedure to use this tool is given next.

- Select the faces of components to which you want to apply the constraint and click on the **Rigid** tool from the **ASSEMBLE** panel in the **DESIGN** tab of the **Ribbon**; refer to Figure-48. The constraint will be applied.

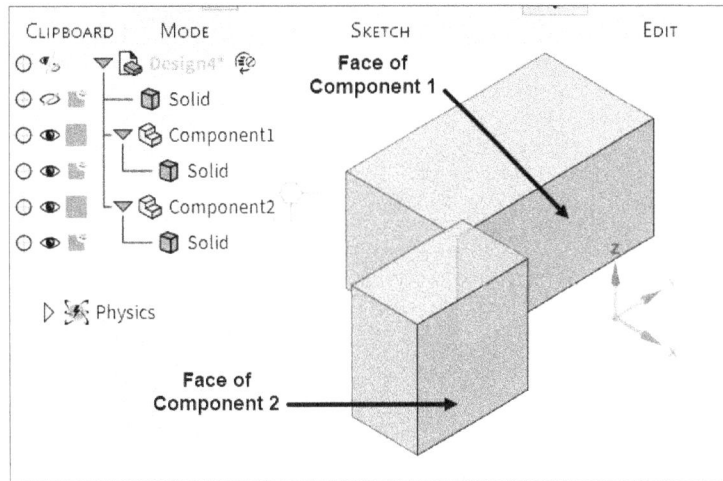

Figure-48. Faces selected for Rigid constraint

Note: If you want to convert a body into a component then right-click on the body from the **Design Tree** and select **Move to New Component** option from the shortcut menu displayed; refer to Figure-49.

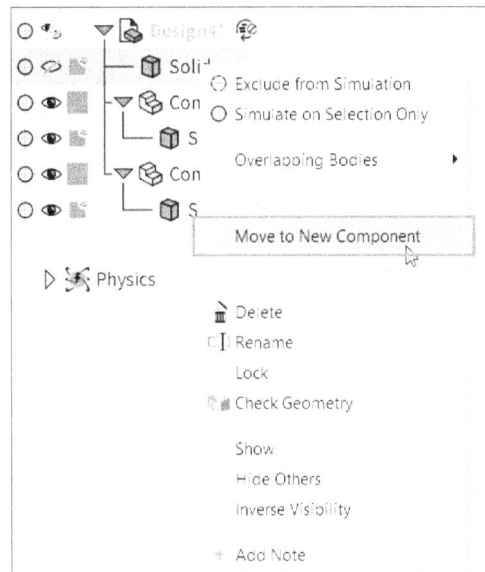

Figure-49. Move to New Component option

Applying Gear Constraint

The **Gear** tool is used to apply no slippage constraint between cylindrical faces of two components. The procedure to use this tool is given next.

• Select cylindrical faces of two components between which you want to apply the gear constraint and click on the **Gear** tool from the **ASSEMBLE** panel in the **DESIGN** tab of the **Ribbon**. The constraint will be applied; refer to Figure-50.

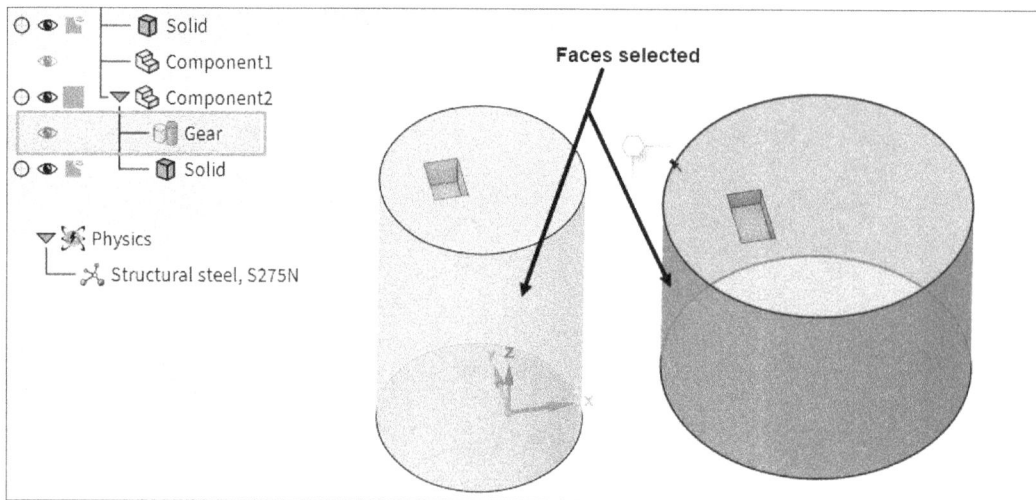

Figure-50. Applying gear relation

Applying Anchor Constraint

Anchor constraint is similar to fixed constraint used in most of the CAD software. This constraint makes selected component fixed in orientation as well as position. To use this tool, select a face of the component to be fixed and click on the **Anchor** tool from the **ASSEMBLE** panel in the **DESIGN** tab of the **Ribbon**. This constraint is generally applied to first component in assembly so that it can be used as reference for placing other components in the assembly.

Note that most of the time, you will import the models of other CAD software and use them to create simulation studies. So, most the design tools of this software will be utilized to modify/simplify the model to perform analysis.

DISPLAY TOOLS

The tools in **DISPLAY** tab of **Ribbon** are used to modify graphics related parameters. You can change colors of different objects in graphics area, you can change shading of objects, you can modify view style, and so on. Various tools and options in this tab are discussed next.

Spinning Operation

The options in **Spin** drop-down of **ORIENT** panel in the **DISPLAY** tab of **Ribbon** are used to modify orientation of model by rotating the graphical view about cursor location or center of gravity point. The options are discussed next.

• Select the **On Center** option from the **Spin** drop-down if you want to spin the model about center of gravity of all objects combined in graphics area. Select the **On Cursor** option from the drop-down if you want to spin the model about current cursor position. On selecting the option, the cursor will change to ☐ to represent that **Spin** tool is active.
• Click at desired location in graphics area and drag the cursor without releasing the mouse button to spin the model.

You can also perform spinning by pressing and holding the middle button of mouse (MMB) while dragging the cursor.

Panning Operation

The **Pan** tool is used to move the view of models in graphics area. Note that moving views is not same as moving objects because in this case, all the objects including references will be moved in the graphics area without relative movements between the objects. The procedure to use **Pan** tool is given next.

*   Click on the **Pan** tool from the **ORIENT** panel in the **DISPLAY** tab of the **Ribbon**. The cursor will change shape to panning cursor.
*   Click at desired location in the graphics area and drag the cursor without releasing left mouse button in desired direction to perform panning operation.

You can also perform panning by holding the **CTRL+MMB** (middle button of mouse) while dragging the cursor.

Zooming Operation

The tools in the **Zoom** drop-down are used to perform different zooming operations like zooming to display all objects in graphics area, zooming in specified region, and so on. The options of this drop-down are discussed next.

*   Click on the **Zoom Extents** tool from the **Zoom** drop-down in the **ORIENT** panel of **DISPLAY** tab in the **Ribbon** if you want to fit all the objects of graphics area in current view. This tool is generally used when objects are far out from current view.
*   Click on the **Zoom Box In** tool from the **Zoom** drop-down to zoom into a specified box region of the model. After selecting this tool, draw a box around the object/ section in the graphics area to zoom in. Enlarged view of object inside box will be displayed; refer to Figure-51.

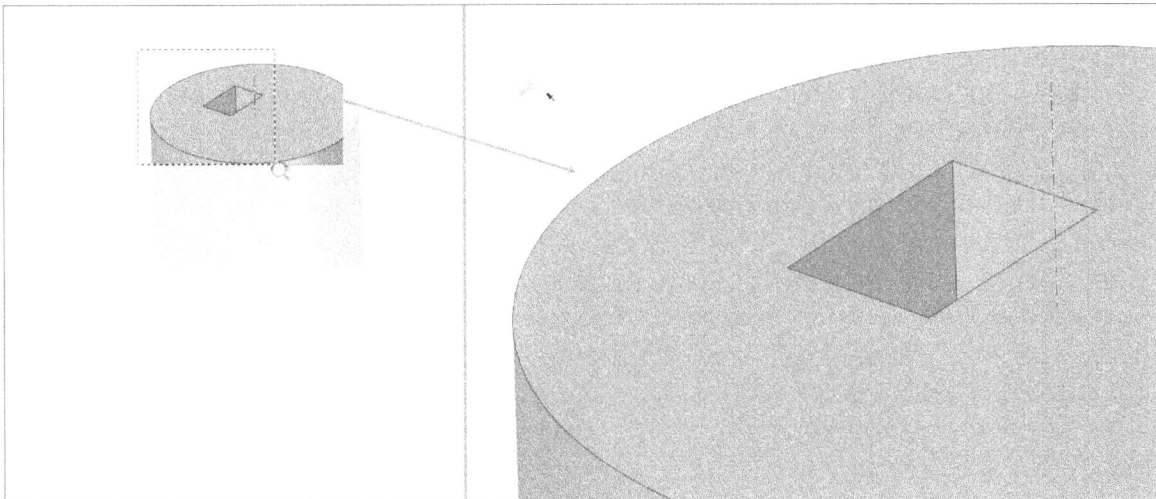

Figure-51. Zoom box in

*   Click on the **Zoom In** tool to enlarge the view of current object in visible graphics area by 50%.
*   Click on the **Zoom Out** tool to diminish the view of current object in visible graphics area by 50%.

You can also perform zoom operations by mouse and hot keys. Scroll the mouse wheel up to zoom out (diminish the view) and scroll the mouse wheel down to zoom in (enlarge the view). You can also use **CTRL**+ **+** key to zoom in and **CTRL** + **-** key to zoom out.

Hold the **SHIFT**+**MMB** (Middle button of mouse) and drag the cursor up to zoom out and down to zoom in.

Changing Colors

The options in the **Color** drop-down are used to change colors of selected faces and bodies; refer to Figure-52. The procedure to use the options in this drop-down is given next.

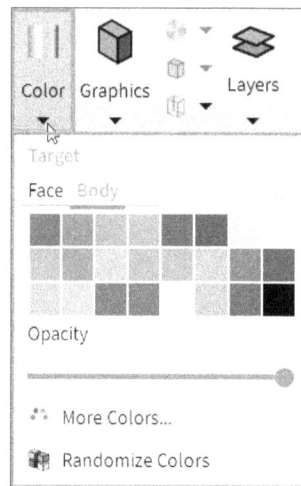

Figure-52. Color drop-down

- Select the face/body from graphics area to which you want to apply the color. Click on the **Color** option in **STYLE** panel of **DISPLAY** tab in the **Ribbon** to display drop-down.
- Select the **Face** tab in the **Target** section of drop-down to apply color to selected face only or select the **Body** tab in the **Target** section of drop-down to apply color to whole body connected with selected face.
- After selecting tab, click on desired color from the color palette to apply respective color. If the color of your choice is not available in the palette then click on the **More Colors** button from the drop-down. The **Select a Color** dialog box will be displayed; refer to Figure-53. Select desired color from the wheel or use the sliders at the right in the dialog box to define color. You can also use **R**, **G**, **B**, or **H**, **S**, **B** edit boxes to define color values. Use the **A** slider/edit box to define transparency value for color. After setting desired values or selecting the color, click on the **OK** button from the dialog box. The defined color will be applied to face/body.

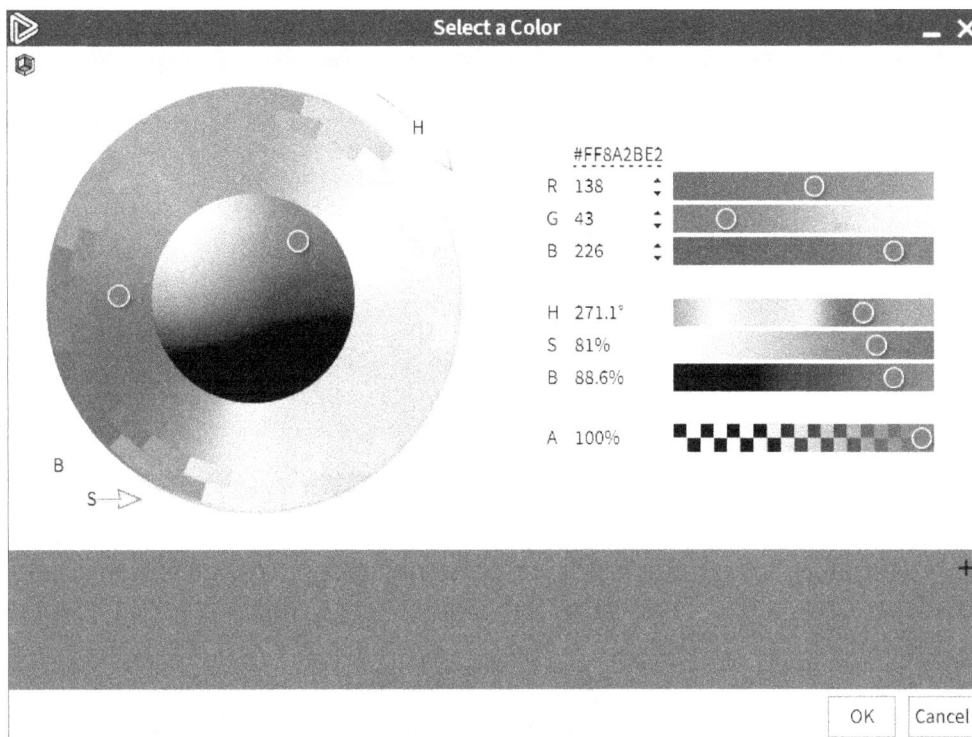

Figure-53. Select a Color dialog box

- Move the **Opacity** slider to make selected face transparent or opaque depending on slider position.
- If you have multiple objects in the graphics area and want to apply different colors to them so that you can easily differentiate then select the objects using window selection and click on the **Randomize Colors** option from the bottom in the **Color** drop-down. System will automatically apply different colors/shades of colors to the objects.

Changing Graphics Style

The options in the **Graphics** drop-down of **STYLE** panel in **DISPLAY** tab of **Ribbon** are used to change the graphics style of objects in the graphics area; refer to Figure-54. Select **Shaded**, **Enhanced Shaded**, **Wireframe**, **Hidden Line**, or **Hidden Line Removed** option from the drop-down to display objects in respective graphics style.

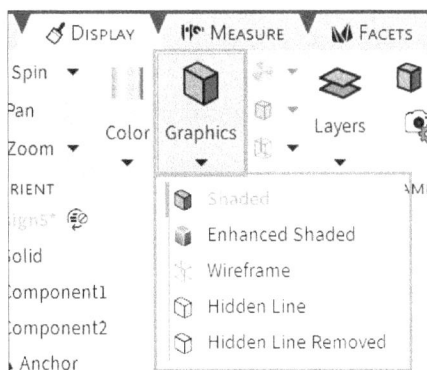

Figure-54. Graphics drop-down

Changing Rendering Styles

The options in the **RENDERING STYLES** drop-down are used to define texture of objects displayed in graphics area; refer to Figure-55. This drop-down is active only after selecting an object from the graphics area. Select desired option from the drop-down to display objects in respective style. For example, select the **Metallic** option from drop-down if you want to display metallic shine in selected objects.

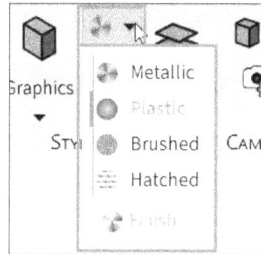

Figure-55. RENDERING STYLE drop-down

STYLE OVERRIDE

The options in the **STYLE OVERRIDE** drop-down are used to define whether selected objects are transparent (see-through) or opaque; refer to Figure-56. Select desired option from drop-down after selecting the object to perform related operation.

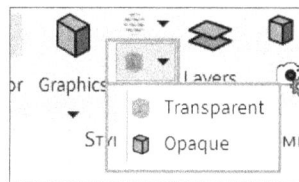

Figure-56. STYLE OVERRIDE drop-down

Edge Display Settings

The options in the **EDGES** drop-down are used to define the types of edges that can be displayed in graphics area; refer to Figure-57. Select/de-select toggle buttons in this drop-down to show/hide respective edges. For example, de-select the **Solid** option if you do not want to show edges on solid bodies in graphics area.

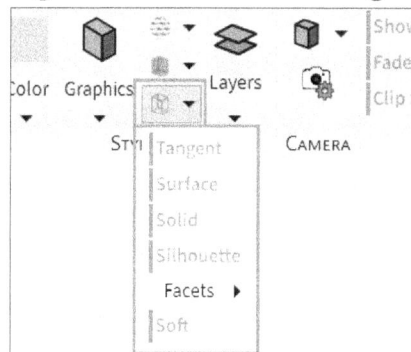

Figure-57. EDGES drop-down

Layer Settings

Layers are used to categorize objects in graphics area based on some common criteria. For example, you want to keep all the surfaces in one category and all the solids in another category so that later you can show/hide them depending on your requirement. In such cases, you can create different layers and then assign objects to them. The procedure to create and manage layers is given next.

• Click on the **Layers** tool from the **STYLE** panel in the **DISPLAY** tab of the **Ribbon**. The **Layers** dialog box will be displayed; refer to Figure-58.

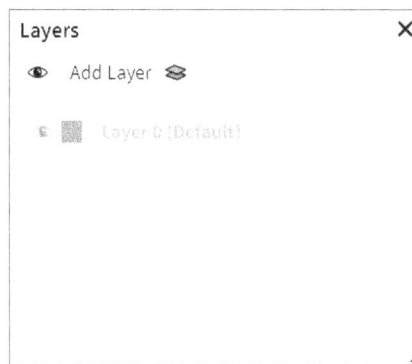

Figure-58. Layers dialog box

- Click on the **Add Layer** option in the dialog box to create a new layer to which you can assign objects. A new layer will be added in the list.
- To assign an object to this new layer, select the object from graphics area and right-click on the new layer created in the **Layer** dialog box. A shortcut menu will be displayed; refer to Figure-59. Select the **Assign to Layer** option from the shortcut menu to assign object to the layer. Now, properties of object like color will be decided by layer properties.

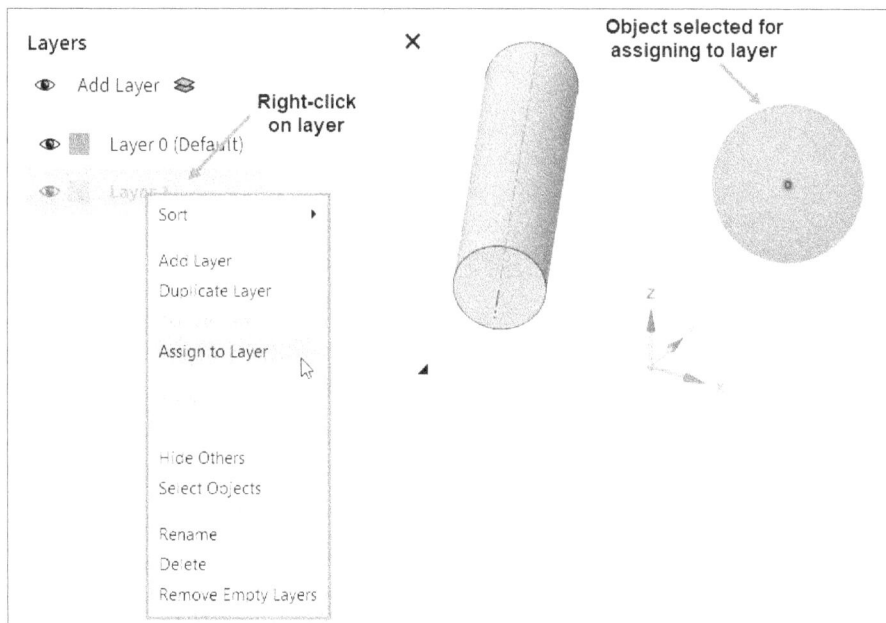

Figure-59. Shortcut menu for layers

- By default, the newly created layer becomes active layer for creating new objects. If you want to use any previous layer as active layer then right-click on that layer and select the **Activate Layer** option from the shortcut menu.
- You can use the other options of shortcut menu as discussed earlier in the book.

Camera Settings

The view of objects in viewport or graphics area is controlled by an imaginary camera through which we, as user, are looking at the objects. So, what you see through camera screen is displayed in your computer display. The options to define the camera settings to control the view are available in the **CAMERA** panel of **DISPLAY** tab in the **Ribbon**. The options of this panel are discussed next.

- Click on the **Orthographic** option from the **CAMERA** drop-down in the **CAMERA** panel of **Ribbon** if you want to display objects of graphics area in orthographic view. Orthographic view is also called orthographic projection as objects are projected perpendicular to view plane from line of sight. All the lines of sight are parallel to each other, resulting in an image where objects do not appear smaller as they move further away. Note that there is no sense of depth in orthographic projections.

- Click on the **Perspective** option from the **CAMERA** drop-down in the **CAMERA** panel of **Ribbon** if you want to display objects using perspective projection. This method aims to create the illusion of depth and distance by showing objects as smaller when they are farther away from the viewer. In a perspective view, objects are drawn with foreshortening, where objects closer to the viewer are larger and more detailed, while those farther to viewer appear smaller and less defined. This effect mimics how objects appear in reality, where depending on distance from viewer, objects to appear smaller or larger. Additionally, parallel lines in a perspective view converge at vanishing points, enhancing the sense of depth in the depiction. Refer to Figure-60 for orthographic vs perspective projection.

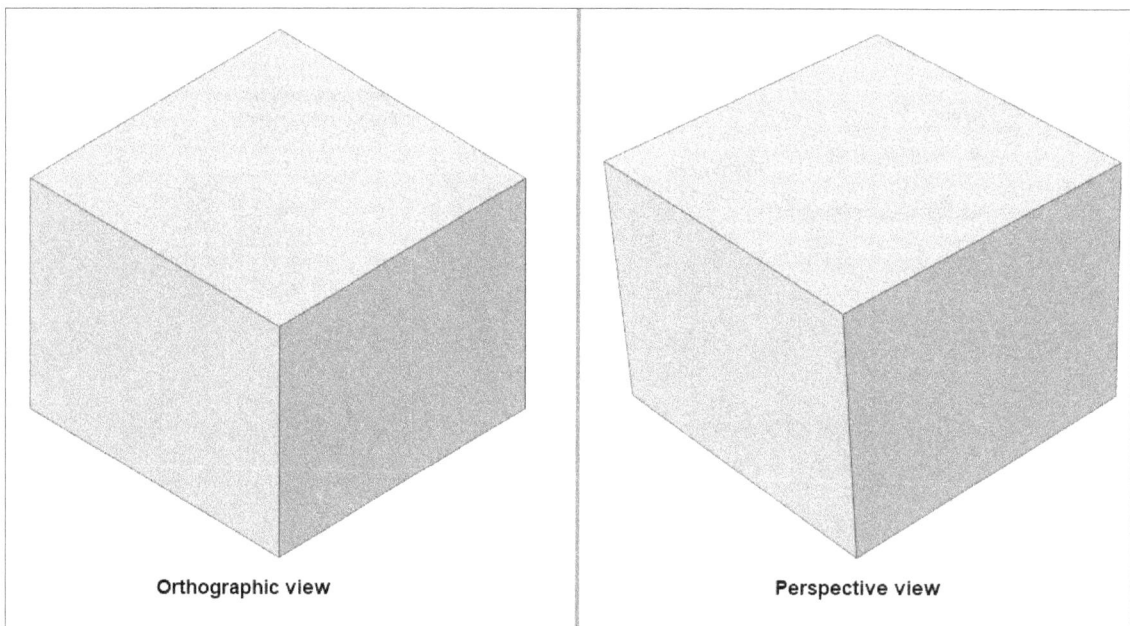

Orthographic view   Perspective view

Figure-60. Orthographic view vs perspective view

- Click on the **SHOW CAMERA OPTIONS** tool from the **CAMERA** panel in the **DISPLAY** tab of **Ribbon** to define size of camera film to be used for viewing objects in the graphics area. The **Width** and **Height** edit boxes will be displayed; refer to Figure-61. Specify desired value to define size of film. Note that in case of Orthographic view, you will be asked to specify film size in length and in case of Perspective view, you will be asked to specify film size in degrees using center of screen as POV (Point of View).

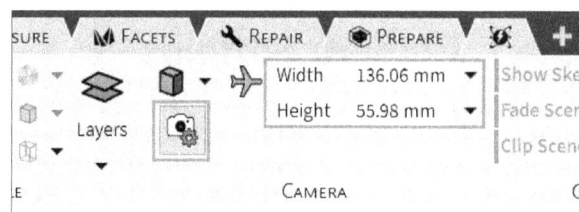

Figure-61. Width and height of camera film

- Click on the **FLY THROUGH** button or press **CTRL**+**SHIFT**+**F** to navigate in graphics area by scrolling as if you are flyout through the object. Click the tool again to exit.

Grid Settings

The options in the **GRID** panel are used to display and manage grids in sketching. These options are discussed next.

- Select the **Show Sketch Grid** toggle button to display grid when performing sketching. Clear this toggle button if you do not want to display grid.
- Select the **Fade Scene** toggle button if you want to make already existing objects faded when creating new sketch.
- Select the **Clip Scene Above Grid** toggle button to make objects above sketching grid plane trimmed.

Showing/Hiding Graphic Elements

The options in the **Show** drop-down of **DISPLAY** panel in the **DISPLAY** tab are used to show/hide various graphic elements of the model; refer to Figure-62. Select desired toggle options from the drop-down to display respective elements. For example, select the **Vertices** option if you want to display all vertices of the solids/surfaces in the graphics area.

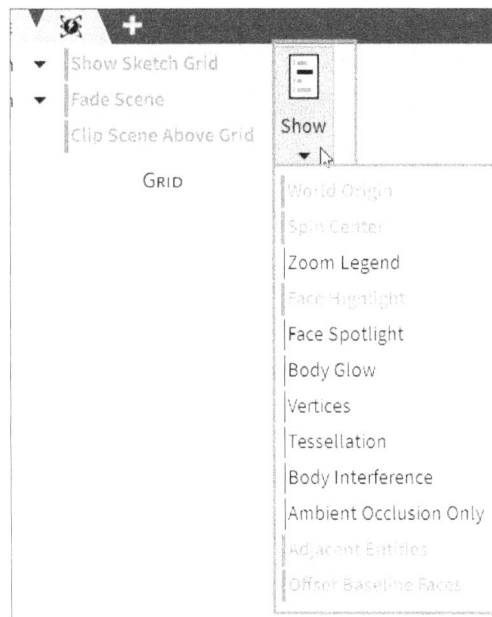

Figure-62. Show drop-down

MEASUREMENT TOOLS

The tools in the **MEASURE** tab of **Ribbon** are used to measure various parameters of model for further modifications or analysis; refer to Figure-63. These tools are discussed next.

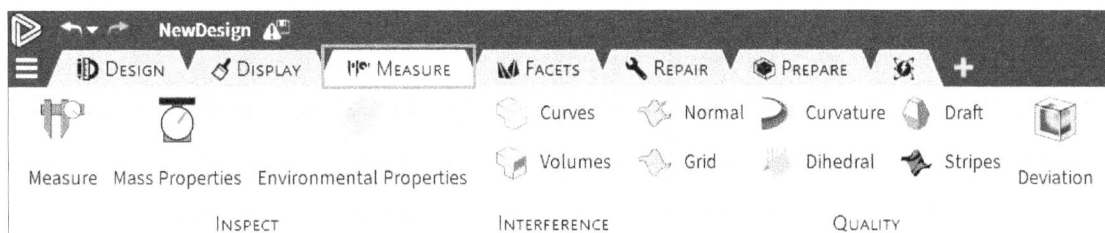

Figure-63. MEASURE tab

Performing Dimension Measurement

The **Measure** tool is used to measure dimensions of objects like length, angle, area, volume, distance, and so on. The procedure to use this tool is given next.

*   Click on the **Measure** tool from the **INSPECT** panel in the **MEASURE** tab of the **Ribbon**. The **MEASURE HUD** toolbar will be displayed; refer to Figure-64 and you will be asked to select objects to be measured.

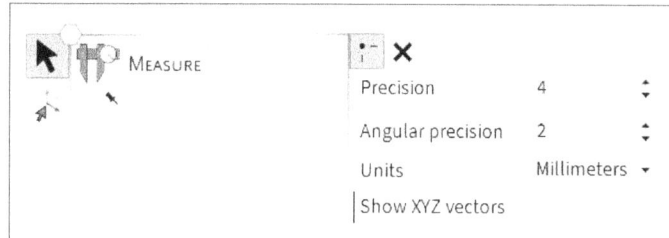

Figure-64. MEASURE HUD toolbar

*   Set desired values in the **Precision** and **Angular precision** spinners to define level of accuracy to be required in measurement of linear and angular dimensions, respectively.
*   Select desired option from the **Units** drop-down to define unit of measurement.
*   Select a flat face to measure its area and perimeter; refer to Figure-65.

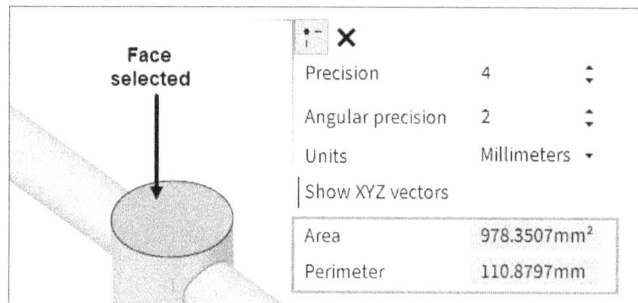

Figure-65. Measuring face parameters

*   Select circular edge of model to measure diameter and perimeter; refer to Figure-66.

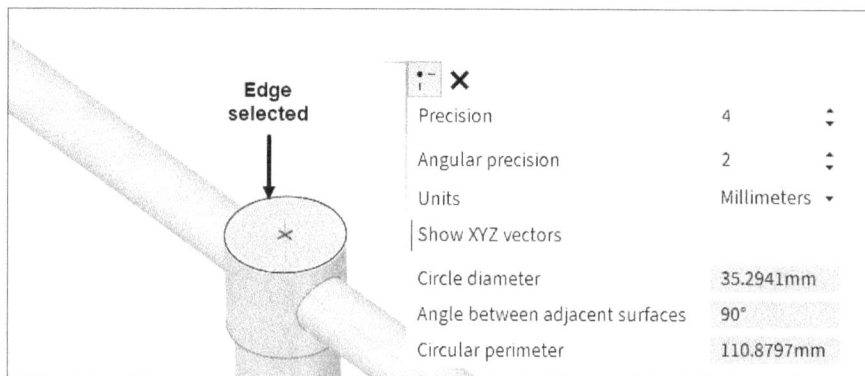

Figure-66. Measuring diameter of edge

*   Select a linear edge of model to measure its length; refer to Figure-67.

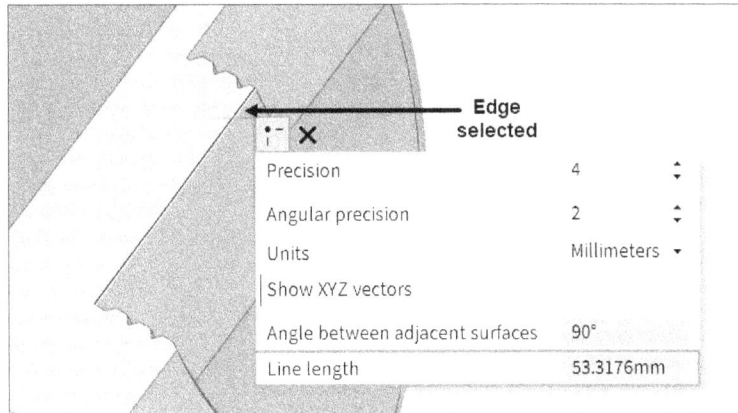

Figure-67. Measuring length of linear edge

- Select two vertices to measure distance between them; refer to Figure-68.

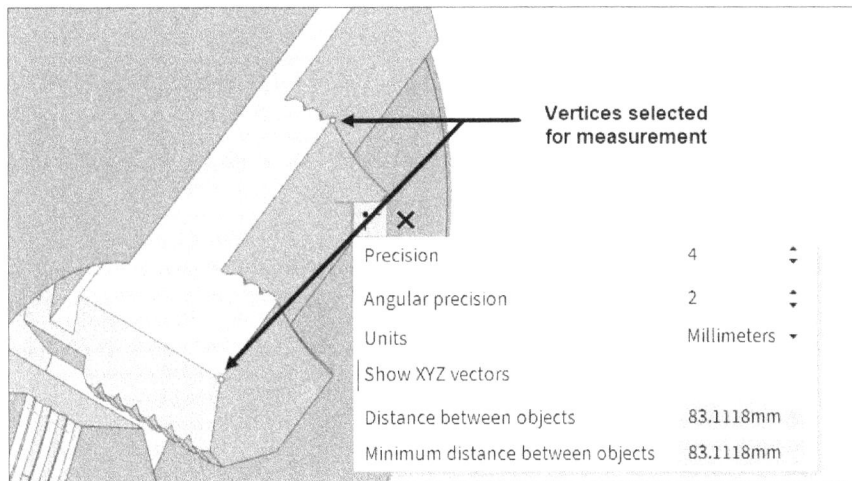

Figure-68. Measuring distance between vertices

- Select the **Show XYZ vectors** toggle button from the **HUD** toolbar if you want to display X, Y, and Z coordinates of vertices as well; refer to Figure-69.

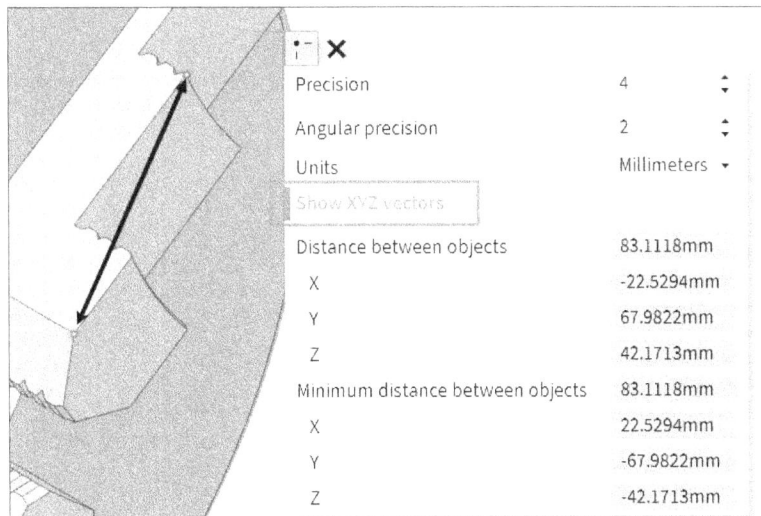

Figure-69. Showing XYZ vectors

- Similarly, you can measure distance and angles between two edges or faces; refer to Figure-70.

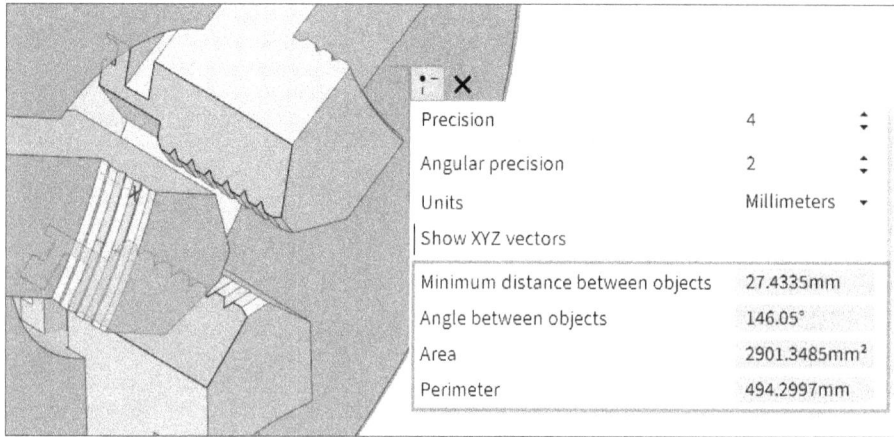

Figure-70. Measuring angle and distance between faces

- Press **ESC** to exit the tool.

Measuring Mass Properties

The **Mass Properties** tool is used to check mass properties of model like mass of selected component, volume, center of mass, and so on. The procedure to use this tool is given next.

- Click on the **Mass Properties** tool from the **INSPECT** panel in the **MEASURE** tab of the **Ribbon**. The **MASS PROPERTIES HUD** toolbar will be displayed; refer to Figure-71 and you will be asked to select the body whose mass is to be measured.

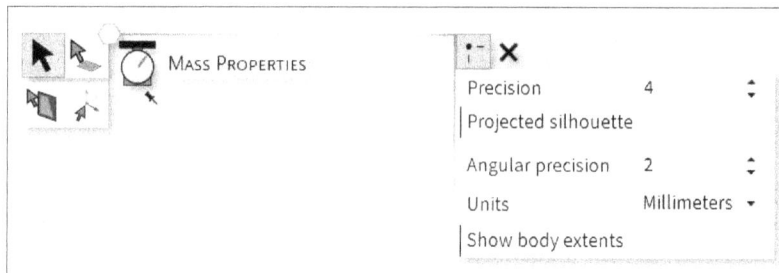

Figure-71. MASS PROPERTIES HUD toolbar

- Select the body whose properties are to be measured. The mass properties will be displayed; refer to Figure-72.

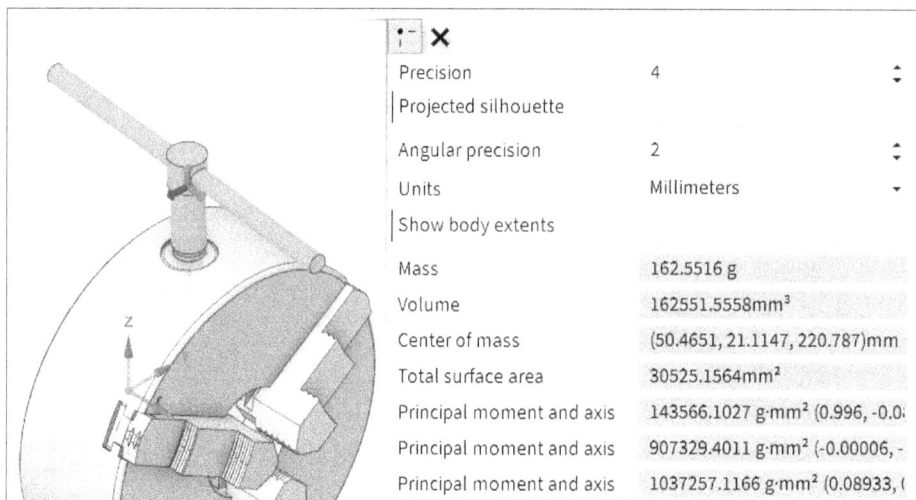

Figure-72. Checking mass properties

- Click on the **Face** selection button ⬚ from left **HUD** toolbar if you want to check area and cross-section properties of selected faces. After activating this selection, click on the face to check properties; refer to Figure-73.

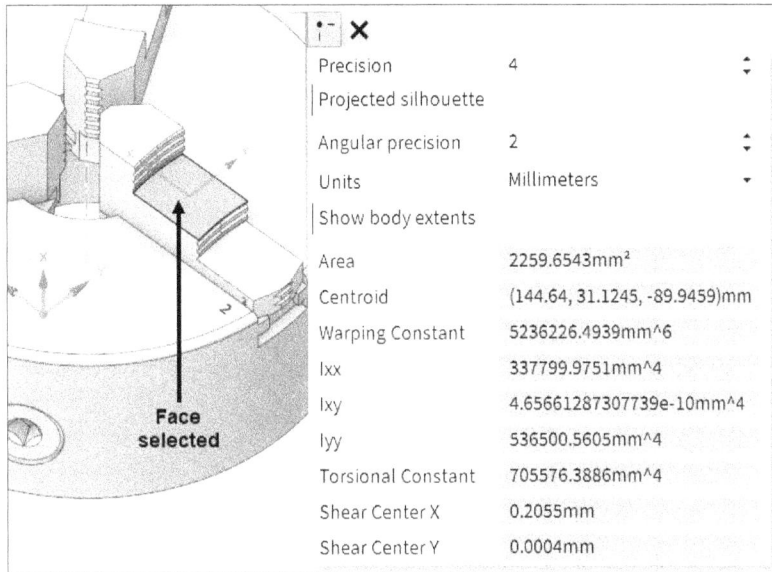

Figure-73. Properties of selected face

- You can also check projected area of a selected body on defined plane by using this tool. To do so, click on the **Plane** selection button ⬚ from the left **HUD** toolbar and select the plane/face to be used as reference for projection. You will be asked to select bodies for measuring projection. Select desired bodies to check the measurement; refer to Figure-74.

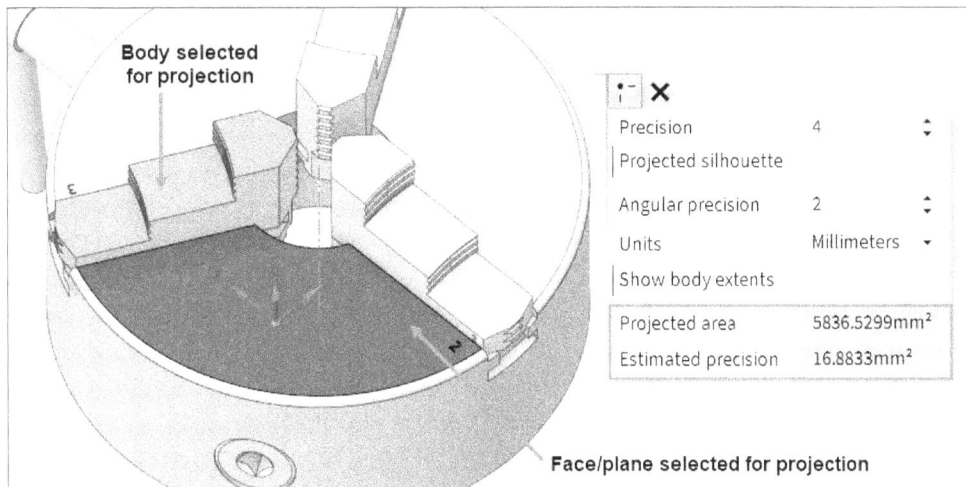

Figure-74. Measuring projection area

- Press **ESC** to exit the tool.

Measuring Environmental Properties

The **Environmental Properties** tool is used to check environmental properties of selected materials in the graphics area. Note that environmental properties are directly connected with material properties defined for the model. The procedure to use this tool is given next.

- Click on the **Environmental Properties** tool from the **INSPECT** panel in the **MEASURE** tab of the **Ribbon**. You will be asked to select the body whose environmental properties are to be displayed.
- Select desired body from the graphics area. The environmental properties will be displayed if assigned in the applied material; refer to Figure-75. Press **ESC** to exit the tool.

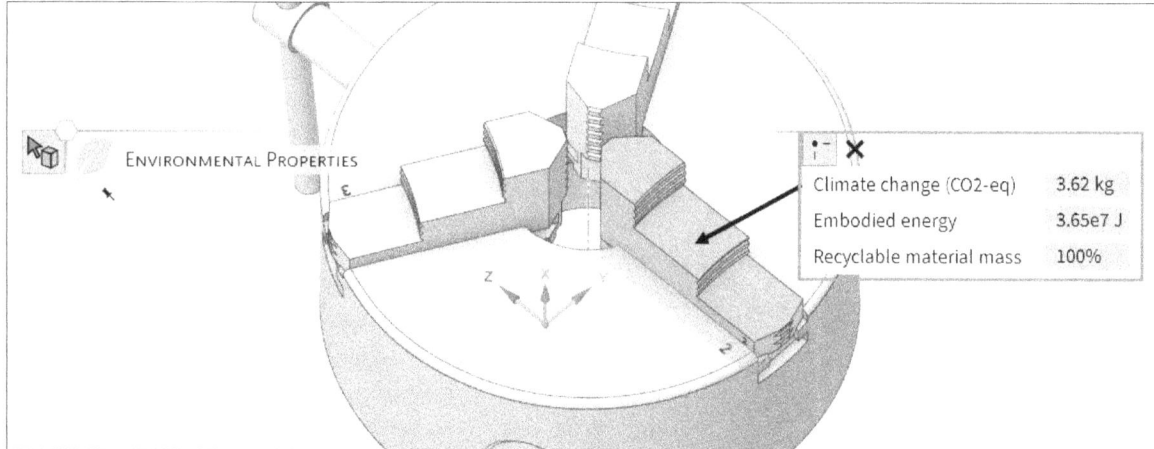

Figure-75. Environmental properties measured

Note: To assign material to a component in the assembly, triple-click on the component to which you want to assign material. An **HUD** toolbar will be displayed with drop-down to select material. Click on the drop-down and select desired material; refer to Figure-76.

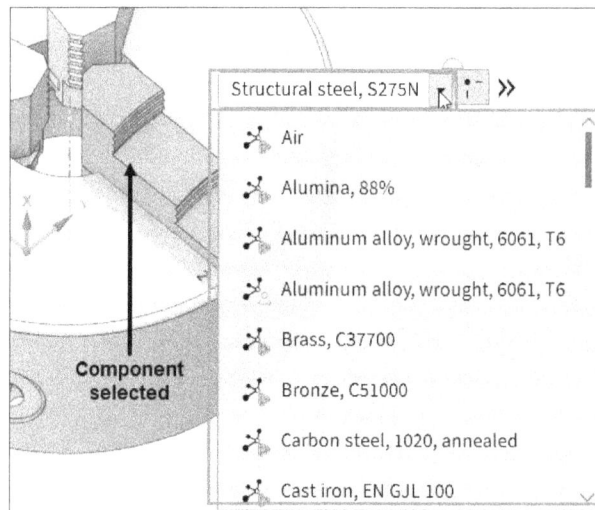

Figure-76. Assigning material for component

Checking Interference Curves

The **Curves** tool of **INTERFERENCE** panel in **Ribbon** is used to check the curves that are common between two selected components. Knowing this detail in advanced allows us to generate better mesh at intersection. The procedure to use this tool is given next.

- Click on the **Curves** tool from the **INTERFERENCE** panel in the **MEASURE** tab of the **Ribbon**. You will be asked to select the objects for checking interference.
- Select two or more objects from the graphics area while holding the **CTRL** key. Interference curves will be displayed; refer to Figure-77.

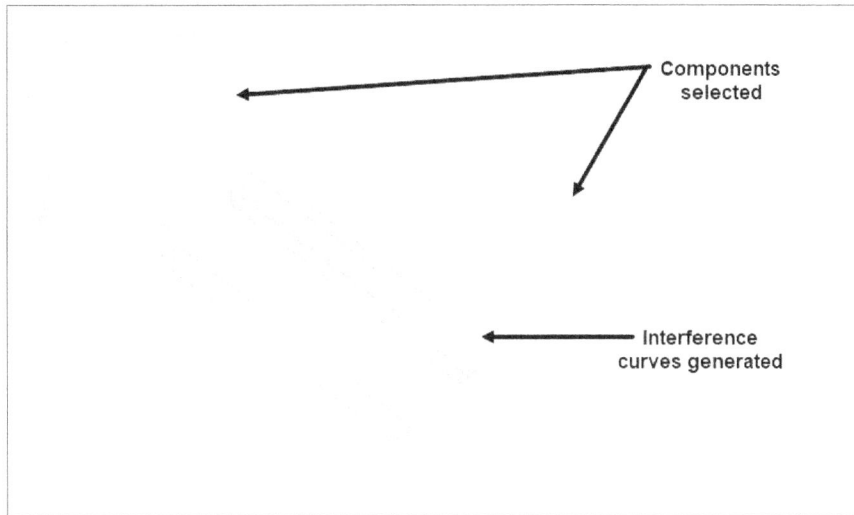

Figure-77. Interference curves generated

- Press **ESC** to exit the tool.

Checking Interference Volume

The **Volumes** tool in the **INTERFERENCE** panel is used to check the region which is overlapped between two selected components. The procedure to use this tool is given next.

- Click on the **Volumes** tool from the **INTERFERENCE** panel in the **MEASURE** tab of the **Ribbon**. You will be asked to select objects for checking interference.
- Select the objects to check interference. The interference volume will be displayed; refer to Figure-78.

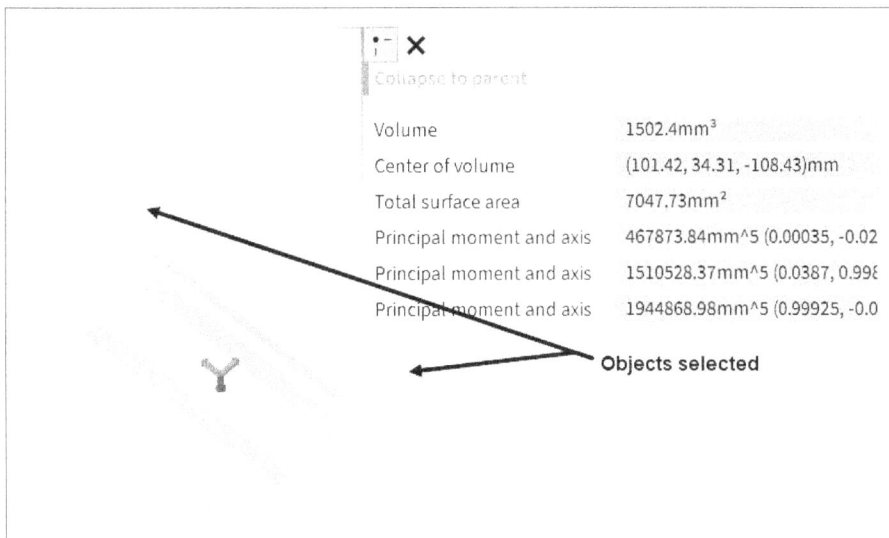

Figure-78. Interference volume

- If you want to convert the interference volume into a solid body then click on the **Select an interfering region** button from left **HUD** toolbar and select the volume to be converted into solid.
- Press **ESC** to exit the tool.

Solid/Surface Quality Checks

The tools in the **QUALITY** panel are used to check various surface quality parameters of selected faces/bodies. Various tools of this panel are discussed next.

- Click on the **Normal** tool from the **QUALITY** panel in the **MEASURE** tab of the **Ribbon** to check normal direction of selected face; refer to Figure-79. By default, arrows are displayed to show normal direction. Select the **Color** option from the right **HUD** toolbar to display color combinations for identifying normal direction. Press **ESC** twice to exit the tool.

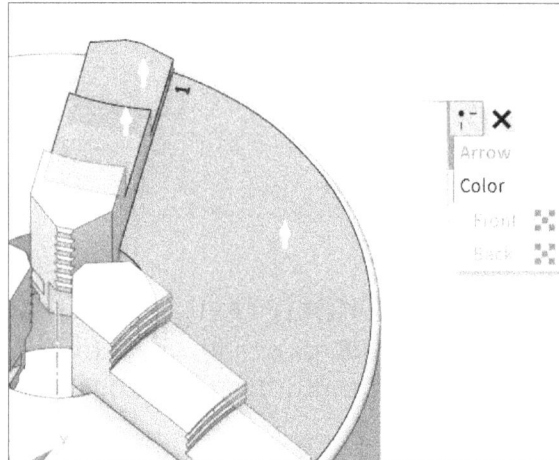

Figure-79. Normal direction displayed on model

- Select the face and click on the **Grid** tool from the **QUALITY** panel in the **MEASURE** tab of the **Ribbon** to display grid lines on selected face; refer to Figure-80.

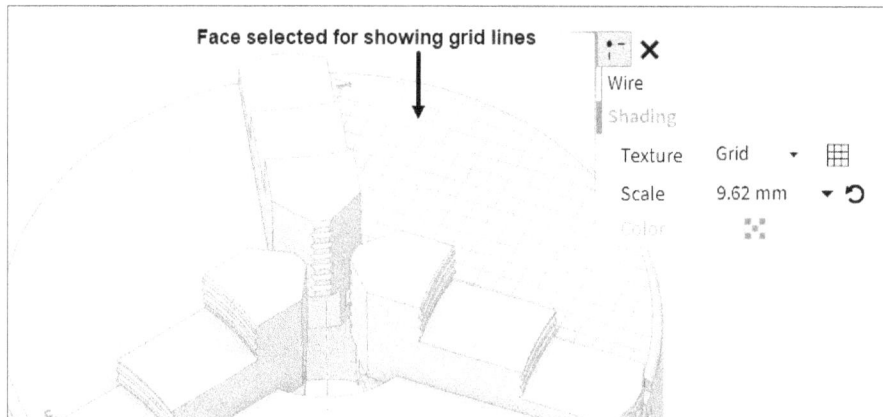

Figure-80. Showing grid lines on face

- Select desired edges/faces and click on the **Curvature** tool from the **QUALITY** panel in the **MEASURE** tab of the **Ribbon**. The curvature comb will be displayed; refer to Figure-81.

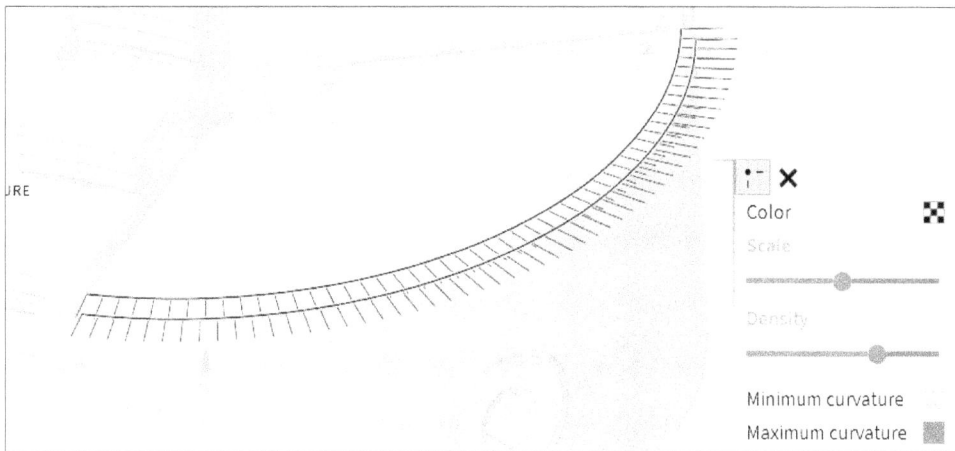

Figure-81. Curvature comb displayed

- Select desired edges and click on the **Dihedral** tool from the **QUALITY** panel in the **Ribbon** to check dihedral angle between faces meeting at selected edge; refer to Figure-82.

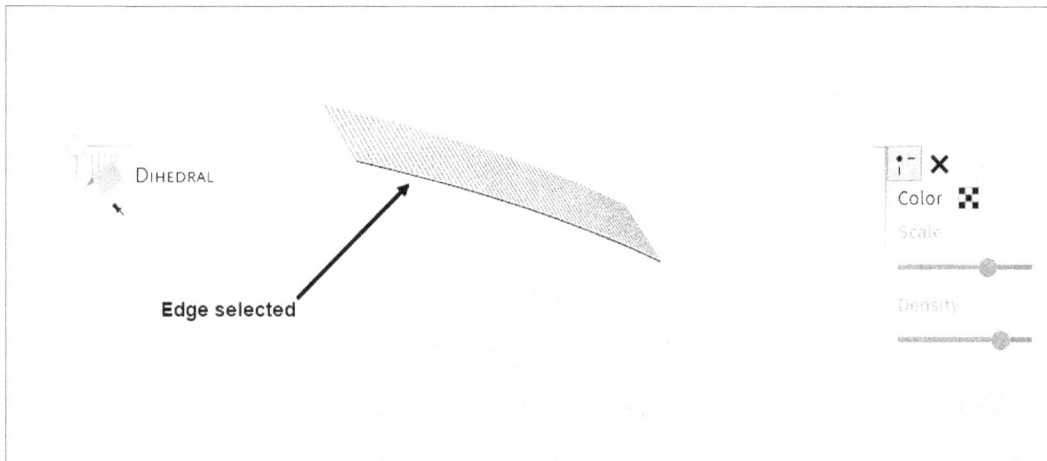

Figure-82. Edge selected for dihedral angle measurement

- Click on the **Draft** tool from **QUALITY** panel in the **MEASURE** tab of the **Ribbon** to check whether selected faces have positive draft or negative draft as compared to value set in the toolbar; refer to Figure-83.

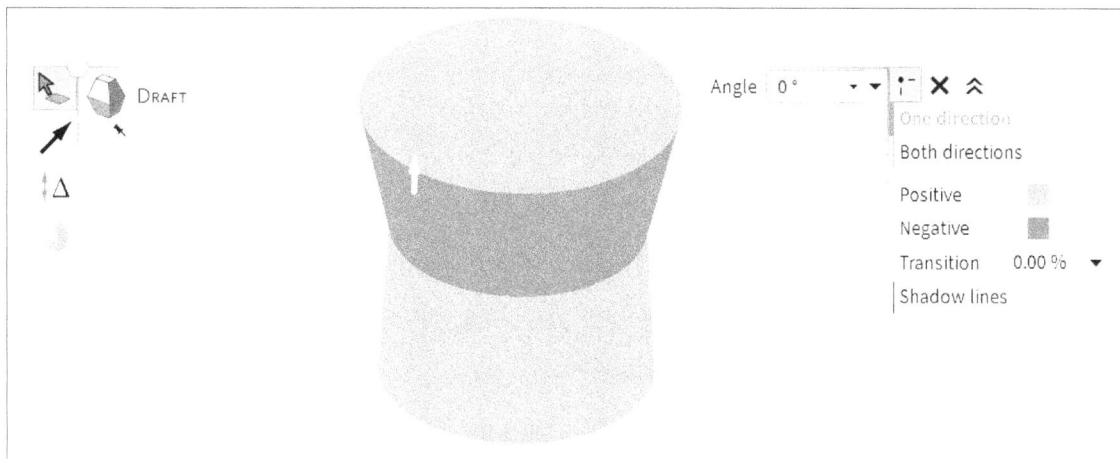

Figure-83. Measuring draft

- Select desired faces and click on the **Stripes** tool from the **QUALITY** panel in the **MEASURE** tab of the **Ribbon** to display surface stripes on the model for checking surface flow; refer to Figure-84.

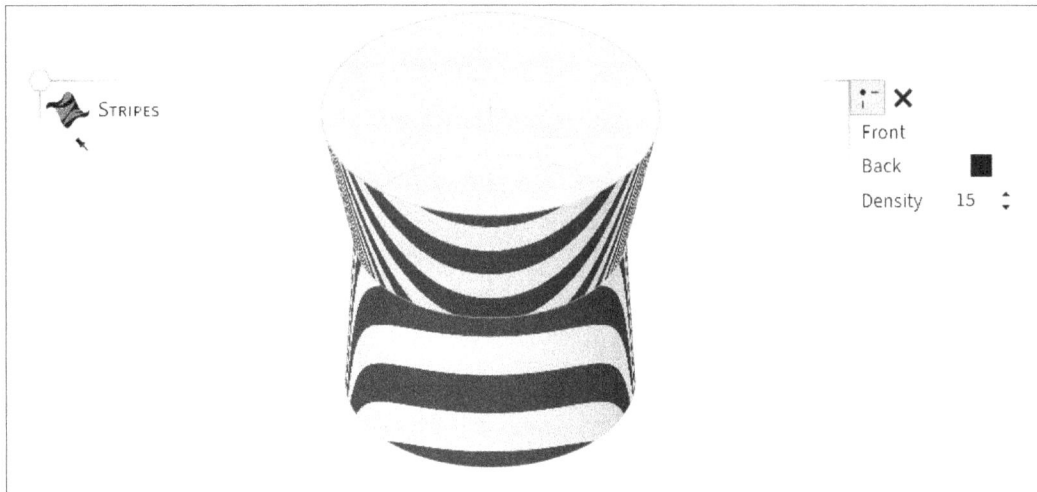

Figure-84. Stripes displayed on model

- Similarly, you can use **Deviation** tool from the **QUALITY** panel in **Ribbon** to find out how much surfaces/faces of selected object are deviating from reference object.

SELF-ASSESSMENT

Q1. What is the primary purpose of the Fill tool?
A. To create solid bodies
B. To fill holes and voids in bodies surrounded by walls
C. To merge two objects into one
D. To replace a selected face with another surface

Q2. Which tool is used to create a solid feature by joining two or more closed loop curves?
A. Fill
B. Blend
C. Replace
D. Split

Q3. What is the function of the Guide curve option in the Blend tool?
A. To define the path for blending multiple curves
B. To create a circular pattern
C. To project curves onto a surface
D. To merge multiple bodies into one

Q4. Which option in the Blend tool is used to create a closed-loop blend, ensuring the first and last sections are connected?
A. Ruled segments
B. Minimal topology
C. Rotational blend
D. Periodic blend

Q5. What does the Replace tool do?
A. Replaces a selected face with another surface in the projected direction
B. Combines two bodies into one
C. Splits a body into multiple parts
D. Extends a selected surface

Q6. Which tool is used to perform Boolean operations like merge, split, and subtract?
A. Replace
B. Blend
C. Combine
D. Fill

Q7. In the Combine tool, what does selecting the "Subtract from target" toggle button do?
A. Removes the cutter body from the target body
B. Merges the target and cutter bodies
C. Keeps the cutter body in the final result
D. Creates imprints on the target body

Q8. Which tool is used to split a body using a face, surface, or edge loop?
A. Fill
B. Split Body
C. Split Face
D. Replace

Q9. What is the purpose of the Project tool?
A. To project faces, surfaces, or curves onto another face/surface
B. To split a face into multiple parts
C. To replace a selected face with another surface
D. To extend faces in a model

Q10. The Plane tool is used to create a datum plane. Which of the following is NOT a valid reference for creating a plane?
A. Face
B. Point
C. Edge
D. Texture

Q11. The Axis tool is used to create a datum axis. How can an axis be created?
A. By selecting a cylindrical face or edge
B. By selecting a texture
C. By selecting an image
D. By selecting a solid body

Q12. Which tool is used to create a coordinate system at a selected reference?
A. Axis
B. Plane
C. Origin
D. Point

Q13. What does the Linear Pattern tool do?
A. Creates multiple copies of an object along X and Y axes
B. Creates circular instances of an object
C. Projects a curve onto a face
D. Merges two bodies into one

Q14. What parameter is required for creating a Circular Pattern?
A. Number of instances
B. Distance between instances
C. Projection angle
D. Blend curvature

Q15. The Fill Pattern tool is used to create multiple instances of an object within a specified boundary. Which option arranges instances in a 2D grid format?
A. Offset
B. Grid
C. Circular
D. Blend

Q16. What is the primary function of the Shell tool?
A. To create a new solid feature
B. To scoop out material and create a thin shell of specified thickness
C. To merge two solid bodies into one
D. To apply a fillet to edges

Q17. Where is the Shell tool located in the software interface?
A. ASSEMBLE panel in the DESIGN tab
B. CREATE panel in the DESIGN tab
C. BODY panel in the DESIGN tab
D. MODIFY panel in the DESIGN tab

Q18. What happens when you apply the Offset tool to two selected faces?
A. The faces get merged into one
B. The faces become bound by an offset relation and move together
C. One face disappears
D. The faces are mirrored

Q19. What does the Mirror tool require to create a mirror copy of an object?
A. A mirror plane
B. A predefined axis
C. A reference point
D. A scaling factor

Q20. What option is selected in the Mirror tool to create a mirror copy of selected features like holes or boss features?
A. Solid(s)/Surface(s) selection button
B. Face(s) selection button
C. Component(s) selection button
D. Offset selection button

Q21. Which tool is used to create a custom surface using specified parameters?
A. Cylinder tool
B. Shell tool
C. Equation tool
D. Offset tool

Q22. In the Equation tool, selecting the "Custom" option from the Curve type drop-down allows you to:
A. Define a parametric user-defined equation
B. Mirror objects
C. Apply a fillet
D. Create a new assembly

Q23. What is the function of the Cylinder tool?
A. To create a thin shell
B. To create a cylindrical solid body
C. To create a conical shape
D. To extrude a sketch

Q24. Which tool is used to apply constraints between bodies in an assembly?
A. Shell tool
B. Mirror tool
C. ASSEMBLE panel tools
D. Equation tool

Q25. What is the purpose of the Tangent tool in the ASSEMBLE panel?
A. To make two selected faces tangent
B. To align two faces
C. To create a mirror relationship
D. To scale an object

Q26. How does the Align tool affect selected faces?
A. It makes the faces tangent
B. It makes the faces aligned in the same plane or axis
C. It rotates the faces
D. It offsets the faces

Q27. What does the Rigid tool do in an assembly?
A. It applies an offset between two faces
B. It keeps the orientation of selected faces fixed with respect to each other
C. It creates a mirror relationship
D. It makes a shell feature

Q28. What is the function of the Gear constraint?
A. It creates a mirror copy of a gear
B. It ensures no slippage between cylindrical faces of two components
C. It applies an offset relation between faces
D. It aligns two faces

Q29. What is the effect of applying an Anchor constraint to a component?
A. It fixes the component in orientation and position
B. It allows the component to move freely
C. It applies an offset between two parts
D. It creates a mirror copy of the component

Q30. Which tab in the Ribbon contains tools to modify graphics-related parameters?
A. HOME
B. DISPLAY
C. VIEW
D. INSPECT

Q31. What does the "On Center" option in the Spin drop-down do?
A. Spins the model about the cursor position
B. Spins the model about the center of gravity of all objects
C. Spins the model about the origin point
D. Spins the model randomly

Q32. How can you perform spinning using the mouse?
A. Hold the left mouse button and drag
B. Hold the right mouse button and drag
C. Hold the middle mouse button (MMB) and drag
D. Hold the SHIFT key and drag

Q33. What does the Pan tool do?
A. Moves individual objects in the graphics area
B. Rotates the model about a fixed axis
C. Moves the view of the model without relative movement between objects
D. Resets the view to the default position

Q34. What is the shortcut for performing the panning operation using the mouse?
A. CTRL + MMB (middle mouse button)
B. ALT + MMB
C. SHIFT + MMB
D. CTRL + SHIFT + MMB

Q35. What does the Zoom Extents tool do?
A. Zooms into a specific region
B. Fits all objects in the graphics area to the current view
C. Zooms in by 50%
D. Zooms out by 50%

Q36. Which method is NOT used to perform zoom operations?
A. Using the mouse scroll wheel
B. Using CTRL + + and CTRL + - keys
C. Holding SHIFT + MMB and dragging the cursor
D. Holding ALT + MMB and dragging the cursor

Q37. How can you apply color to an entire object instead of just a face?
A. Select the Face tab in the Color drop-down
B. Select the Body tab in the Color drop-down
C. Select the Object tab in the Color drop-down
D. It is not possible to apply color to an entire object

Q38. What is the function of the Opacity slider in the Color drop-down?
A. Adjusts the brightness of the object
B. Adjusts the texture of the object
C. Adjusts the transparency of the object
D. Adjusts the reflection of the object

Q39. Which tool automatically assigns different colors to multiple objects for easy differentiation?
A. Opacity
B. Randomize Colors
C. Texture
D. Edge Display

Q40. Which of the following is NOT a graphics style option in the Graphics drop-down?
A. Shaded
B. Wireframe
C. Hidden Line
D. Textured

Q41. What does the STYLE OVERRIDE drop-down control?
A. Object transparency or opacity
B. Object color settings
C. Object position in the graphics area
D. Object material properties

Q42. How can layers be used in the graphics area?
A. To delete unwanted objects
B. To categorize objects for easy visibility control
C. To merge multiple objects into one
D. To apply different colors to objects

Q43. What is the function of the Orthographic option in the CAMERA panel?
A. Displays objects with depth perception
B. Displays objects in an orthographic projection without perspective
C. Hides objects in the graphics area
D. Increases the size of the camera view

Q44. How does the Perspective option affect the view?
A. Objects appear smaller when they are farther away
B. Objects appear the same size regardless of distance
C. Objects disappear when zoomed out
D. Objects are displayed in wireframe mode

Q45. Which shortcut key combination activates the Fly Through feature?
A. CTRL + SHIFT + F
B. ALT + SHIFT + F
C. CTRL + ALT + F
D. SHIFT + F

Q46. What does the Show Sketch Grid toggle button do?
A. Displays a grid when sketching
B. Hides all objects in the graphics area
C. Rotates the model automatically
D. Moves objects to a new layer

Q47. Which measurement tool is used to check mass properties of a model?
A. Measure tool
B. Mass Properties tool
C. Environmental Properties tool
D. Interference Volume tool

Q48. How do you measure the distance between two vertices?
A. Use the Measure tool and select two vertices
B. Use the Pan tool and select two vertices
C. Use the Zoom tool and select two vertices
D. Use the Mass Properties tool and select two vertices

Q49. What does the Environmental Properties tool check?
A. The environmental impact of the model
B. The material properties related to the environment
C. The interference between objects
D. The colors assigned to objects

Q50. How can you check interference curves between two components?
A. Use the Measure tool
B. Use the Mass Properties tool
C. Use the Curves tool in the INTERFERENCE panel
D. Use the Zoom Extents tool

Chapter 3

Ansys Discovery-Design III

Topics Covered

The major topics covered in this chapter are:

- *Repairing Tools*
- *Introduction to Facets*
- *Performing Cleanup Operations*
- *Organizing Facets*
- *Modifying Facets*
- *Adjusting and Inspecting Facets*
- *Performing Reverse Engineering*

REPAIRING TOOLS

The tools available in **REPAIR** tab of **Ribbon** are used to modify and repair objects imported in the software from other CAD applications; refer to Figure-1. These tools fix issues in the imported models that can undermine the accuracy of analysis in software like cleaning gaps and overlaps in geometry, merging faces, simplifying model, and so on.

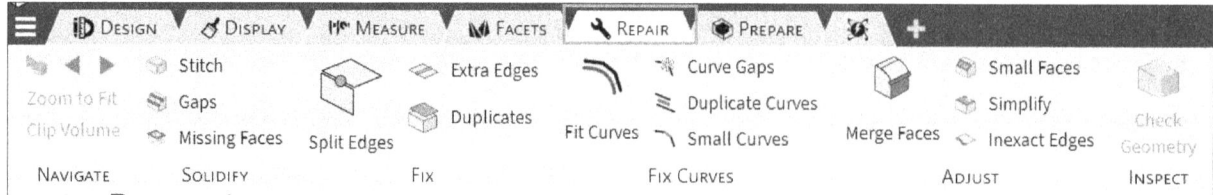

Figure-1. REPAIR tab

Stitching Faces

The **Stitch** tool is used to combine surfaces to a single body. The procedure to use this tool is given next.

* Click on the **Stitch** tool from the **SOLIDIFY** panel in the **REPAIR** tab of the **Ribbon**. The **STITCH HUD** toolbar will be displayed and surfaces eligible for stitching will be highlighted automatically; refer to Figure-2.

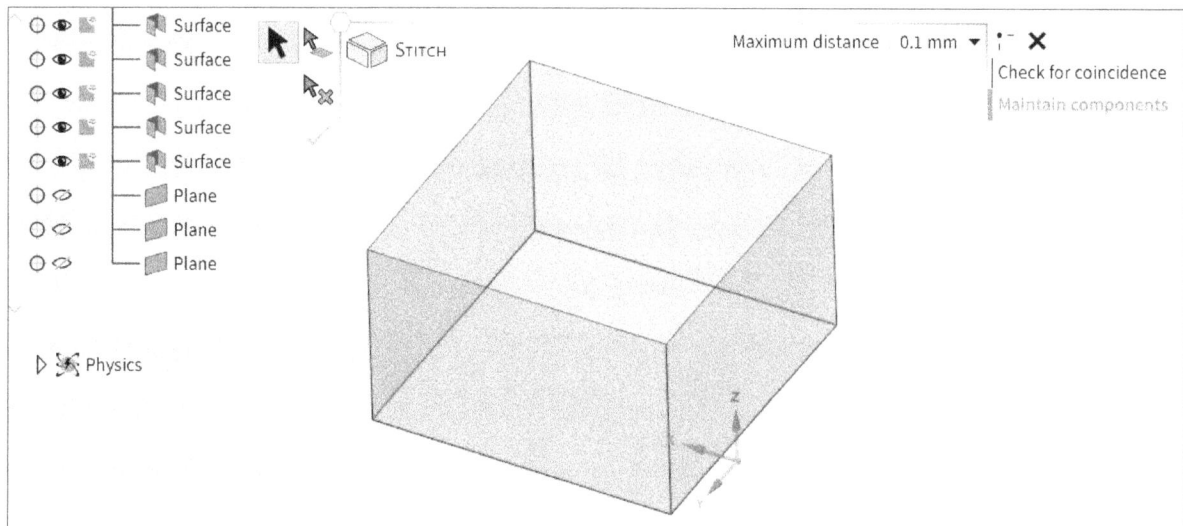

Figure-2. STITCH HUD Toolbar

* Specify desired value in the **Maximum distance** edit box to define the gap size upto which system can ignore gap between two surface edges and stitch surfaces together.
* Select the **Check for coincidence** toggle option from the right **HUD** toolbar if you want the software to check for coincident faces before applying stitching and ignore multiple copies of same face.
* Select the **Maintain components** toggle option from right **HUD** toolbar if you want to keep the component same after joining multiple surfaces into single surface/body.
* By default, the surface which will be affected by the tool are highlighted. If you want to manually define faces/surfaces that are to be stitched then click on the **Select geometry** button ⟍ from the left **HUD** toolbar and select desired faces/ surface that can be stitched (they are within specified gap tolerance).

- If you want to remove any automatically selected section from the model so that stitching is not performed for that section then click on the **Exclude Selection** button from left **HUD** toolbar and select the highlighted edges; refer to Figure-3.

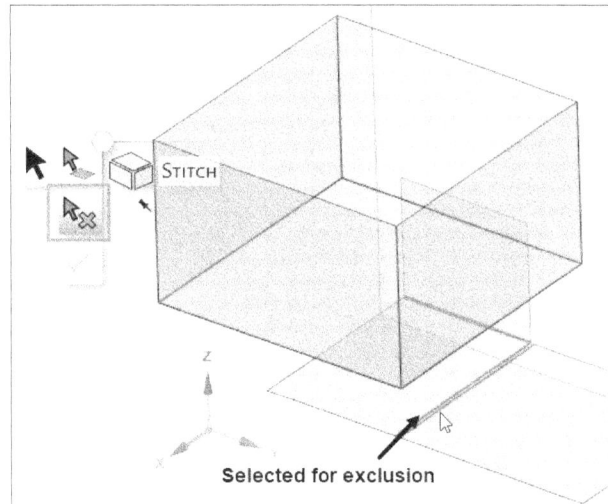

Figure-3. Excluding object from selection

- After setting desired parameters, click on the **OK** button to perform stitching.

Identifying and Filling Gaps

The **Gaps** tool is used to find and fill gaps in surfaces of imported object. The procedure to use this tool is given next.

- Click on the **Gaps** tool from the **SOLIDIFY** panel in the **REPAIR** tab of the **Ribbon**. The **Gaps HUD** toolbar will be displayed and sections of model with gaps will be highlighted.
- Set desired parameters as discussed earlier and click on the **OK** button to fill the gaps.

Similarly, you can use the **Missing Faces** tool from the **SOLIDIFY** panel in the **Ribbon** to find and fix missing faces from the imported body.

Detecting and Fixing Split Edges

The **Split Edges** tool is used to find edges in model that should be a single entity instead of multiple small segments of edge. On clicking this tool from **Fix** panel in the **REPAIR** tab of **Ribbon**, edges that are split in the imported model will be highlighted; refer to Figure-4. Select the highlighted dots to fix split edges and click on the **OK** button from the left **HUD** toolbar.

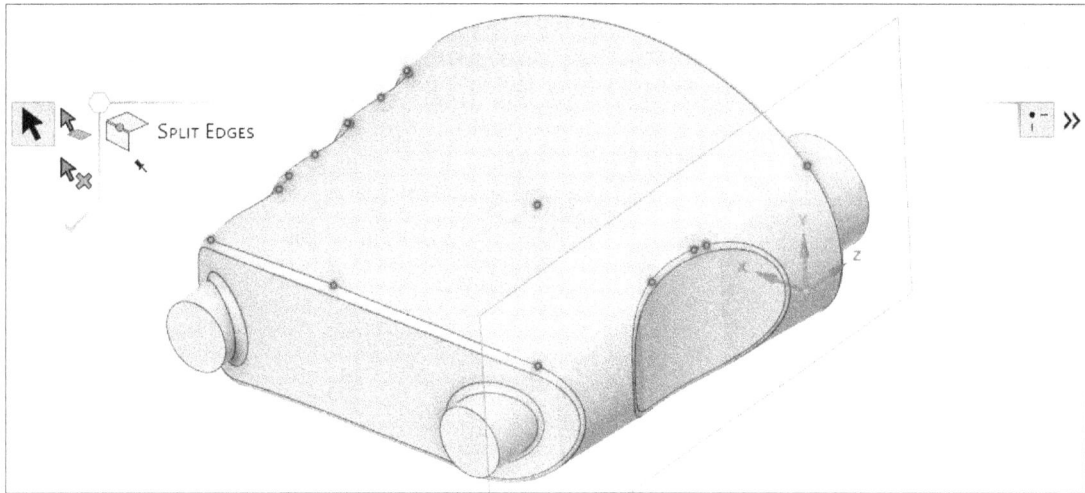

Figure-4. Split edges

Similarly, you can use **Extra Edges** tool from the **FIX** panel in the **Ribbon** to trim extra portions of edges in imported model and **Duplicates** tool from the **FIX** panel in the **Ribbon** to remove duplicate/overlapping faces from the model. In the same way, you can use other tools in the **REPAIR** tab of the **Ribbon** to fix problems related to various entities in the imported model.

INTRODUCTION TO FACETS

Facets are small segments of faces of the model which can be used for generating mesh and performing analyses. Using facets allows us to simplify the complex geometries for performing analyses and then combine the results of individual facets for generating final result of analyses. The tools to create and manage facets are available in the **FACETS** tab of the **Ribbon**; refer to Figure-5. Various tools of this tab are discussed next.

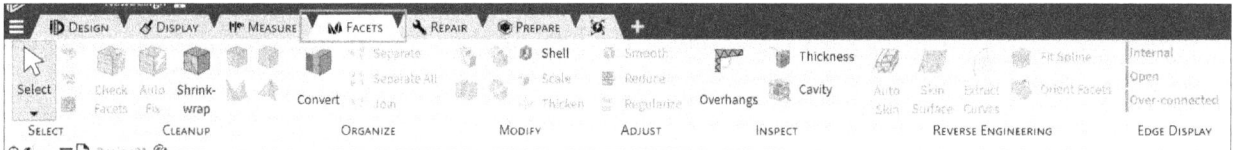

Figure-5. FACETS tab

Converting Solids/Surfaces to Facets

The **Convert** tool is used to convert selected bodies into facets for further modifications. The procedure to use this tool is given next.

- Click on the **Convert** tool from the **ORGANIZE** panel in the **FACETS** tab of the **Ribbon**. The **CONVERT HUD** toolbar will be displayed; refer to Figure-6 and you will be asked to select the bodies to be converted to facets.

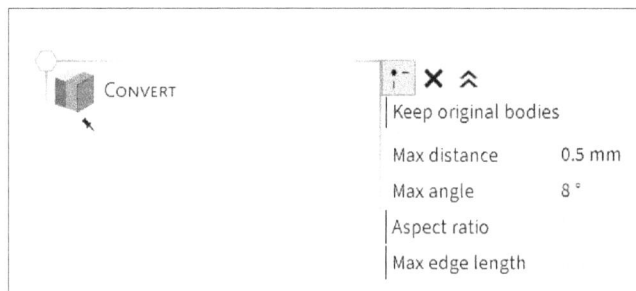

Figure-6. Convert HUD toolbar

- Set desired values in the **Max distance** and **Max angle** edit boxes of the **HUD** toolbar to define maximum allowed distance from model edge and angle of facet edges.
- Select the **Aspect ratio** toggle button to create facets having length to width ratio near specified value.
- Select the **Max edge length** toggle button to define the maximum value upto which facet edges can have length.
- Select the **Keep original bodies** toggle button to keep the original body as well after creating the facet body.
- After setting desired parameters, click on the body to be converted into facet body; refer to Figure-7.

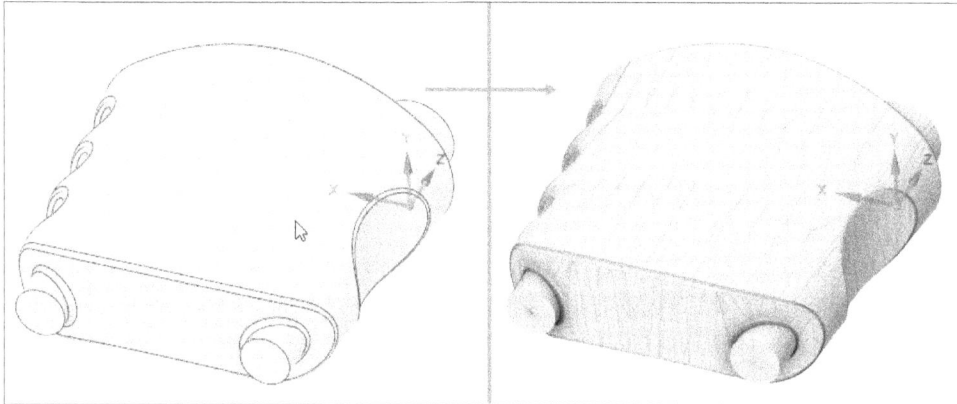

Figure-7. Coverting solid to facets

Checking Facets

The **Check Facets** tool is used to check quality of facets and highlight areas with problems. The procedure to use this tool is given next.

- Click on the **Check Facets** tool from the **CLEANUP** panel in the **FACETS** tab of the **Ribbon**. The **CHECK FACETS HUD** toolbar will be displayed and you will be asked to select the facet body for checking quality.
- Select desired facet body from graphics area. The **Check Facets** dialog box will be displayed with list of errors, if any; refer to Figure-8.
- Select the errors to check them in graphics area and click on the **Close** button to exit the dialog box.

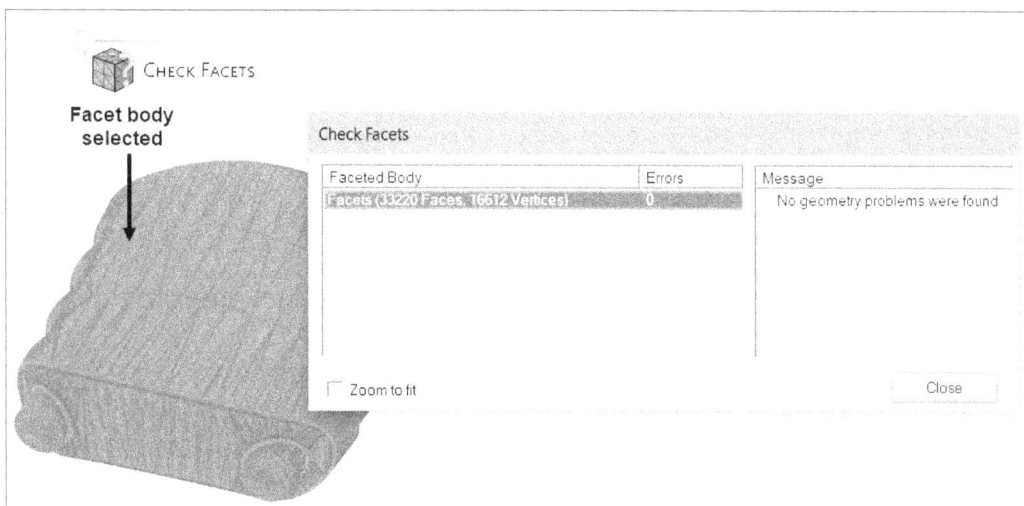

Figure-8. Check Facets dialog box

Automatically Fixing Facets

The **Auto Fix** tool is used to automatically fix defective facets, if possible. The procedure to use this tool is given next.

*   Click on the **Auto Fix** tool from the **CLEANUP** panel in the **FACETS** tab of the **Ribbon**. The **AUTO FIX HUD** toolbar will be displayed; refer to Figure-9.

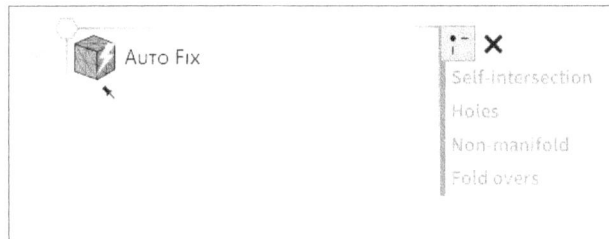

Figure-9. AUTO FIX HUD Toolbar

*   Select the **Self-intersection** toggle button if you want to repair self-intersection of facets.
*   Select the **Holes** toggle button to automatically fill unintentional holes in the facets.
*   Select the **Non-manifold** toggle button to fix non-manifold edges. Non-manifold edges are those which have breaks in continuity or they are connected to multiple faces in unintentional ways.
*   Select the **Fold overs** toggle button to remove and repair non meaningful folded sections of the facets.
*   After setting desired parameters, click on the body to be fixed, and click on the **OK** button from left **HUD** toolbar to apply changes. Press **ESC** twice to exit the tool.

Creating Shrinkwrapped Body

The **Shrinkwrap** tool is used to create a shrunk copy of selected body with specified shrinkage value. The procedure to use this tool is given next.

*   Click on the **Shrinkwrap** tool from the **CLEANUP** panel in the **FACETS** tab of the **Ribbon**. The **SHRINKWRAP HUD** toolbar will be displayed; refer to Figure-10 and you will be asked to select body to be used as reference for creating shrink wrap body.

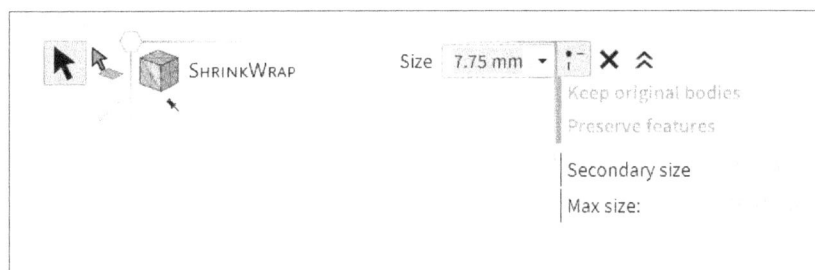

Figure-10. SHRINKWRAP HUD Toolbar

*   Specify desired value in the **Size** edit box to define offset distance by which new body will be shrunk as compared to selected body.
*   Set the other parameters as discussed earlier and select the body from graphics area to create shrunk wrap body.
*   After selecting body, click on the **OK** button from the **HUD** toolbar. The body will be created; refer to Figure-11.

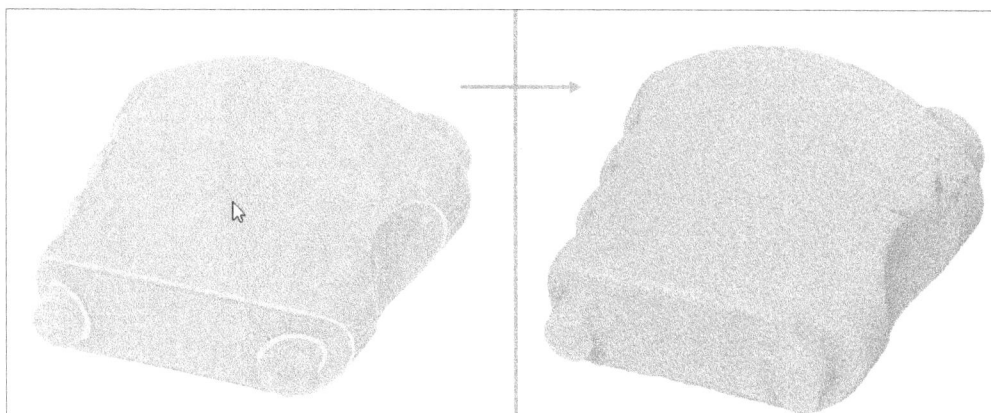

Figure-11. Shrinkwrap body generated

• Press **ESC** to exit the tool.

Repairing and Fixing Facet Defects

There are various tools available in the **CLEANUP** panel of the **Ribbon** to perform repair and fixing of defects in facets. These tools perform the operation automatically after selecting body and the procedures to use these tools are same as discussed earlier. The functions of these tools are discussed next.

• Use the **HOLES** tool from the **CLEANUP** panel in the **FACETS** tab of the **Ribbon** to automatically fix unwanted holes in the facet body.
• Use the **FIX SHARPS** tool from the **CLEANUP** panel in the **FACETS** tab of the **Ribbon** to automatically fix sharp edges connected abruptly with abnormally high vertices.
• Use the **INTERSECTIONS** tool from the **CLEANUP** panel in the **FACETS** tab of the **Ribbon** to automatically fix self-intersecting and overlapping edges.
• Use the **OVER-CONNECTED** tool from the **CLEANUP** panel in the **FACETS** tab of the **Ribbon** to automatically fix edges that are overly connected with multiple vertices of the facet body due to previous operations or conversions.

Separating Facet Bodies

The **Separate** tool is used to separate multiple facet bodies combined due to conversion or other pre-operations. The procedure to use this tool is given next.

• Click on the **Separate** tool from the **ORGANIZE** panel in the **FACETS** tab of the **Ribbon**. The **SEPARATE HUD** toolbar will be displayed and you will be asked to select the objects to be created as separate facet body.
• Select desired objects to create facet bodies; refer to Figure-12.
• Press **ESC** to exit the tool.

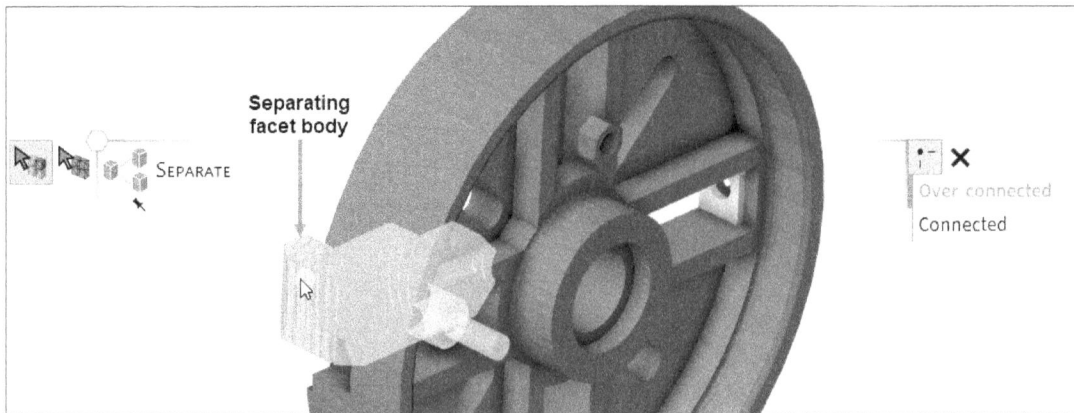

Figure-12. Creating separate facet body

Separating All Facet Bodies

The **Separate All** tool is used to separate all the facet bodies of current imported or converted body. Click on the **Separate All** tool from the **ORGANIZE** panel in the **FACETS** tab of the **Ribbon**.

Joining Facet Bodies to Single Body

The **Join** tool is used to combine multiple selected facet bodies into one single body. It can also attempt to stitch those facet bodies together, if possible. The procedure to use this tool is given next.

*   Click on the **Join** tool from the **ORGANIZE** panel in the **FACETS** tab of the **Ribbon**. The **JOIN HUD** toolbar will be displayed and you will be asked to select bodies to be joined together.
*   Select all the bodies to be joined one by one.
*   If you want to select one target body to which all the other facet bodies should be joined then select the **Target Faceted Body** selection button 🔖 from the **HUD** toolbar and select desired body. All the other facet bodies will join with selected target body.

Merging Selected Facet Bodies

The **MERGE** tool is used to join small faces at intersection/common edge of two selected bodies to reduce complexity in mesh and form larger facets. The procedure to use this tool is given next.

*   Click on the **MERGE** tool from the **MODIFY** panel in the **FACETS** tab of the **Ribbon**. The **MERGE HUD** toolbar will be displayed and you will be asked to select the bodies.
*   Select desired facet bodies that are to be merged.

Splitting Selected Facet Body

The **SPLIT** tool is used to split selected facet body using a plane as cutter object. The procedure to use this tool is given next.

*   Click on the **SPLIT** tool from the **MODIFY** panel in the **FACETS** tab of the **Ribbon**. The **MERGE HUD** toolbar will be displayed and you will be asked to select the body to be split.

- Select the **No Cap** option from right **HUD** toolbar if you do not want to create covering cap on facet at splitting faces. Select the **Cap** option from the **HUD** toolbar to create cap facets at the splitting site. Select the **Cap with internal points** option to close all the open sections and generate points for newly created cap facets.
- Select the body from graphics area and then select a plane passing through selected body to split it. The split bodies will be created; refer to Figure-13.

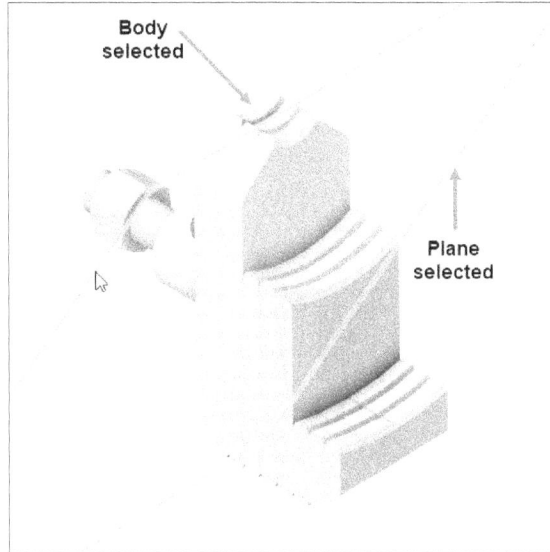

Figure-13. Selection for Split

Subtracting from Facet Body

The **SUBTRACT** tool is used to remove selected tool body from main body. The procedure to use this tool is given next.

- Click on the **SUBTRACT** tool from the **MODIFY** panel in the **FACETS** tab of the **Ribbon**. The **SUBTRACT HUD** toolbar will be displayed and you will be asked to select target body from which other body will be subtracted.
- Select the **Keep original bodies** and **Keep cutter** toggle buttons to keep respective bodies after performing the operation. By default, these toggle buttons are not active and only the result body remains after performing the operation.
- Select desired body from which other body will be subtracted. The subtract feature will be created; refer to Figure-14.

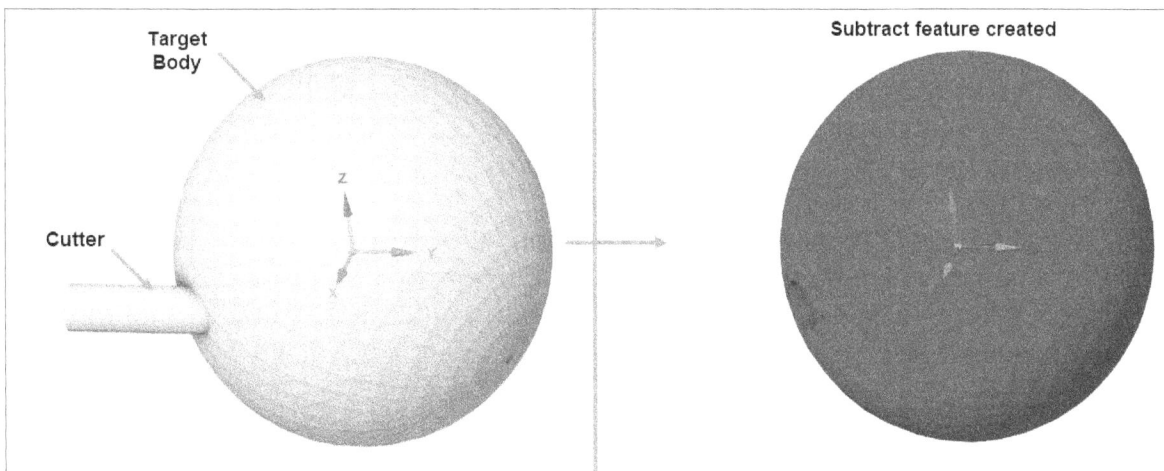

Figure-14. Subtracting facets

Performing Intersection Operation

The **INTERSECT** tool is used to generate body from section common to two selected bodies. The procedure to use this tool is given next.

- Click on the **INTERSECT** tool from the **MODIFY** panel in the **FACETS** tab of the **Ribbon**. The **INTERSECT HUD** toolbar will be displayed and you will be asked to select intersecting facet bodies.
- Select two bodies to be used for generating intersection body. The feature will be created; refer to Figure-15. Press **ESC** to exit the tool.

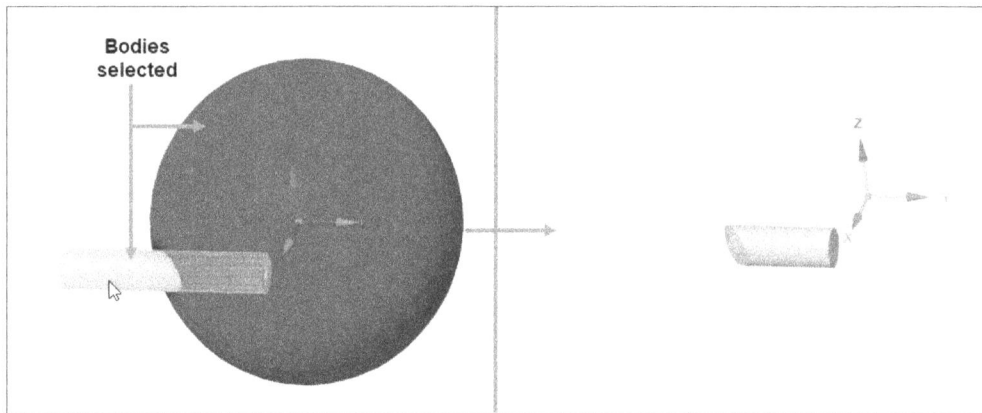

Figure-15. Generating Intersection body

Creating Shell Feature

The **Shell** tool is used to create thin walled facet body from selected solid facet body. The procedure to use this tool is given next.

- Click on the **Shell** tool from the **MODIFY** panel in the **FACETS** tab of the **Ribbon**. The **SHELL HUD** toolbar will be displayed and you will be asked to select the body for applying shell operation.
- Select the **Inside** toggle button from the right **HUD** toolbar to create shell feature with thickness specified inside the walls. Select the **Outside** toggle button from the right **HUD** toolbar to create shell feature with thickness specified outside the walls. Select the **Keep original bodies** toggle button to keep original bodies after performing the operation.
- Select the **Basic Infill** toggle button to define the pattern of filling, if needed in the created shell feature. The options in toolbar will be displayed as shown in Figure-16.
- Set desired shapes and parameters for the infill, if needed.
- Specify desired value in the **Thickness** edit box to define thickness of shell walls.
- Click on the facet body on which shell operation is to be applied. Preview of shell feature will be displayed; refer to Figure-17.
- Click on the **Select Outer face(s) to open** selection button from the left **HUD** toolbar and select the face(s) to be removed from the shell feature. After selecting faces, click on the **OK** button from **HUD** toolbar and exit the tool. The shell feature will be created; refer to Figure-18.

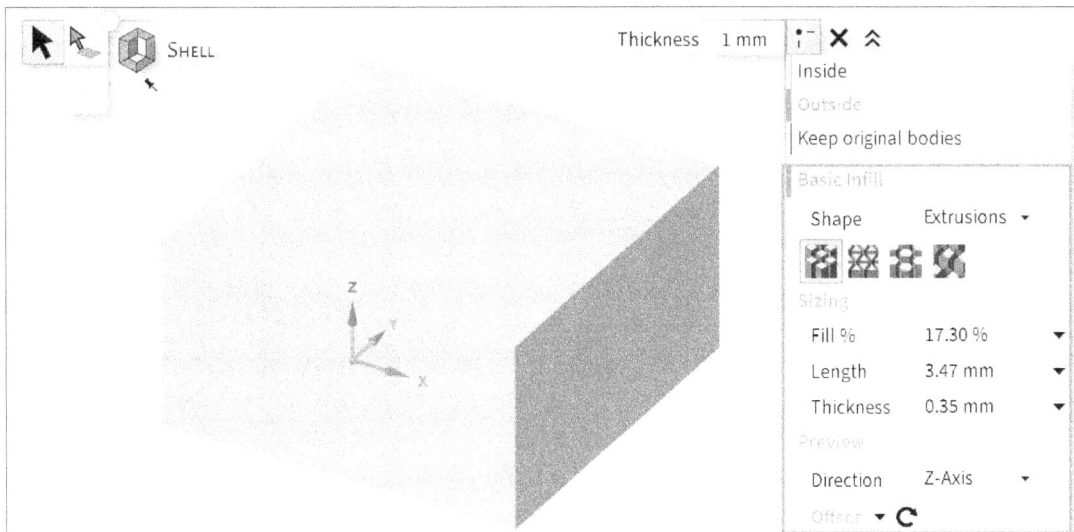

Figure-16. Basic Infill options

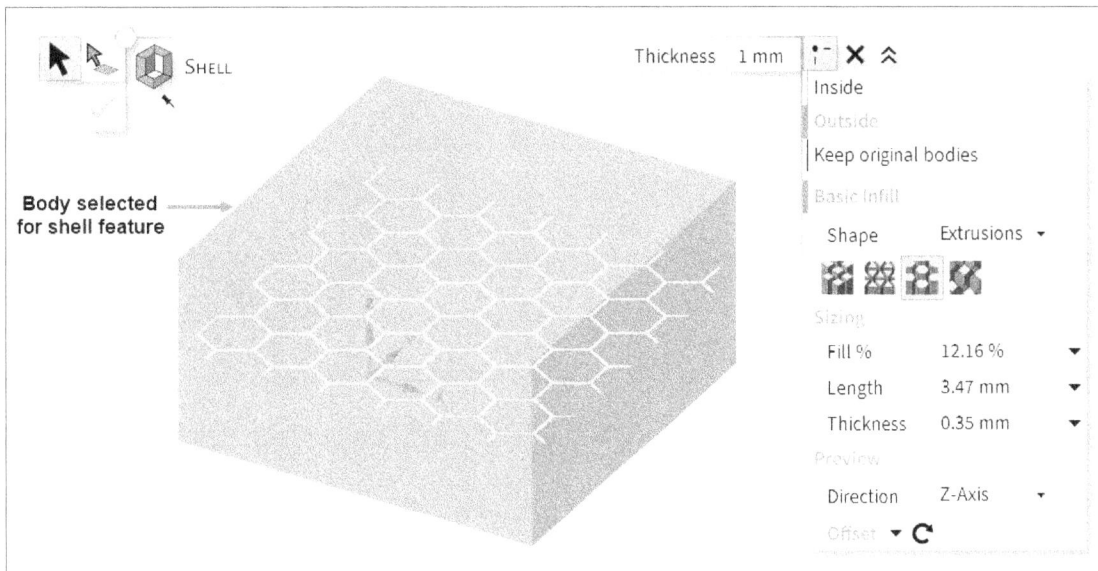

Figure-17. Body selected for shell feature

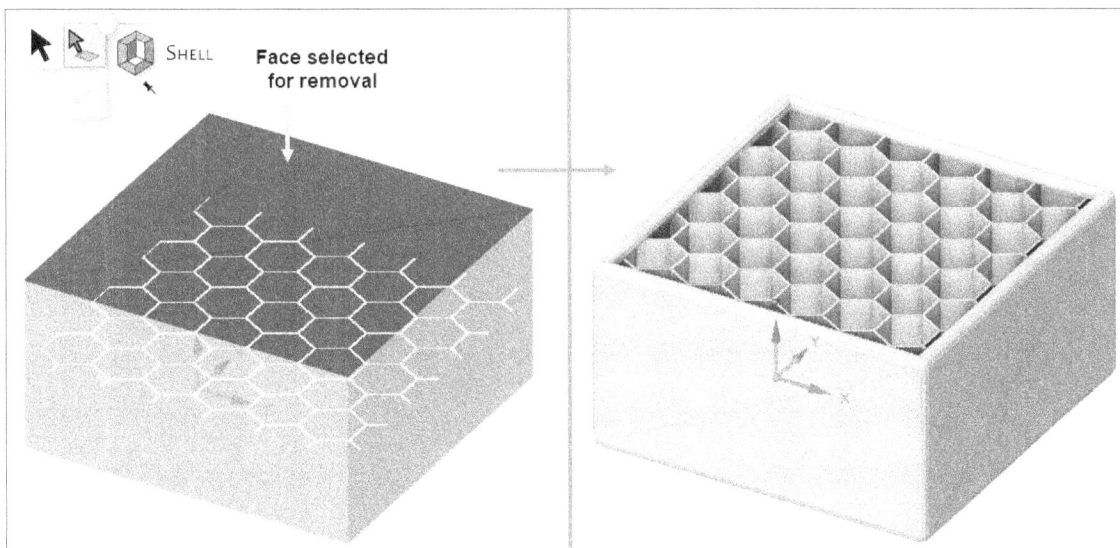

Figure-18. Creating shell facet feature

Using Scale

The **Scale** tool is used to increase/decrease the size of body by specified scale factor. The procedure to use this tool is given next.

*   Click on the **Scale** tool from the **MODIFY** panel in the **FACETS** tab of the **Ribbon**. The **SCALE HUD** toolbar will be displayed and you will be asked to select the body to be scaled.
*   By default, all three toggle buttons are selected in the right **HUD** toolbar. Clear desired toggle button if you do not want to scale the model in respective direction.
*   After setting desired parameters, click on the body to be scaled. You will be asked to specify anchor point with respect to which scale operation is being performed.
*   Click at desired location on the model to define anchor point and then drag it to scale selected body; refer to Figure-19.

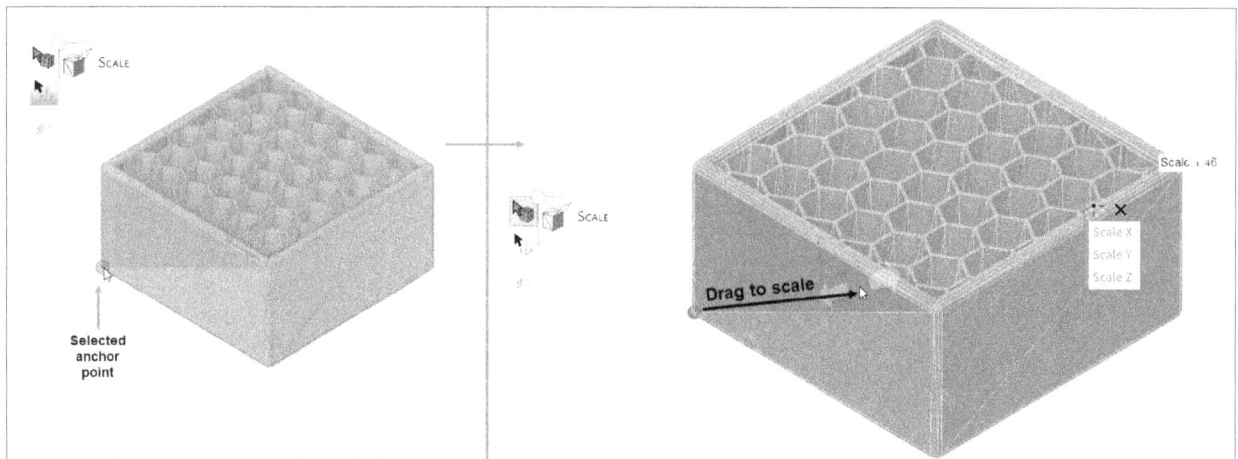

Figure-19. Scaling facet body

*   Press **ESC** to exit the tool.

Using Thicken Operation

The **Thicken** tool is used to apply thickness to selected facet face/surface. The procedure to use this tool is given next.

*   Click on the **Thicken** tool from the **MODIFY** panel in the **FACETS** tab of the **Ribbon**. The **THICKEN HUD** toolbar will be displayed.
*   Specify desired value in the **Thickness** edit box to define thickness of solid feature created from the facet.
*   Set the other parameters as discussed earlier and click on the facet surface feature to be thickened. The feature will be created; refer to Figure-20.

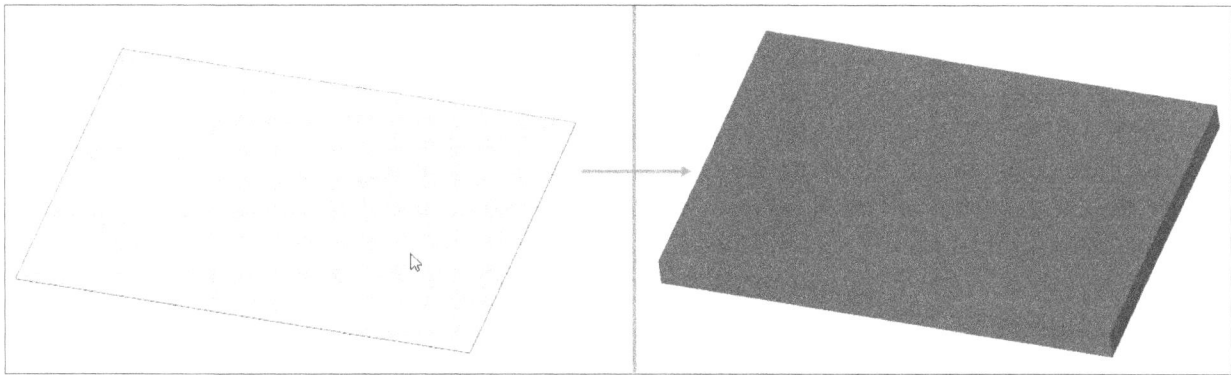

Figure-20. Applying thickness to facet

Performing Facet Smoothening Operation

The **Smooth** tool is used to smoothen selected facets of the model so that there are no peaks in the facets. The procedure to use this tool is given next.

- Click on the **Smooth** tool from the **ADJUST** panel in the **FACETS** tab of the **Ribbon**. The **SMOOTH HUD** toolbar will be displayed.
- Specify desired value in the **Angle threshold** edit box to define angle value upto which facet edges are allowed.
- Select the **Flatten peaks** toggle button from the right **HUD** toolbar to flatten sharp peaks in the facets; refer to Figure-21.

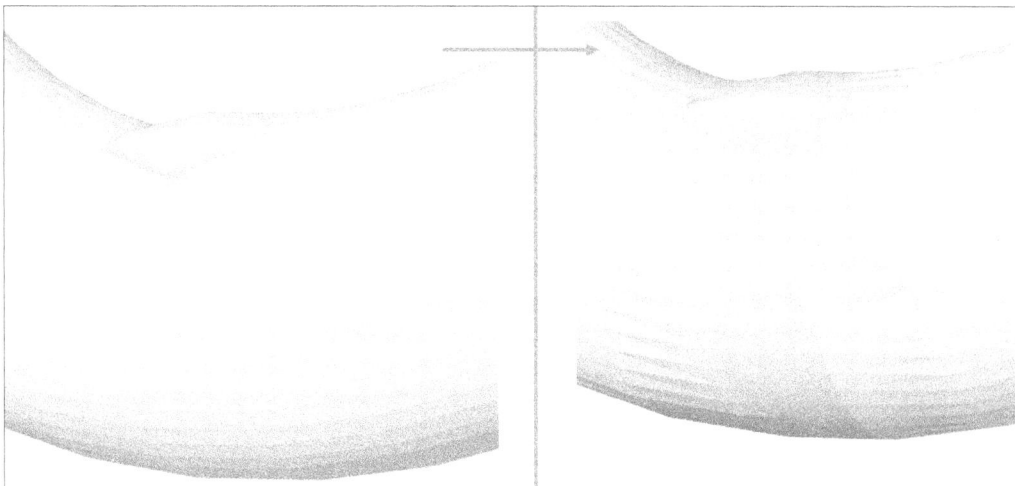

Figure-21. Applying flatten peaks smoothening operation

- Select the **Add facets** toggle button to add more facets in sharp peaks section of facet to smoothen the transition.
- Select the **Volume aware** toggle button if you want to smoothen the facets according to curvature intent of the model while applying smoothening operation.
- After setting desired parameters, select the region to be smoothen and click on the **OK** button from the toolbar. The operation will be applied.

Reducing Facets Count

The **Reduce** tool is used to decrease number of facets in selected region by specified percentage. The procedure to use this tool is given next.

- Select desired region of facets where you want to reduce number of facets and click on the **Reduce** tool from **ADJUST** panel in the **FACETS** tab of the **Ribbon**. The **REDUCE HUD** toolbar will be displayed.
- Specify desired value in the **Triangle reduction** edit box to define number of triangles to be reduced from selected facets in percentage.
- Set desired value in the **Maximum deviation** edit box to define distance between triangle edges of facet and original model edge.
- After setting desired parameters, click on the **OK** button from the HUD toolbar to apply operation. Press **ESC** to exit the tool.

Tip: Triple click on a facet of model to select all the facets of that model.

Regularizing Facets for Improved Mesh

The **Regularize** tool is used to make facet meshing uniform in the model. The procedure to use this tool is given next.

- Click on the **Regularize** tool from the **ADJUST** panel in the **FACETS** tab of the **Ribbon**. The **REGULARIZE HUD** toolbar will be displayed and you will be asked to select facets to be regularized by specified size parameters.
- Set desired values in the right **HUD** toolbar like maximum edge length of facets, threshold angle, and so on to define shape and size parameters for facets.
- Select the object(s) from the graphics area and click on the **OK** button from left **HUD** toolbar to perform the operation; refer to Figure-22. Press **ESC** to exit the tool

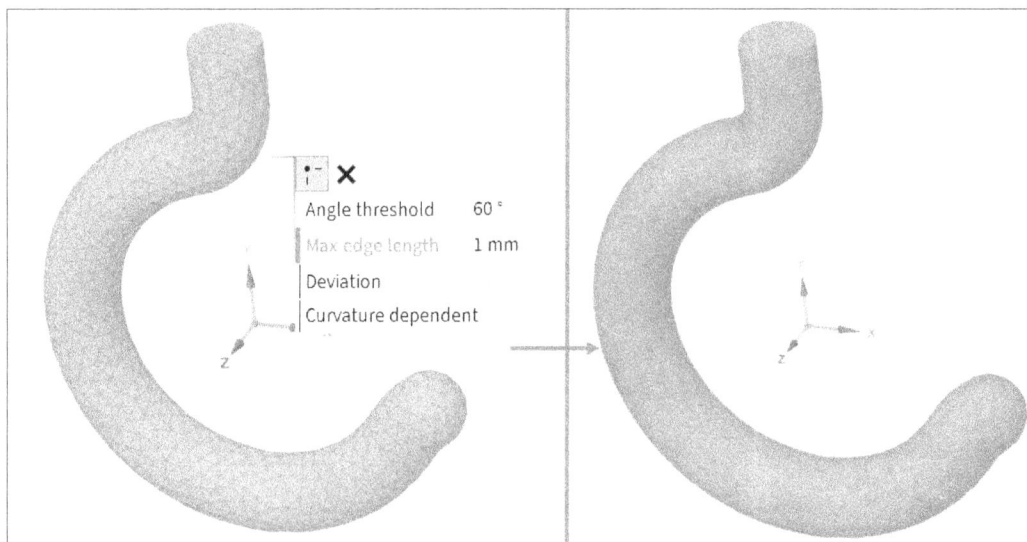

Figure-22. Regularizing facets

Inspecting Overhangs

The **Overhangs** tool is used to find out sections of model that would need a support when performing 3D printing of the model. The procedure to use this tool is given next.

- Click on the **Overhangs** tool from the **INSPECT** panel in the **FACETS** tab of the **Ribbon**. The **OVERHANGS HUD** toolbar will be displayed; refer to Figure-23.

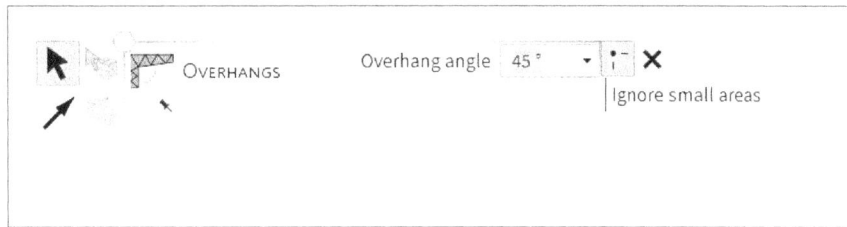

Figure-23. OVERHANGS HUD toolbar

- Set desired parameters in the **Overhang angle** edit box to define angle value above which faces will be considered overhang.
- Select the **Ignore small areas** toggle button to specify size of overhanging areas that can be ignored by the tool when performing check.
- After setting desired parameters, select the model to check overhanging areas; refer to Figure-24.

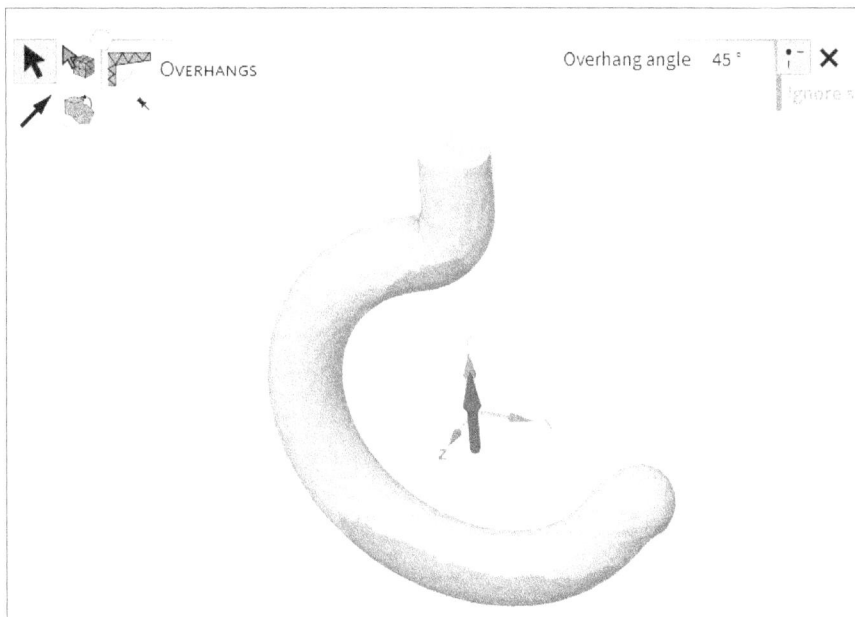

Figure-24. Overhanging areas in model

- Click on the **Direction** button ⬈ from the left **HUD** toolbar to select a line or axis reference for defining analysis direction for the tool and then select desired edge/line/axis.
- Click on the **Minimize overhang area** button from the left **HUD** toolbar to reorient the model so that minimum overhang area is generated.
- Click on the **Convert highlighted items to selection** button from the **HUD** toolbar to add current highlighted overhang area in selection.
- Press **ESC** to exit the tool.

Checking Thickness of Facets

The **Thickness** tool is used to check minimum thickness regions on the model. If there is any region in model where thickness of facets is less than specified value then those regions will be highlighted in red color and the regions passing this check will be displayed in green color. The procedure to use this tool is given next.

- Click on the **Thickness** tool from the **INSPECT** panel in the **FACETS** tab of the **Ribbon**. The **THICKNESS HUD** toolbar will be displayed.

- Set desired value in the **Minimum thickness** edit box to define the thickness of facets. The facets with thickness below this value will be highlighted in red color in graphics area.
- After setting desired parameters, click on the facet body. The regions passing and failing check will be highlighted in different colors; refer to Figure-25. Press **ESC** to exit the tool.

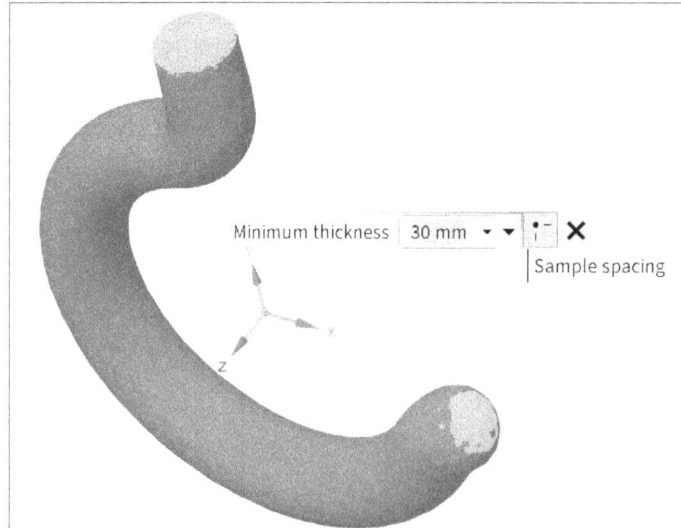

Figure-25. Checking thickness

Checking Cavities

The **Cavity** tool is used to find out empty regions trapped by facets in the body. The procedure to use this tool is given next.

- Click on the **Cavity** tool from the **INSPECT** panel in the **FACETS** tab of the **Ribbon**. The **CAVITY HUD** toolbar will be displayed and you will be asked to select the faceted body for performing check.
- If you want to ignore a specified amount of gap in model then select the **Max opening** toggle button and specify the size of gap that can be ignored by tool when performing checks.
- After setting desired parameters, click on the model to perform checks. Press **ESC** to exit the tool.

Generating CAD Body from Facet Body

The **Auto Skin** tool is used to convert selected facet body into a solid body. The procedure to use this tool is given next.

- Click on the **Auto Skin** tool from the **REVERSE ENGINEERING** panel in the **FACETS** tab of the **Ribbon**. The **AUTO SKIN HUD** toolbar will be displayed and you will be asked to select faceted body.
- Select desired faceted body from the graphics area and click on the **OK** button from the **HUD** toolbar. The body will be converted to patched solid body; refer to Figure-26.

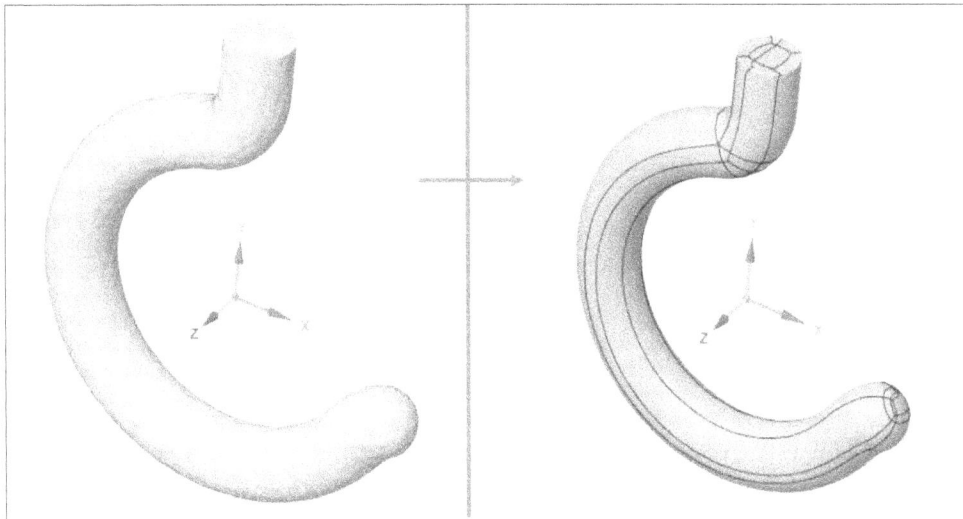

Figure-26. Patched body created

- Press **ESC** to exit the tool. Note that you can show/hide both patched CAD body and faceted body from the **Design Tree**.

Generating Surface from Facet Body

The **Skin Surface** tool is used to create a surface by using specified closed boundary. The procedure to use this tool is given next.

- Click on the **Skin Surface** tool from the **REVERSE ENGINEERING** panel in the **FACETS** tab of the **Ribbon**. The **SKIN SURFACE HUD** toolbar will be displayed.
- Single click in the face of faceted body at desired location to specify start point of surface boundary and then click at desired locations to specify intermediatory points of the boundary curve; refer to Figure-27.

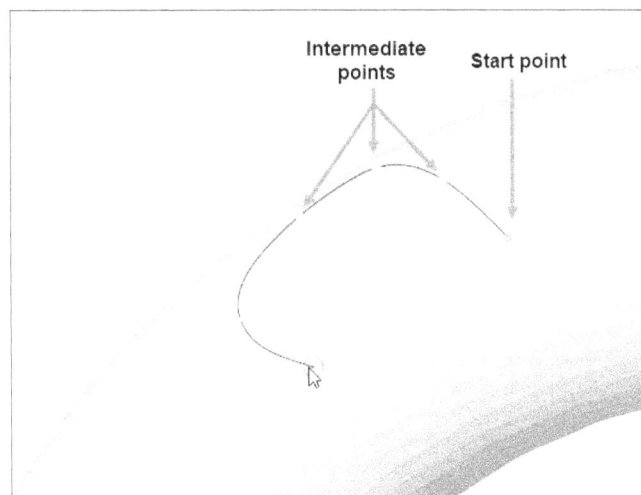

Figure-27. Specifying start point and intermediate points

- Double-click at desired location to specify a corner point; refer to Figure-28. Note that you need to specify at least two corner points by double-clicking before you can create a closed loop boundary for surface.

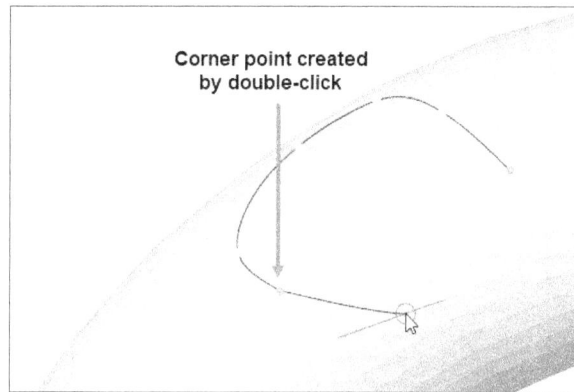

Figure-28. Creating corner point

- After creating intermediate corner points, click at the start point to create closed loop boundary curve; refer to Figure-29. The surface boundary will be generated.

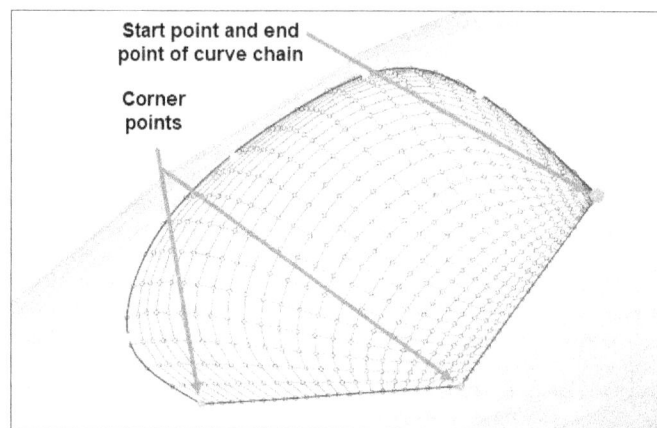

Figure-29. Creating surface boundary

- Set desired value in the **Samples** edit box of right **HUD** toolbar to define number of points to be taken on the control curves to draw surface.
- Select the **Full preview** toggle button to check full preview with specified number of points on the curve.
- Set the other parameters as discussed earlier and click on the **OK** button from the right **HUD** toolbar. The surface will be generated; refer to Figure-30. You can hide rest of the model from **Design Tree** to check the surface. Press **ESC** to exit the tool.

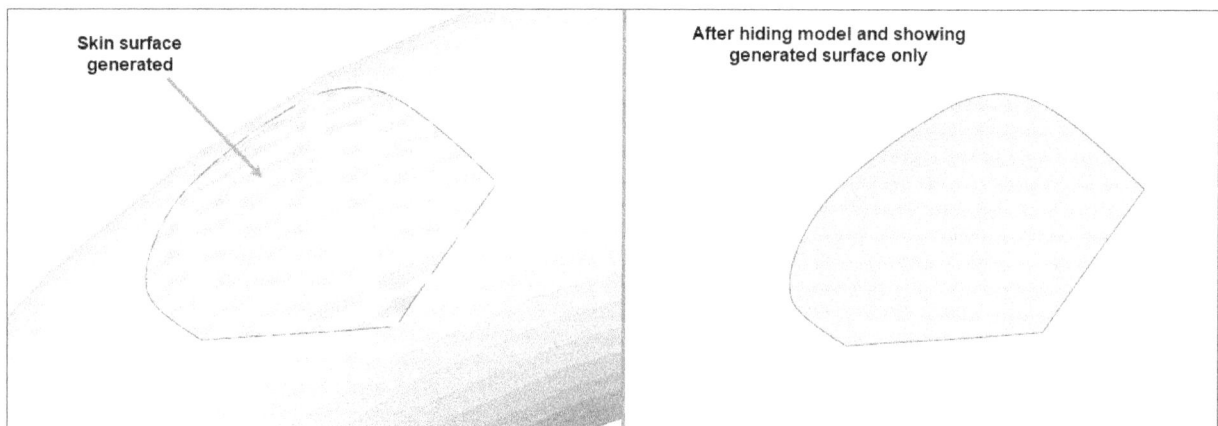

Figure-30. Surface generated

Generating Cross-section Curves from Faceted Body

The **Extract Curves** tool is used to generate curves at the intersection of a plane and faceted body. So, you need to make sure that a plane intersecting with the body at desired orientation is already created before using this tool. The procedure to use this tool is given next.

- Click on the **Extract Curves** tool from the **REVERSE ENGINEERING** panel in the **FACETS** tab of the **Ribbon**. The **EXTRACT CURVES HUD** toolbar will be displayed; refer to Figure-31.

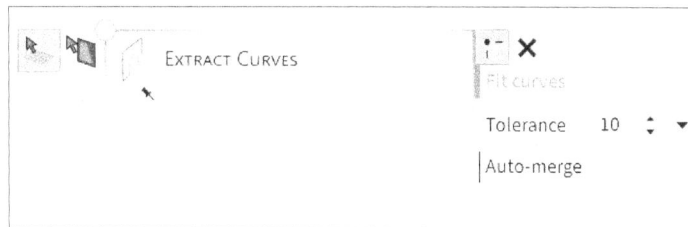

Figure-31. EXTRACT CURVES HUD Toolbar

- Select the **Fit curves** toggle button to make extracted curve follow curvature of faceted body.
- Specify desired value in the **Tolerance** spinner to define how closely the extracted curves will follow curvature of faceted body. Lower value means lesser deviation from the curvature.
- Select the **Auto-merge** toggle button from right **HUD** toolbar to automatically join segments of extracted curve to form a spline.
- Select the plane intersecting with faceted body to generate curves and click on the **OK** button from the left **HUD** toolbar. The curves will be generated at intersection; refer to Figure-32. If there are multiple intersections between plane and faceted body and you want to specifically define which curves to be extracted then click on the **Select curves** button from **HUD** toolbar after selecting plane and select desired curves.

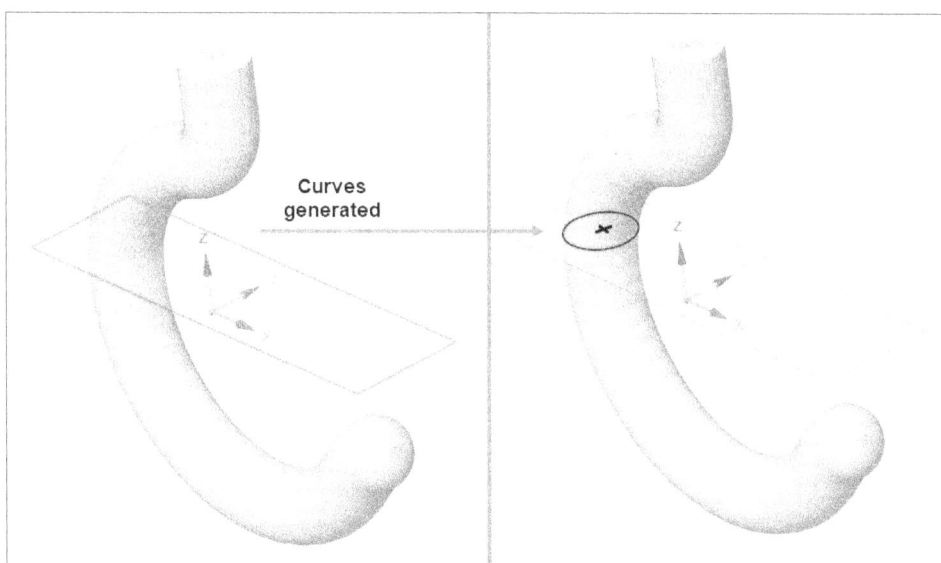

Figure-32. Extracting curves

- Press **ESC** to exit the tool.

Creating Spline Bound Surface

The **Fit Spline** tool is used to create surface around selected section of faceted body using spline curve bounding selected facets. The procedure to use this tool is given next.

• Select desired facets from the faceted body and click on the **Fit Spline** tool from the **REVERSE ENGINEERING** panel in the **FACETS** tab of the **Ribbon**. The spline surface will be created; refer to Figure-33.

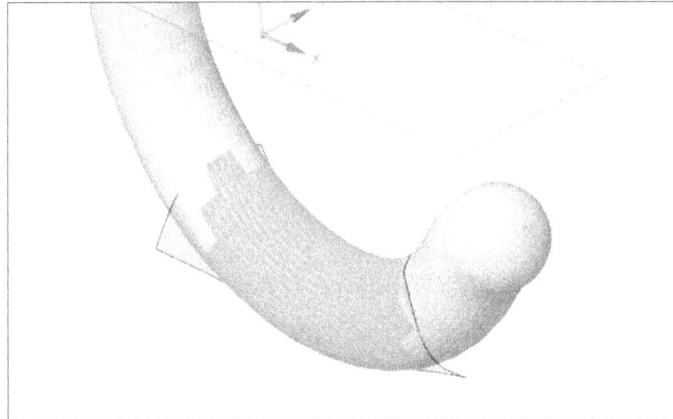

Figure-33. Surface generated by Fit Spline

• Press **ESC** to exit the tool.

Re-orienting Facets

The **Orient Facets** tool is used to re-orient the faceted body using selected facet. The procedure to use this tool is given next.

• Click on the **Orient Facets** tool from the **REVERSE ENGINEERING** panel in the **FACETS** tab of the **Ribbon**. You will be asked to select facet of faceted model to be used for orientation.
• Hover the cursor on a facet to check possible orientation plane and select the facet when you are satisfied with orientation. The model will be reoriented.

Edge Display Options

The options in the **EDGE DISPLAY** panel of **FACETS** tab are used to show/hide various elements of the facet edge. Various options in this panel are discussed next.

• Select the **Internal** toggle button from the panel to display edges generated at the intersection of multiple facets within the outer boundaries of the body; refer to Figure-34.

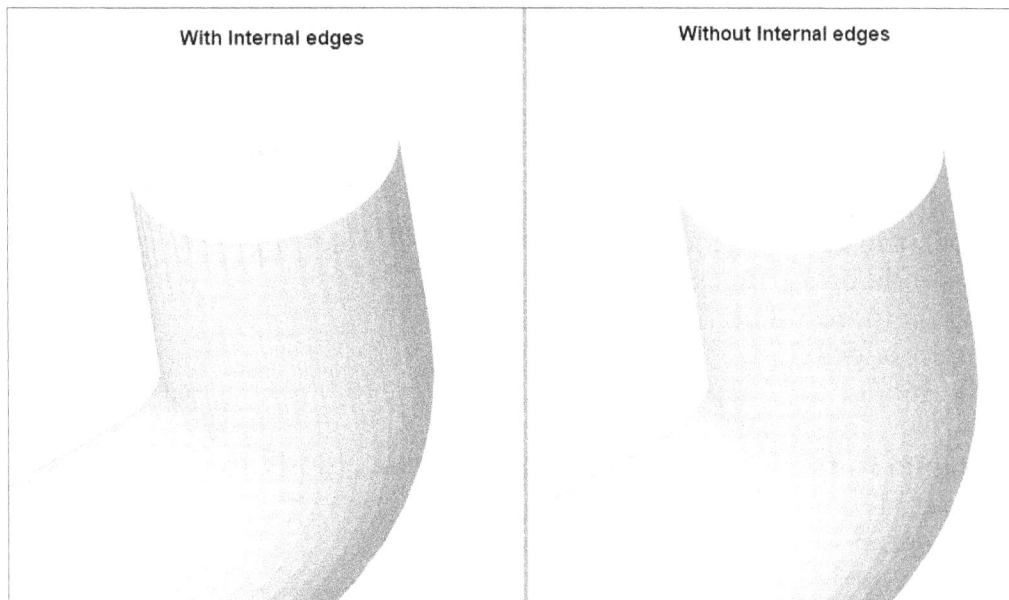

Figure-34. Model with and without internal edges

*   Select the **Open** toggle button from the **EDGE DISPLAY** panel in **Ribbon** to show open edges of the faceted body.
*   Select the **Over-connected** toggle button to highlight over-connected edges of the faceted body.

SELF-ASSESSMENT

Q1. Which panel in the REPAIR tab contains the Stitch tool?
A. FIX panel
B. SOLIDIFY panel
C. ORGANIZE panel
D. MODIFY panel

Q2. What is the function of the Stitch tool?
A. To combine surfaces into a single body
B. To check and repair facets
C. To split a facet body
D. To remove extra edges

Q3. What does the "Check for coincidence" toggle option do in the Stitch tool?
A. It removes duplicate edges
B. It ensures faces do not overlap before stitching
C. It simplifies model geometry
D. It converts solid bodies into facet bodies

Q4. Which tool is used to find and fill gaps in surfaces of imported objects?
A. Stitch
B. Gaps
C. Convert
D. Separate

Q5. How can split edges in an imported model be detected?
A. Using the Check Facets tool
B. Using the Split Edges tool
C. Using the Stitch tool
D. Using the Shrinkwrap tool

Q6. What does the Convert tool do?
A. Converts solids or surfaces into facets
B. Repairs gaps in surfaces
C. Detects split edges
D. Scales the model

Q7. Which panel in the FACETS tab contains the Auto Fix tool?
A. MODIFY panel
B. CLEANUP panel
C. ORGANIZE panel
D. FIX panel

Q8. What does the Auto Fix tool do?
A. Removes duplicate faces
B. Automatically repairs defective facets
C. Converts models into mesh
D. Splits facet bodies

Q9. The Shrinkwrap tool creates a shrunk copy of a selected body by:
A. Merging its facets
B. Applying a shell feature
C. Using a specified shrinkage value
D. Splitting the facet body

Q10. Which tool is used to join multiple facet bodies into a single body?
A. Join
B. Merge
C. Convert
D. Scale

Q11. What is the purpose of the Split tool?
A. To divide a facet body using a plane
B. To separate all facet bodies
C. To fix extra edges
D. To detect and fix gaps

Q12. The Subtract tool removes:
A. Selected tool body from the main body
B. Small facets from the mesh
C. Overlapping edges
D. Duplicate surfaces

Q13. What does the Shell tool do?
A. Converts solid bodies into facets
B. Creates a thin-walled body from a solid facet body
C. Detects duplicate faces
D. Identifies and fills gaps

Q14. The Thicken tool applies:
A. Additional facets to a model
B. Thickness to selected facet surfaces
C. Smoothing to facet edges
D. Shrinkwrapping to a body

Q15. Which of the following tools is used to smoothen peaks in a facet body?
A. Stitch
B. Thicken
C. Smooth
D. Check Facets

Q16. Which tool is used to identify model sections that require support for 3D printing?
A. Thickness
B. Overhangs
C. Cavity
D. Auto Skin

Q17. What is the purpose of the Overhang angle edit box in the Overhangs tool?
A. To set the minimum thickness of facets
B. To define the angle above which faces are considered overhang
C. To specify the number of support structures needed
D. To define the depth of cavities

Q18. Which button in the Overhangs tool is used to reorient the model for minimizing overhang areas?
A. Convert highlighted items to selection
B. Minimize overhang area
C. Ignore small areas
D. Direction

Q19. What does the Thickness tool highlight when the thickness is below the specified value?
A. Blue
B. Yellow
C. Green
D. Red

Q20. What is the function of the Cavity tool?
A. To check minimum thickness of facets
B. To detect trapped empty regions in a faceted body
C. To create a spline-bound surface
D. To extract intersection curves

Q21. Which tool is used to convert a faceted body into a solid body?
A. Auto Skin
B. Skin Surface
C. Extract Curves
D. Fit Spline

Q22. What must be specified before using the Skin Surface tool?
A. A faceted body
B. A closed boundary
C. A thickness value
D. A minimum overhang angle

Q23. Which tool generates intersection curves between a plane and a faceted body?
A. Fit Spline
B. Extract Curves
C. Orient Facets
D. Cavity

Q24. What does the Auto-merge toggle button in the Extract Curves tool do?
A. Joins segments of extracted curves to form a spline
B. Increases the overhang angle
C. Highlights minimum thickness areas
D. Removes small cavity openings

Q25. Which tool is used to create a surface by bounding selected facets with a spline?
A. Auto Skin
B. Extract Curves
C. Fit Spline
D. Thickness

Q26. What is the purpose of the Orient Facets tool?
A. To create a CAD body from a faceted body
B. To check facet thickness
C. To adjust the orientation of a faceted body
D. To extract intersection curves

Q27. What does the Internal toggle button in the Edge Display panel do?
A. Highlights internal edges within the model
B. Displays over-connected edges
C. Shows thickness variations
D. Highlights cavity openings

Q28. What is the function of the Over-connected toggle button in the Edge Display panel?
A. Highlights edges that have more connections than expected
B. Displays internal edges only
C. Shows open edges of the faceted body
D. Removes small overhang areas

Chapter 4

Ansys Discovery-
Preparing Model

Topics Covered

The major topics covered in this chapter are:

- *Beams and Shell Creation Tools*
- *Adding Bolts*
- *Applying Weld Joints*
- *Generating Volumes and other Features*
- *Removing Features*
- *Identifying Features from Imported Model*
- *Sharing Topology*

INTRODUCTION

In Ansys Discovery, you will import various complex models to perform analyses in your career. Most of the time, there will be features in the model that would not be necessary for performing analyses like chamfers, rounds on edges, features placed for aesthetics, and so on. There are also instances where you can replace the objects in 3D model by structural members like tubes, channels, angles, etc. to simplify the analysis and reduce solution time. The tools to perform these tasks for preparing model are available in the **PREPARE** tab of the **Ribbon**; refer to Figure-1. Various tools of this tab are discussed next.

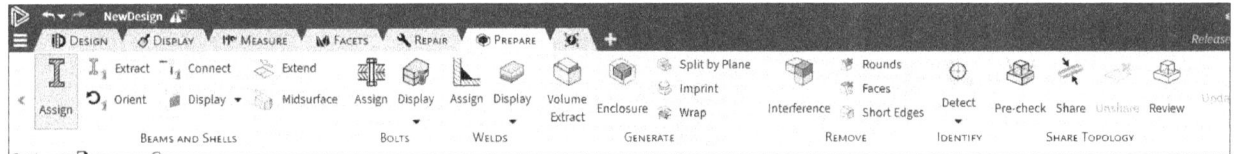

Figure-1. PREPARE tab in Ribbon

BEAMS AND SHELL CREATION TOOLS

The tools of **BEAMS AND SHELLS** panel are used to assign and generate structural features from the model to simplify the analysis. You can use these tools to convert tubes, rods, bars, and so on to beam elements. Various tools of this panel are discussed next.

Extracting Beams from Solid Model

The **Extract** tool is used to automatically extract beam elements from the solid model which has uniform cross-section shaped like beam with/without taper faces. The procedure to use this tool is given next.

- Click on the **Extract** tool from the **BEAMS AND SHELLS** panel in the **PREPARE** tab of the **Ribbon**. The **EXTRACT HUD** toolbar will be displayed and you will be asked to select solid objects from which beams are to be extracted.
- Select the objects that have been created as beams/columns in CAD software and you want to extract them as beams; refer to Figure-2. On extraction, beams and columns will change to line elements.

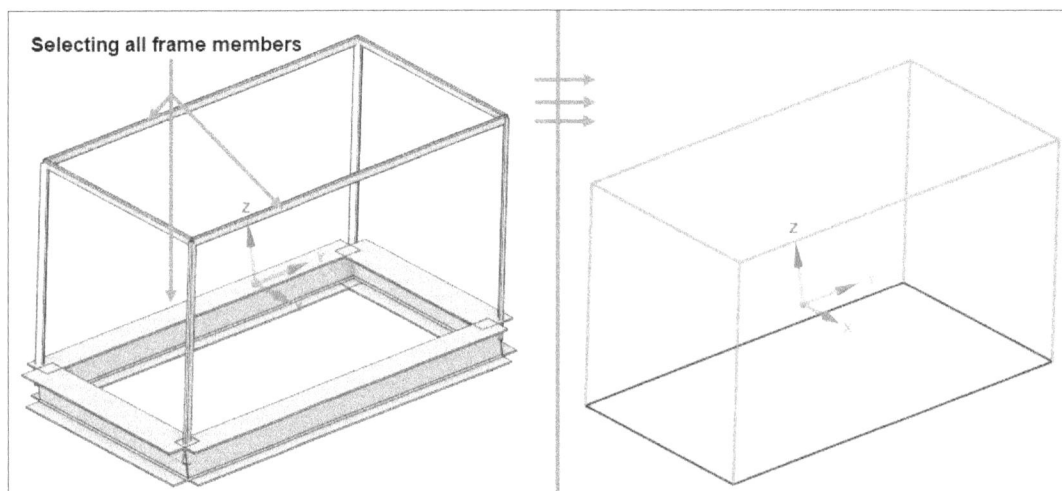

Figure-2. Extracting frame members

- Press **ESC** to exit the tool.

Assigning Beam to Curves

The **Assign** tool is used to assign a specific type of beam element to selected curve (sketch curves or extracted curves). The procedure to use this tool is given next.

- Click on the **Assign** tool from the **BEAMS AND SHELLS** panel in the **PREPARE** tab of the **Ribbon**. The **ASSIGN BEAMS HUD** toolbar will be displayed and you will be asked to select line segments to be assigned as beams.

- Hold **CTRL** key while selecting segments to perform multiple selection and select desired line segments (sketched or extracted) from the drawing area. After selecting line segments, select desired beam type from the drop-down in **HUD** toolbar; refer to Figure-3. The beam type will be assigned to selected segments and detailed properties of the beam will be displayed; refer to Figure-4. You can check the properties of created beams using relevant categories in displayed information.

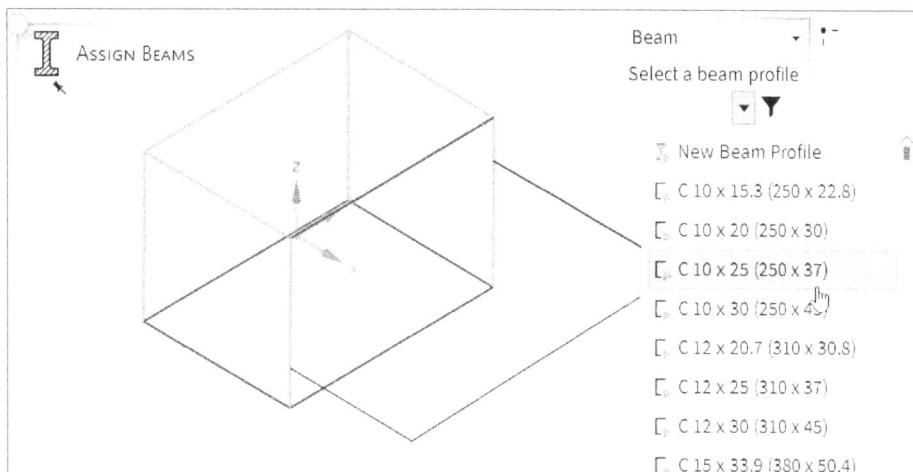

Figure-3. Assigning beams

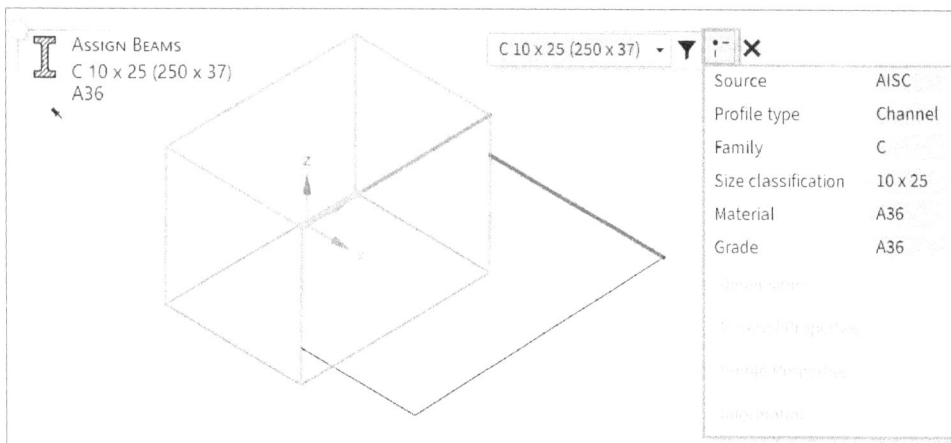

Figure-4. Details of selected beam

- You can also create and use custom profiles for the beams by selecting the **New Beam Profile** option from the drop-down in **HUD** toolbar. Set desired parameters in the displayed menu to define properties of new profile. You can the new profile in the same way as discussed earlier. Press **ESC** twice to exit the tool.

Showing/Hiding 3D Profile of Beams

The options in the **Display** drop-down are used to show/hide 3D profiles of beams; refer to Figure-5. Select the **Display 3D Profiles** option from the drop-down to display beam elements in 3D shapes and select the **Hide 3D Profiles** option if you want to display beam elements as line objects; refer to Figure-6.

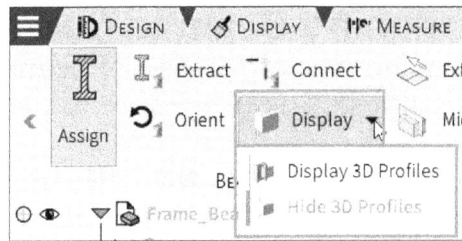

Figure-5. Display drop-down

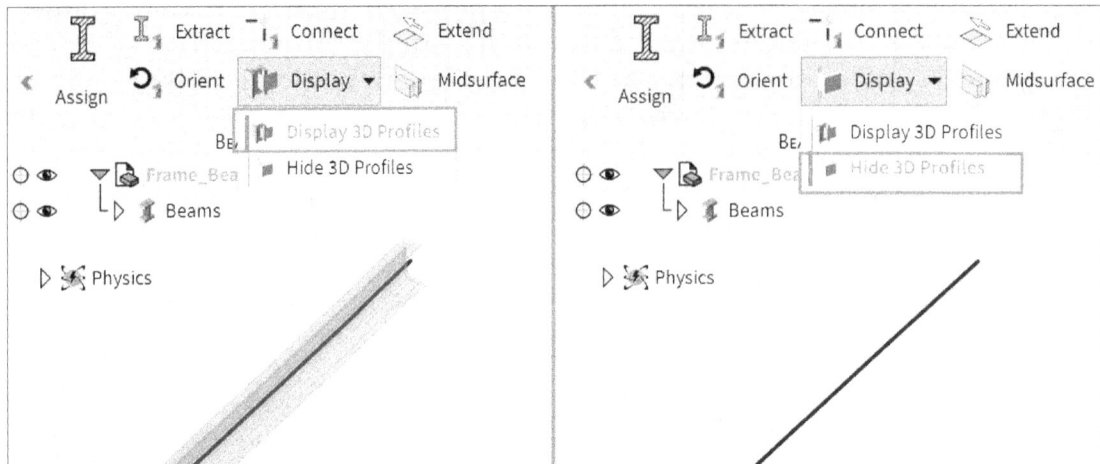

Figure-6. Showing hiding 3D profiles

Orienting Beams

The **Orient** tool is used to change the orientation of 3D profile of selected beam(s). This operation is performed when orientation of your beams is not as required by design or one of the beam element is facing different direction then intended. The procedure to use this tool is given next.

• Click on the **Orient** tool from the **BEAMS AND SHELLS** panel in the **PREPARE** tab of the **Ribbon** and select the beam(s) to be reoriented. The handles to change orientation of beam(s) are displayed; refer to Figure-7.

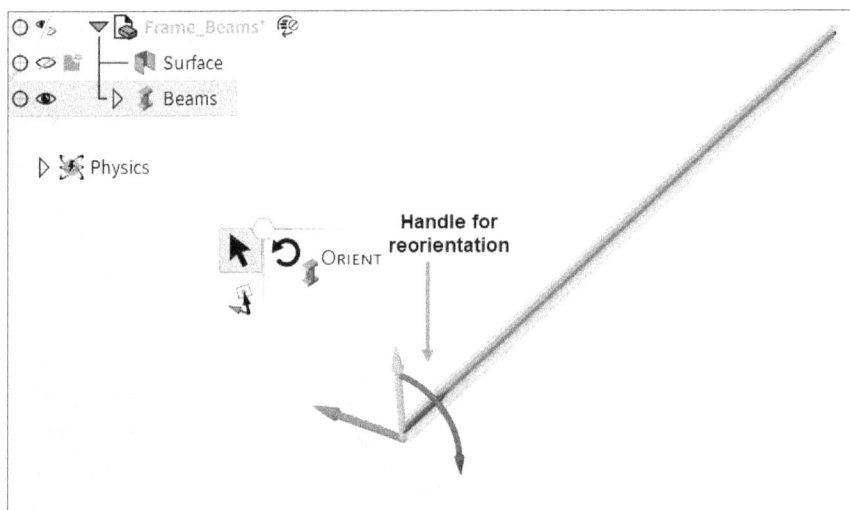

Figure-7. Reorientation handle for beam

• Use the drag handles to change the orientation of beams.

Connecting Gaps in Frame

The **Connect** tool is used to connect small gaps left in the wire (line sketch) frame model of beams when importing from other software. The procedure to use this tool is given next.

- Click on the **Connect** tool from the **BEAMS AND SHELLS** panel in the **PREPARE** tab of the **Ribbon** if you have beam elements in the model that have unintentional gaps. The **CONNECT HUD** toolbar will be displayed and you will be asked to select gaps to be connected; refer to Figure-8.

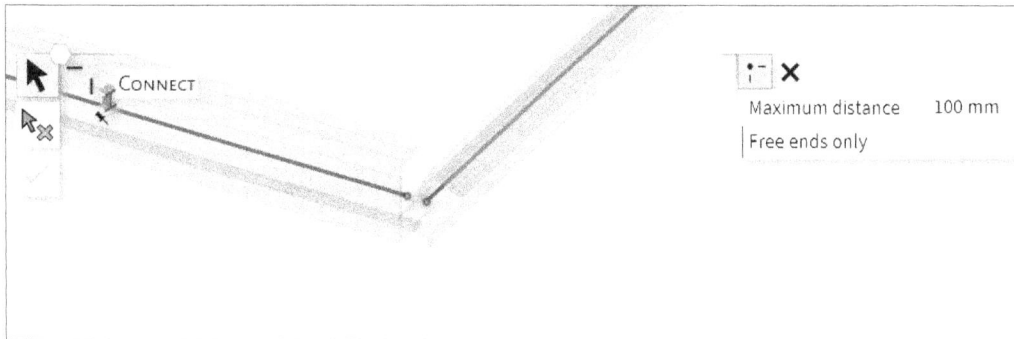

Figure-8. CONNECT HUD Toolbar

- Specify desired value in the **Maximum distance** edit box to define the maximum gap distance that will be highlighted for using **Connect** tool. If gap is more than the specified value then software will consider it as intentional gap and will not connect it.
- Select the **Free ends only** toggle button from the **HUD** toolbar if you want to connect only free end points of beams which are not connected to any other beams already.
- After setting desired parameters, select the highlighted end points to connect them. The selected end points of beams will be connected. If you want to connect multiple gap points then use window selection to select all the point sets and click on the **OK** button from the **CONNECT HUD** toolbar.

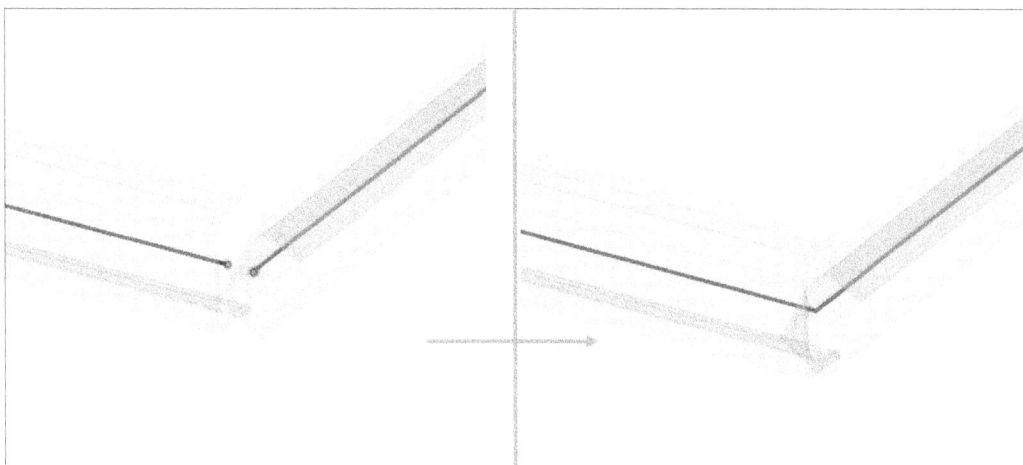

Figure-9. Connecting beams

- If you want to mark a highlighted set of points as problem area that should not be connected when using window selection for connecting multiple points then click on the **Select problem areas** selection button ![icon] from the **HUD** toolbar and select such end points set; refer to Figure-10. Click again on the set of points if you have selected them unintentionally.

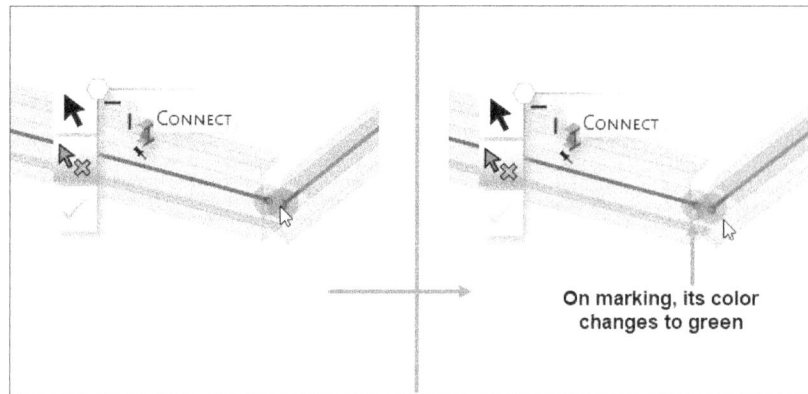

On marking, its color
changes to green

Figure-10. Marking problem area

- Press **ESC** to exit the tool.

Generating Mid Surface

Midsurfaces are generating from the solid model to simplify the analytical model. Ultimately, this analytical model will be used for performing various analyses. As the name suggests, midsurfaces are the surfaces created at the center of each wall of the model and these combined together surfaces show the design intent of model. The procedure to generate midsurfaces is given next.

- Click on the **Midsurface** tool from the **BEAMS AND SHELLS** panel in the **PREPARE** tab of the **Ribbon**. The **MIDSURFACE HUD** toolbar will be displayed; refer to Figure-11.
- Select the **Use selected faces** option from the right **HUD** toolbar if you want to select two offset faces and midsurface will be created at center between these two faces.

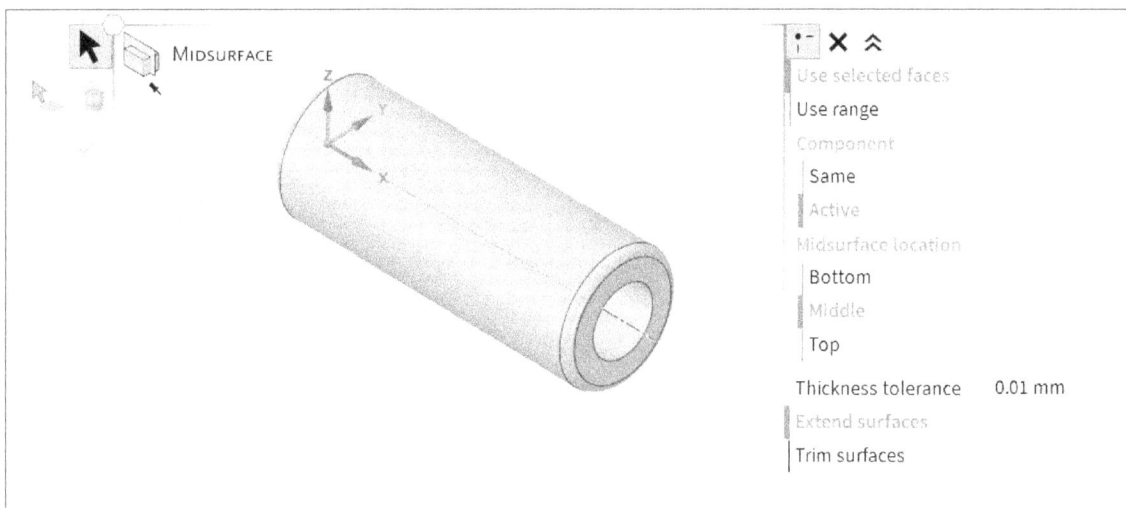

Figure-11. MIDSURFACE HUD Toolbar

- Select the **Use range** option if your model is too large or too complex to select faces individually and you want to define a maximum and minimum range for thickness of walls in the model. The sections of model which fall within this range will be selected automatically and midsurface will be created at the center of each pair of faces.
- Select the **Same** option from **Component** section in the right **HUD** toolbar if you want to add newly created midsurface(s) as part of selected component(s). Select the **Active** option from this section if you want to add midsurface(s) as new feature in the current active component of **Design Tree**.

- Select desired option from the **Midsurface location** section of the **HUD** toolbar to define the location of midsurface with respect to selected face set of solid. For example, select the **Top** option if you want to create midsurface at the top of selected face of solid.
- Specify desired value in the **Thickness tolerance** edit box to define maximum error allowed when detecting thickness of model for creating midsurface.
- Select the **Extend surfaces** toggle button from **HUD** toolbar to automatically extend midsurfaces to nearby surface closing unintended gaps.
- Select the **Trim surfaces** toggle button from the toolbar if you want to automatically trim midsurfaces intersecting with other nearby surfaces.
- After setting desired parameters, click on the outer and inner faces of model if **Use selected faces** toggle button is active in **HUD** toolbar and click on the **OK** button to created the mid surface; refer to Figure-12.

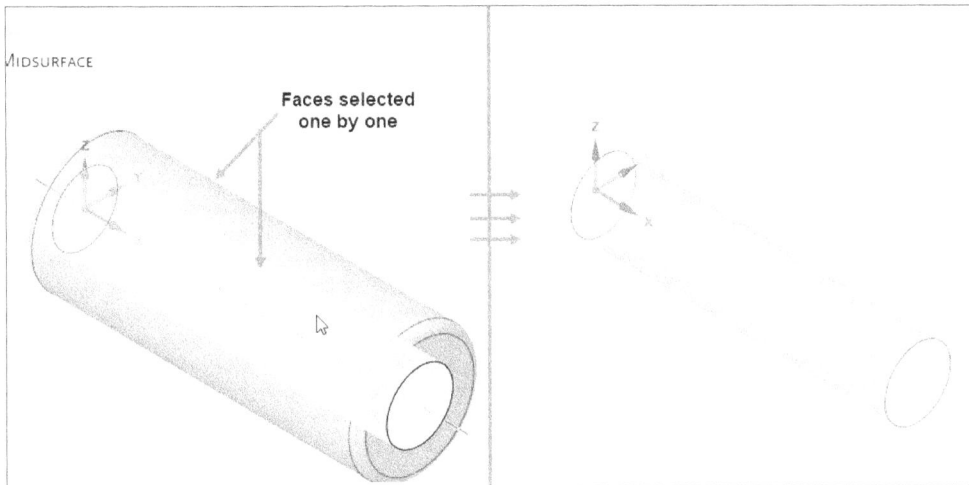

Figure-12. Generating midsurface

Extending/Trimming Surfaces

The **Extend** tool is used to extend selected surface edge to intersecting bodies. You can also use this tool to trim a portion of selected surface. The procedure to use this tool is given next.

- Click on the **Extend** tool from the **BEAMS AND SHELLS** panel in the **PREPARE** tab of the **Ribbon**. The **EXTEND HUD** toolbar will be displayed and edges of surfaces that can be extended to nearby face/surface will be highlighted in red color; refer to Figure-13.

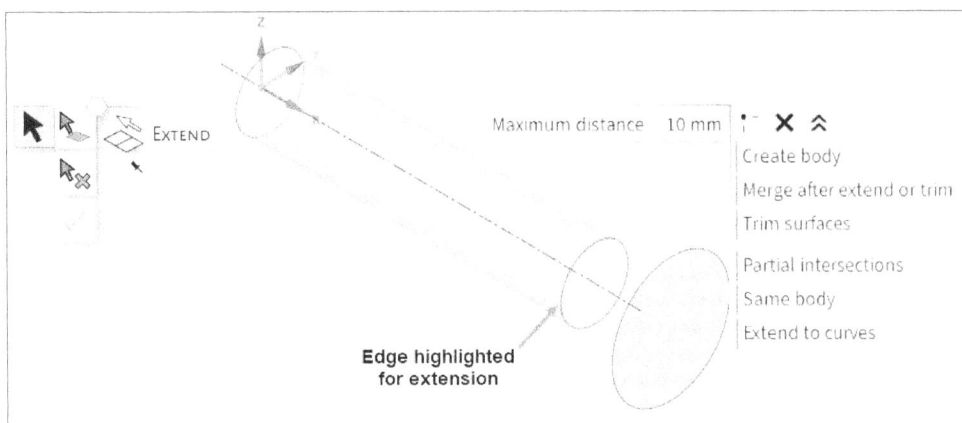

Figure-13. EXTEND HUD toolbar

- Select the highlighted edge to extend it to nearest face/surface. Preview of extension will be displayed; refer to Figure-14.

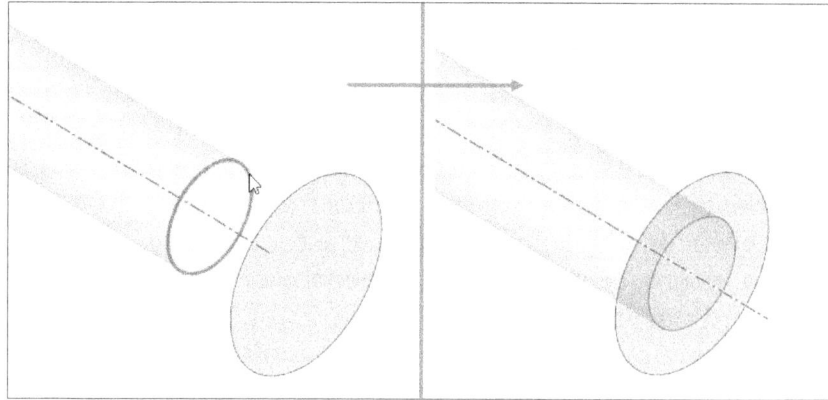

Figure-14. Preview of extension

- Similarly, select the **Trim surfaces** option from the right **HUD** toolbar if you want to trim a surface to nearest reference face/surface and select the highlighted edges; refer to Figure-15.

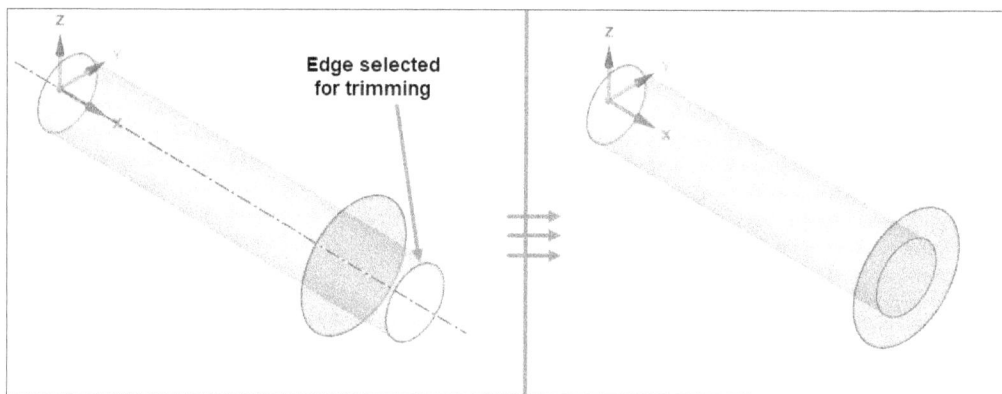

Figure-15. Trimming surface

- Select the **Create body** toggle option from the toolbar if you want to create a new body which joins two surfaces by extension. This will create **WeldSurface** bodies in **Design**; refer to Figure-16.

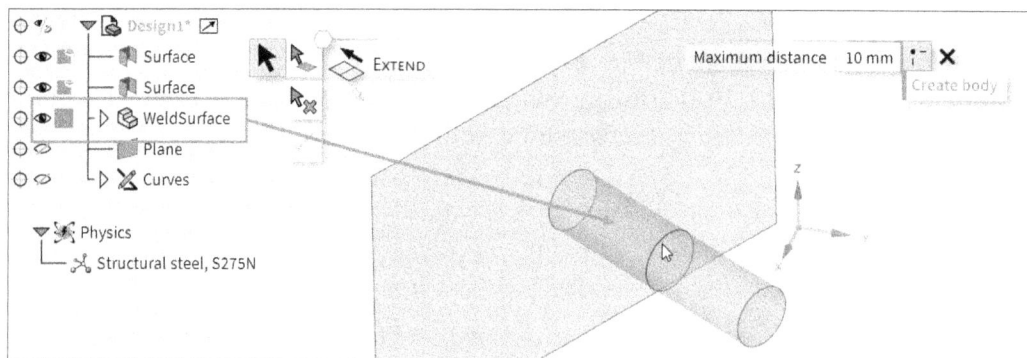

Figure-16. WeldSurface created

- Select the **Merge after extend or trim** toggle option from the shortcut menu if you want to merge the surfaces after performing trimming or extension operation.
- Select the **Partial intersections** toggle option if you want to include faces/surfaces that cause partial intersection with selected surfaces for trimming or extension operation; refer to Figure-17.

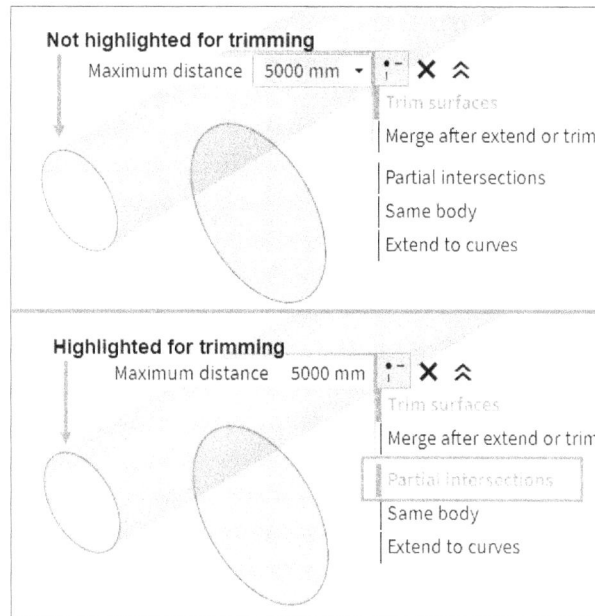

Figure-17. Partial intersections option

- Select the **Same body** toggle option from the **HUD** toolbar if you want to use a face/edge of same body to perform trimming or extension.
- Select the **Extend to curves** toggle option from toolbar if you want to include curves along with surfaces/faces to be used as reference for trimming/extension.
- After setting desired parameters, click on the **OK** button to apply the operation. Press **ESC** to exit the tool.

ASSIGNING BOLTS

When performing analyses, you will find assemblies where two or more components are bolted together using various fasteners. Most of the time in such assemblies, your focus area will not be the bolted site but when you will be generating mesh in such cases, system will automatically allot resources for meshing the bolts and nuts. Even worse, system will generate fine mesh for these bolts. You can avoid this situation by assigning bolts in Ansys Discovery. The procedure to assign bolts is given next.

- Click on the **Assign** tool from the **BOLTS** panel in the **PREPARE** tab of the **Ribbon**. The **ASSIGN BOLTS HUD** toolbar will be displayed; refer to Figure-18 and you will be asked to select coaxial revolved faces for assigning bolt connection. Note that there should be two or more components in the drawing area with coaxial holes to which you can assign the bolts.

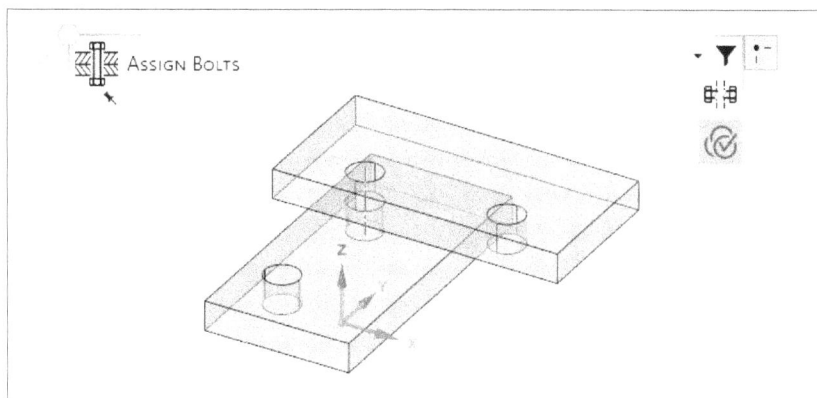

Figure-18. ASSIGN BOLTS HUD toolbar

- Select desired hole/hole set from the model to which you want to assign bolt. The options to assign bolt type will be displayed; refer to Figure-19.

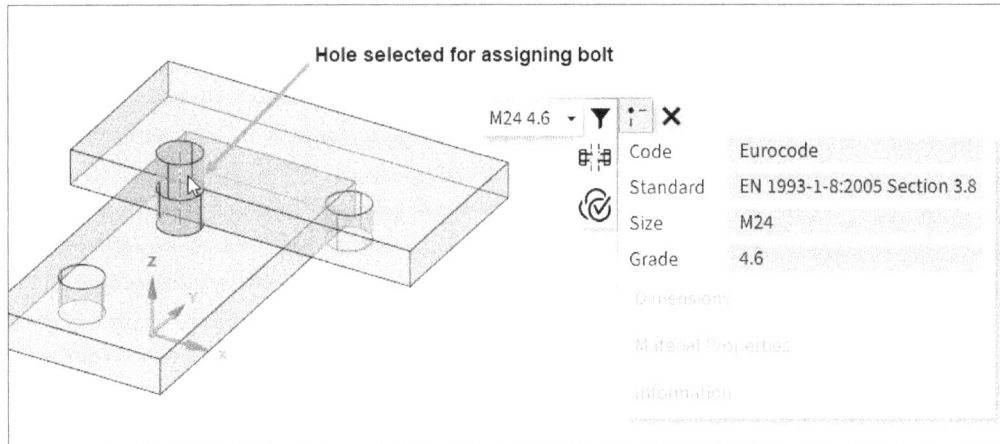

Figure-19. Hole selected for assigning bolt

- Select desired option from the drop-down list in right **HUD** toolbar to define bolt size and click on the **OK** button from the toolbar. Press **ESC** to exit the tool.

To display 3D bolts, select the **Display 3D Bolts** option from the **Display** drop-down in the **BOLTS** panel of the **Ribbon** as discussed earlier.

ASSIGNING WELDS

The **Assign** tool of the **WELDS** panel in **Ribbon** is used to apply welded connection at selected edges common to two or more faces. The procedure to use this tool is given next.

- Click on the **Assign** tool from the **WELDS** panel in the **PREPARE** tab of the **Ribbon**. The **FILLET WELD HUD** toolbar will be displayed and you will be asked to select faces/bodies/edges on which weld connection is to be applied.
- Select desired geometries (hold **CTRL** key for multiple selection) to define edge for creating fillet weld; refer to Figure-20. The edge to be welded will be highlighted in pink color.

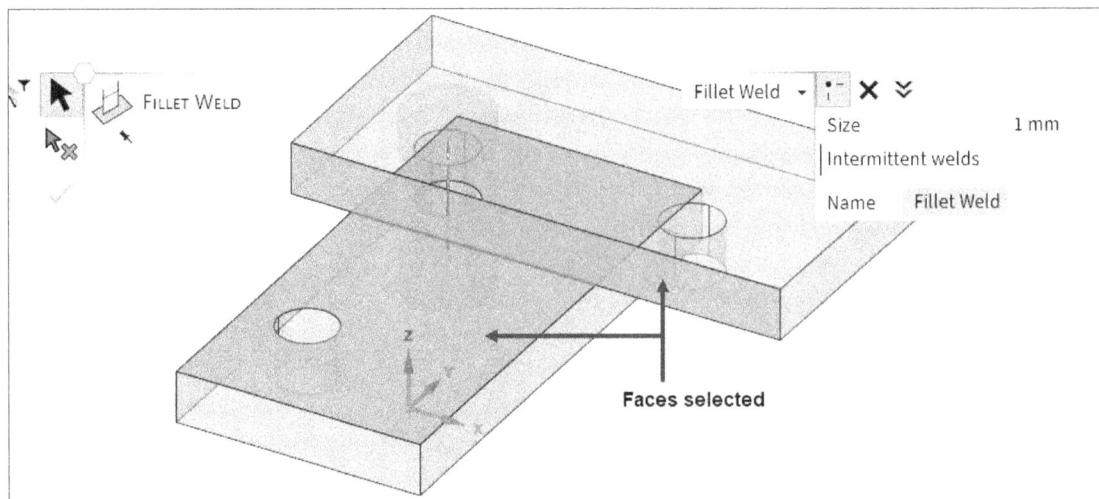

Figure-20. Selection for Weld Assignment

- Specify desired size value of fillet weld in the **Size** edit box.

- Set the other parameters as discussed earlier and click on the **OK** button. The weld will be created. Press **ESC** to exit the tool

GEOMETRY GENERATION TOOLS

The tools in the **GENERATE** panel of **Ribbon** are used to generate different types of geometries like extracted volumes, enclosure volumes, and so on. Various tools of this panel are discussed next.

Extracting Volumes

The **Volume Extract** tool is used to extract volume from region bounded by other geometries. This option is generally used when performing Computational Fluid Dynamics analyses because we will need fluid volumes to perform flow analyses. The procedure to use this tool is given next.

- Click on the **Volume Extract** tool from the **GENERATE** panel in the **PREPARE** tab of the **Ribbon**. The **VOLUME EXTRACT HUD** toolbar will be displayed; refer to Figure-21 and you will be asked to select bounding faces & a seed face.

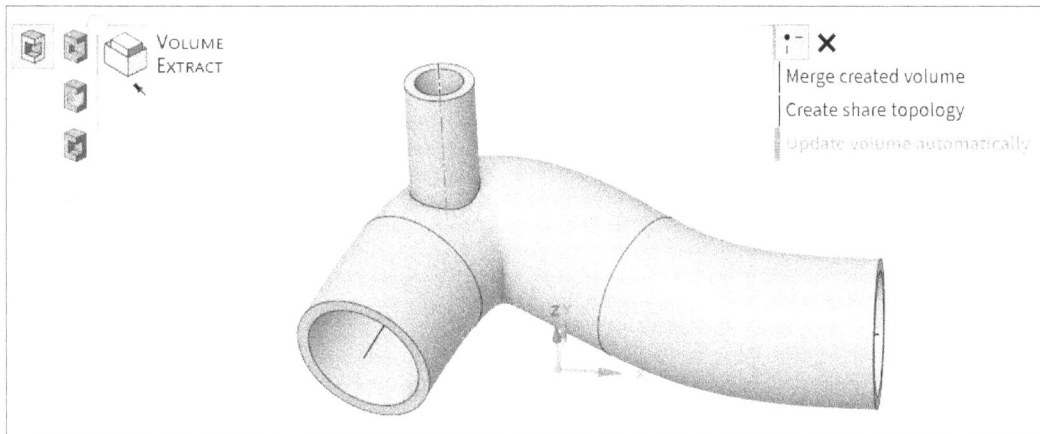

Figure-21. VOLUME EXTRACT HUD toolbar

- Select the faces of model that are to be used as inlet and outlet ports for CFD analysis; refer to Figure-22.

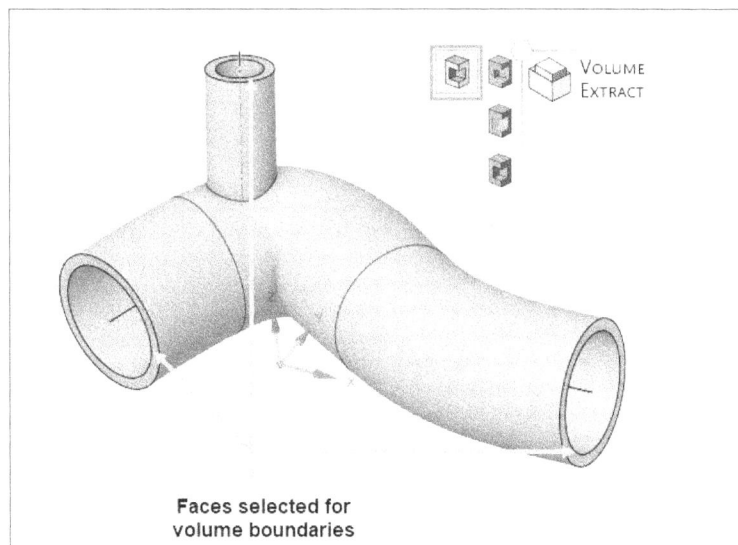

Figure-22. Faces selected for volume extract

- Click on the **Select Seed Face** selection button from the **HUD** toolbar to define the starting face of volume to be extracted and select desired internal volume face from the model; refer to Figure-23.

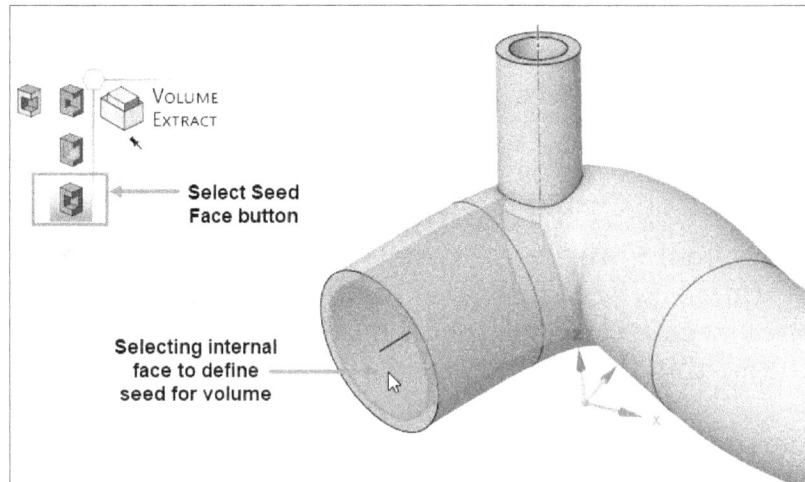

Figure-23. Selecting seed face

- Select the **Preview inside faces** toggle button from the right **HUD** toolbar to check preview of volume to be extracted; refer to Figure-24. Using the preview slider, you can check step by step formation of volume. Note that each section shown in this animation is a separate volume.

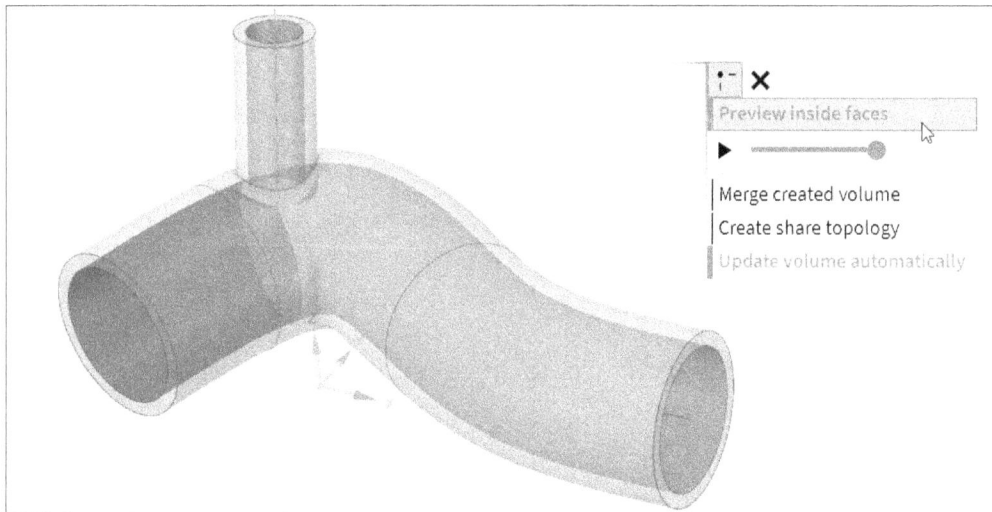

Figure-24. Preview of volume extract

- You can also generate the volume by selecting the edge loops at inlet and outlet faces of the model. To do so, click on the **Select loops of edges** selection button from the **HUD** toolbar and select desired edge loops; refer to Figure-25.

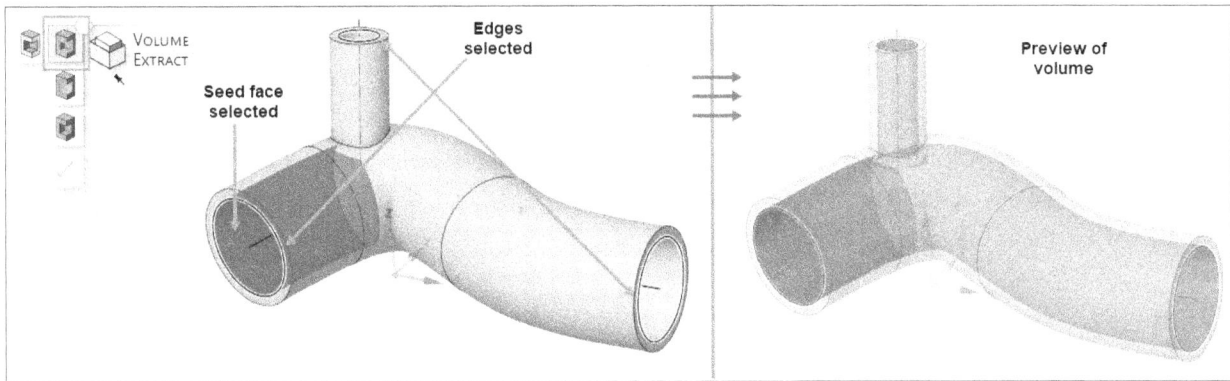

Figure-25. Volume extracted using edges

- If you want to define capping surfaces for volume then click on the **Select optional capping surfaces** selection button and select desired faces. Note that this step is optional and system automatically creates the capping, if not defined explicitly.
- Select the **Merge created volume** toggle button if you want to combine the generated volume with outer shell.
- Select the **Imprint capping edges** toggle button if you want to keep the capping faces unchanged even after merging with outer shell.
- Select the **Create share topology** toggle button if you want to allow interaction between solid body of outer shell with fluid volume extracted via this tool. Selecting this option is important if you want to check changes in solid shell due to the changes in fluid volume.
- Select the **Update volume automatically** toggle button if you want to allow changes in fluid domain generated by this tool when the main body is modified.
- After setting desired parameters, click on the **OK** button from the **HUD** toolbar. The volume feature will be created and added in the **Design Tree**. Press **ESC** to exit the tool.

Generating Enclosure

The **Enclosure** tool is used to create an enclosure volume surrounding selected objects. This enclosure volume is generally used for performing external Computational Fluid Dynamics (CFD) analyses. The enclosure volume is used as fluid medium to perform analyses like Aerodynamics study. The procedure to use this tool is given next.

- Click on the **Enclosure** tool from the **GENERATE** panel in the **PREPARE** tab of the **Ribbon**. The **ENCLOSURE HUD** toolbar will be displayed; refer to Figure-26.

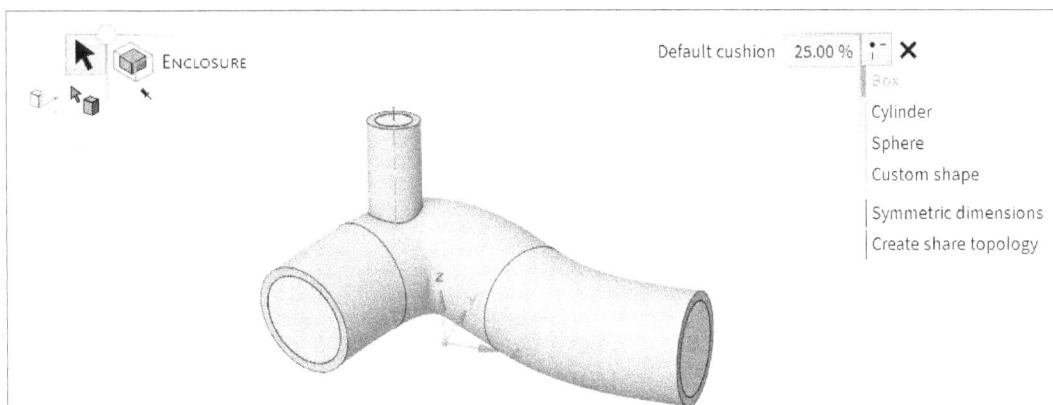

Figure-26. ENCLOSURE HUD toolbar

- Select **Box**, **Cylinder**, or **Sphere** toggle option from the right **HUD** toolbar to use respective shape for enclosure and then select the body to be enclosed.
- If you want to use a custom shape for defining enclosure then after selecting the body to be enclosed, select the **Custom shape** toggle option from the right **HUD** toolbar and click on the **Select a custom shape** selection button from left **HUD** toolbar. You will be asked to select the body. Select desired model body to create the custom enclosure; refer to Figure-27.

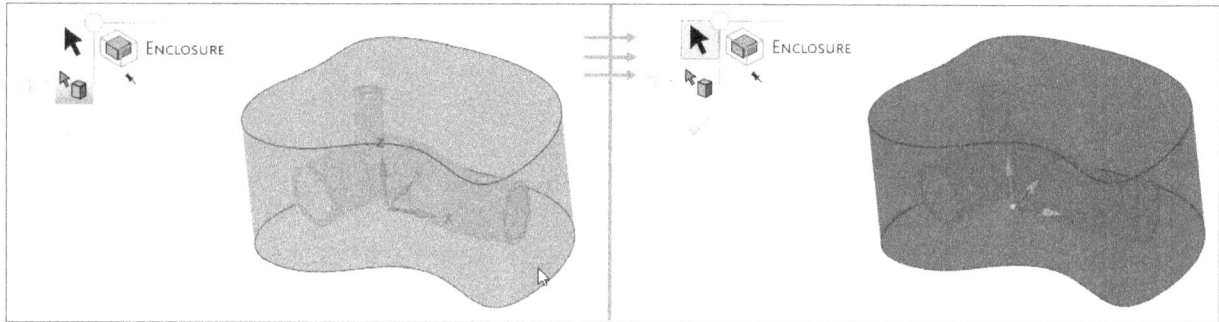

Figure-27. Using custom enclosure

- Set the other parameters as discussed earlier and click on the **OK** button. The enclosure volume will be created. Press **ESC** to exit the tool.

Splitting Bodies by Planes or Planar Faces

The **Split by Plane** tool is used to split selected bodies by specified plane. The procedure to use this tool is given next.

- Click on the **Split by Plane** tool from the **GENERATE** panel in the **PREPARE** tab of the **Ribbon**. The **SPLIT BY PLANE HUD** toolbar will be displayed.
- Select desired body to be split by plane/face. You will be asked to select a face/plane.
- Select desired face/plane from the model. The splitting will be performed; refer to Figure-28.

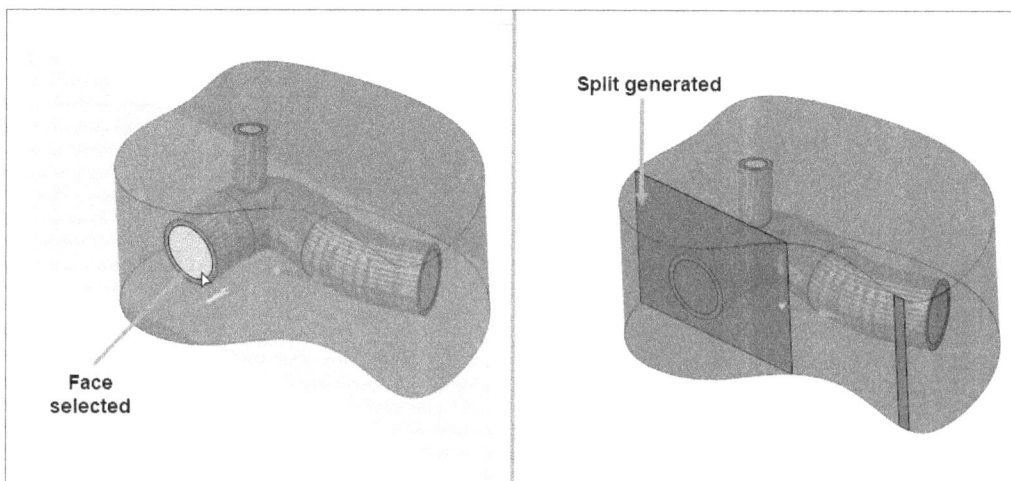

Figure-28. Splitting performed on model

- Select desired region of the split body to be removed.
- Select the **Merge when done** option from the right **HUD** toolbar if you want to combine split body with connected solid/surface features.
- Press **ESC** to exit the tool.

Imprinting

Imprinting is used to connect various sections of model for mesh generation. Generally, there can be instances where your model may seem connected visually after importing from other software but when you generate mesh for the complete model, different sections of mesh model will be acting as separate bodies, hence not transferring effects of loads properly. The procedure to perform imprinting is given next.

- Click on the **Imprint** tool from the **GENERATE** panel in the **PREPARE** tab of the **Ribbon**. The **IMPRINT HUD** toolbar will be displayed and you will be asked to select points for objects to be imprinted; refer to Figure-29.

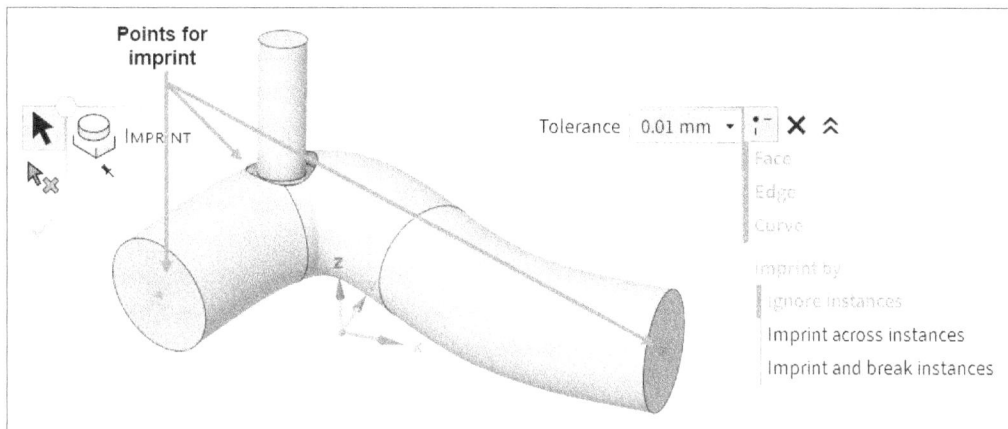

Figure-29. IMPRINT HUD Toolbar and imprint points

- Select desired points from the model or click on the **OK** button from the **HUD** toolbar to select all the points for applying imprinting operation. Press **ESC** to exit the tool.

Performing Wrap Operation

The **Wrap** tool is used to wrap selected curves, faces, or solids around curve faces. The procedure to use this tool is given next.

- Click on the **Wrap** tool from the **GENERATE** panel in the **PREPARE** tab of the **Ribbon**. The **WRAP HUD** toolbar will be displayed and you will be asked to select the body around which other body will be wrapped.
- Select desired curved body. You will be asked to select faces/edges/curves to be wrapped.
- Select the face of other body to be wrapped on the main body. Preview of wrap operation will be displayed; refer to Figure-30.
- Select the **Imprint as edges** toggle option from the right **HUD** toolbar if you want to generate only projection edges of wrapped face on main body otherwise full body will be wrapped. You should select this toggle option when you want to later perform splitting of main body. You should not select this toggle option when you want to create embossed or engraved design on curved faces of main body.
- Select the **Delete source geometry** toggle option from **HUD** toolbar if you do not keep the geometry being wrapped after creating wrap feature.
- After setting desired parameters, click on the **OK** button. The wrap feature will be generated; refer to Figure-31. Press **ESC** to exit the tool.

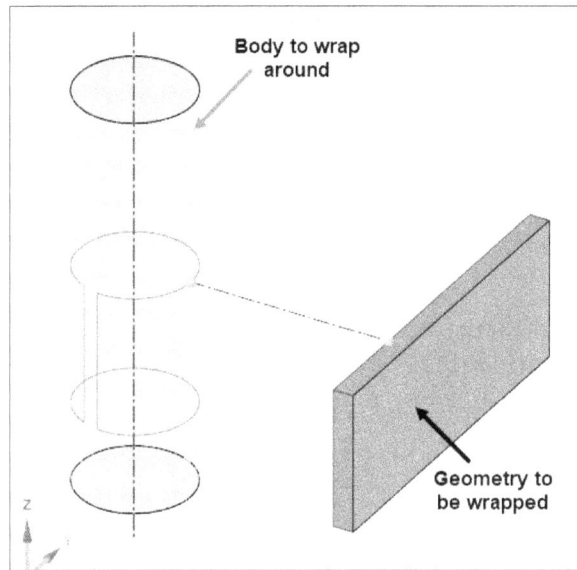

Figure-30. Selection for wrap operation

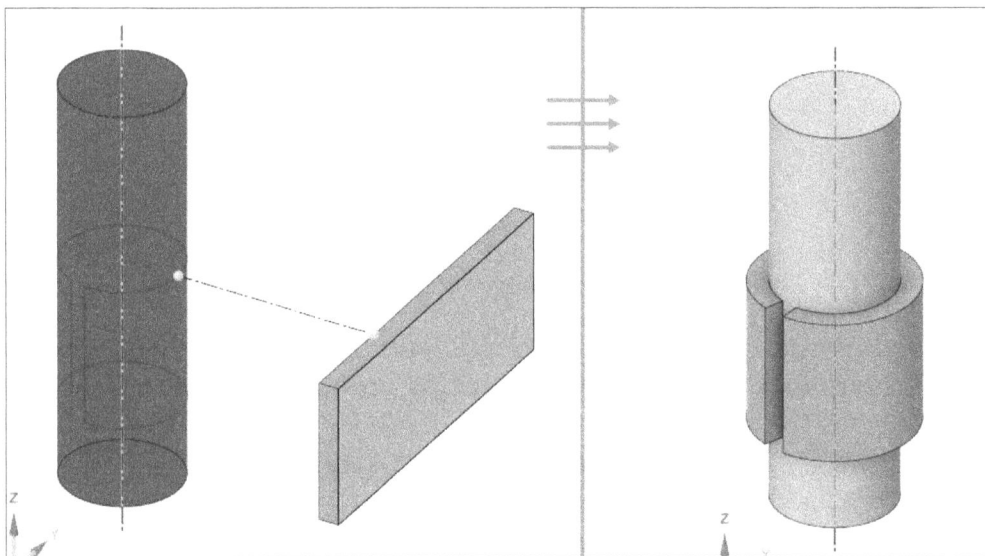

Figure-31. Creating wrap feature

FEATURE REMOVAL TOOLS

The tools in the **REMOVE** panel of **Ribbon** are used to perform various checks like interference check, round check, short edge check, etc. and remove them; refer to Figure-32. Various tools of this panel are discussed next.

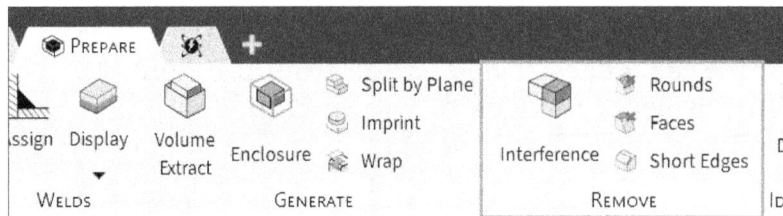

Figure-32. REMOVE panel

Checking and Removing Interference

The **Interference** tool is used to check for bodies that are overlapping and fix them by removing overlapping portions. The procedure to use this tool is given next.

- Click on the **Interference** tool from the **REMOVE** panel in the **PREPARE** tab of the **Ribbon**. The **INTERFERENCE HUD** toolbar will be displayed and overlapping bodies of the model will be highlighted; refer to Figure-33.

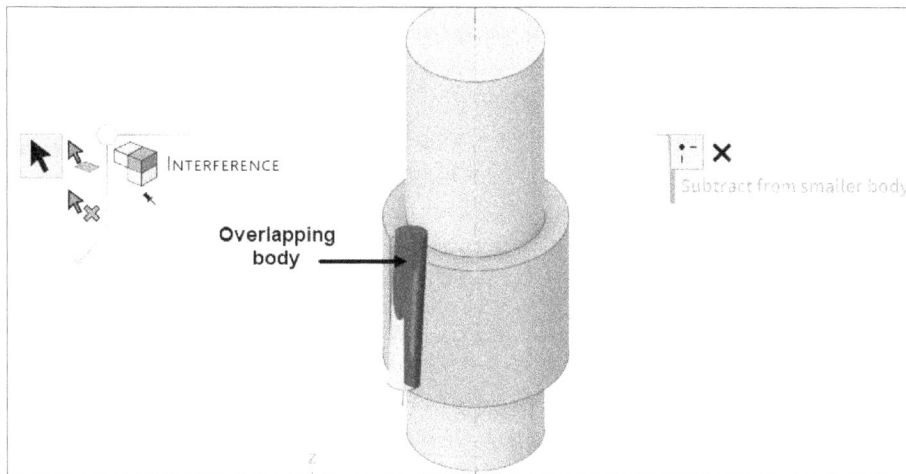

Figure-33. Overlapping bodies

- Select the **Subtract from smaller body** toggle option if you want to delete overlapping region of smaller body otherwise this region will be deleted from larger body.
- Set the other parameters and selections as discussed earlier and click on the **OK** button to apply the operation. Press **ESC** to exit the tool.

Removing Rounds

The **Rounds** tool in **REMOVE** panel of **Ribbon** is used to delete selected rounds from the model. The procedure to use this tool is given next.

- Click on the **Rounds** tool from the **REMOVE** panel in the **PREPARE** tab of the **Ribbon**. The **ROUNDS HUD** toolbar will be displayed and you will be asked to select rounds to be removed.
- Select the **Auto-shrink fill area** toggle option from the right **HUD** toolbar if the rounds to be removed are on holes and voids, and you want to reduce the expanded boundaries of these features. Generally, when you apply rounds on holes and voids then their boundaries get expanded. So, if you want to compensate this expansion when removing rounds then you should select this option.
- Specify desired value in the **Cap width** edit box of right **HUD** toolbar to define the limit upto which rounds related to selected edge will be removed; refer to Figure-34.

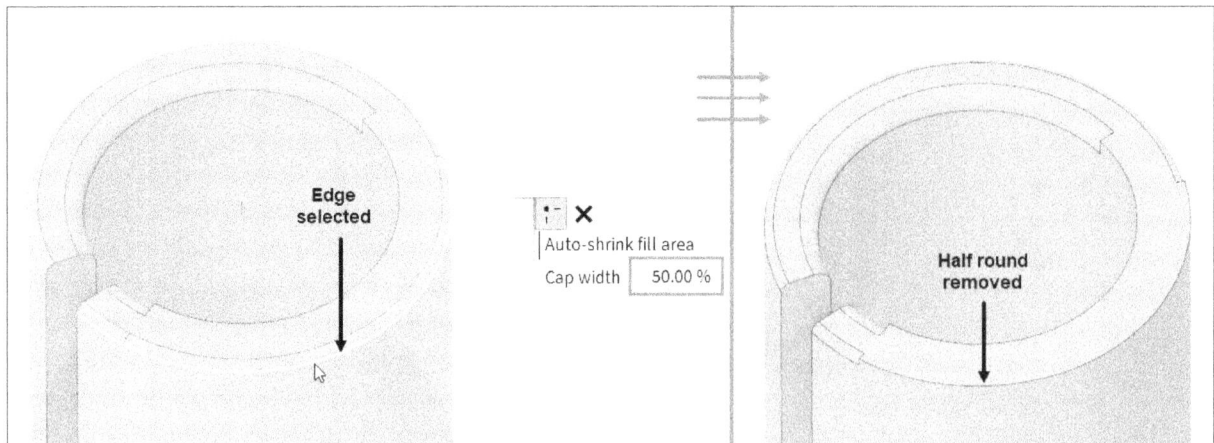

Figure-34. Removing half round

- If you want to remove full round then select the curved faces of rounds (hold **CTRL** key for multiple selection) and click on the **OK** button from the **HUD** toolbar.
- After removing desired rounds, press **ESC** to exit the tool.

Removing Faces

Faces are removed from the model to fill voids/gaps. When simplifying geometry for analysis, you will find cavities and holes in the model that do not affect the result of analyses as much as they create meshing problems and increase processing time. In such cases, it is better to remove such features by deleting their faces. The procedure to do so is given next.

- Click on the **Faces** tool from the **REMOVE** panel in the **PREPARE** tab of the **Ribbon**. The **FACES HUD** toolbar will be displayed and you will be asked to select the faces to be removed.
- Select desired faces from the model to be removed and click on the **OK** button from the **HUD** toolbar. The faces will be removed; refer to Figure-35.
- Press **ESC** to exit the tool.

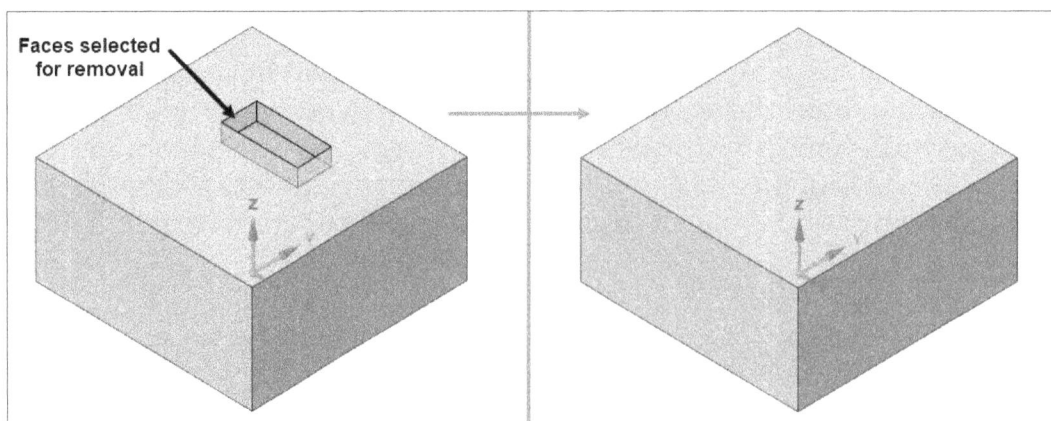

Figure-35. Removing faces

Removing Short Edges

The **Short Edges** tool is sued to find and remove short edges from the model. The procedure to use this tool is given next.

- Click on the **Short Edges** tool from the **REMOVE** panel in **PREPARE** tab of the **Ribbon**. The **SHORT EDGES HUD** toolbar will be displayed and edges shorter than specified maximum length will be highlighted; refer to Figure-36.

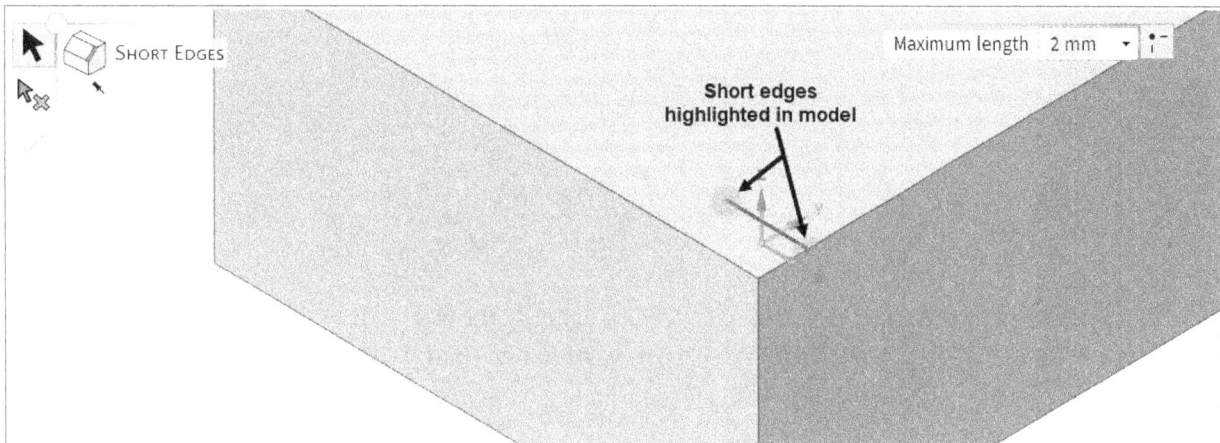

Figure-36. Short edges highlighted in model

- Set desired value in the **Maximum length** edit box to define the size upto which edges will be highlighted by this tool.
- Select the highlighted edges to remove them; refer to Figure-37. Press **ESC** to exit the tool.

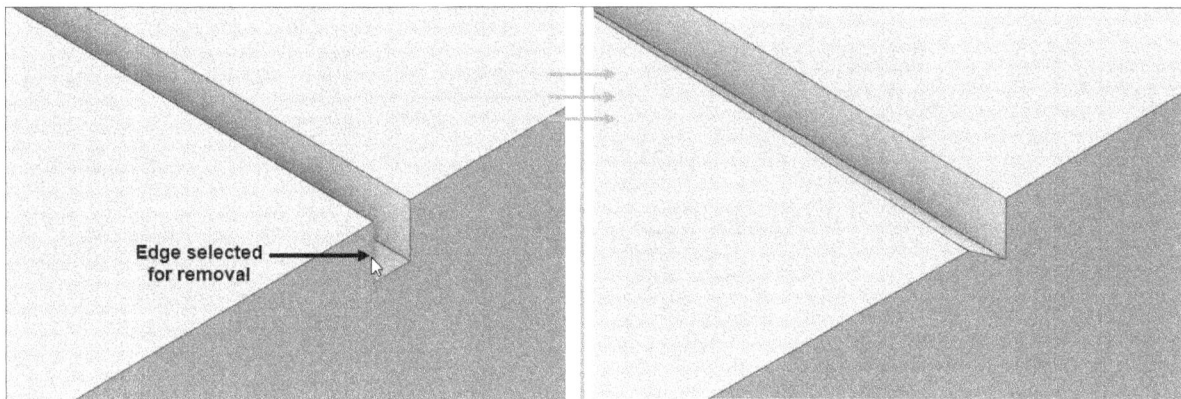

Figure-37. Removing edge

Note that you can also remove various features from the model by selecting them and pressing **DELETE** key.

Detecting Features

The tools in the **Detect** drop-down are used to identify various features of the model that can be defeatured for simplifying the model; refer to Figure-38.

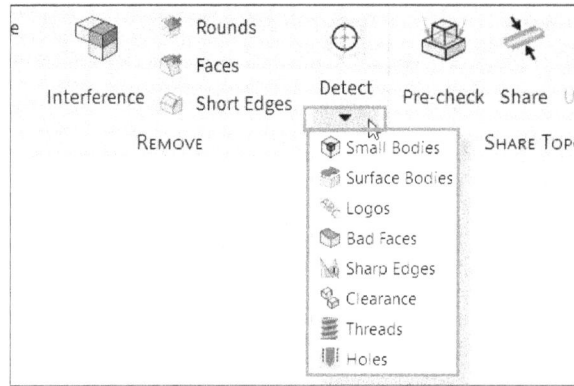

Figure-38. Detect drop-down

Select desired tool from the drop-down to highlight features falling under specified conditions. We will discuss **Holes** tool in this case and you can apply same concept for other tools in the drop-down.

Detecting Holes

*   Select the **Holes** tool from the drop-down and specify desired value in the **Max diameter** edit box. The holes which have diameter lower than specified value will be highlighted; refer to Figure-39.

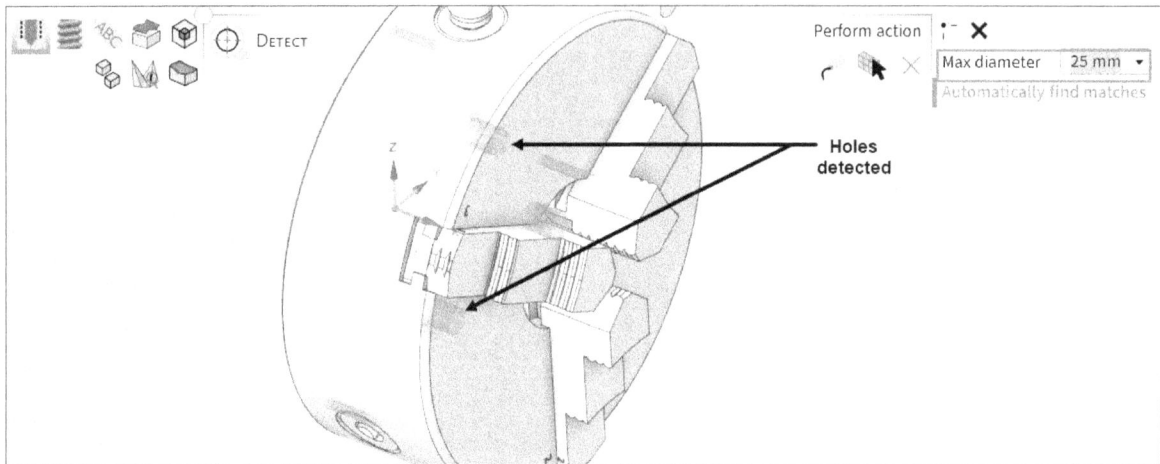

Figure-39. Holes detected in model

*   Click on the **Convert detected area to named selection** button from the right **HUD** toolbar to save automatically highlighted sections as named selection.
*   Click on the **Convert detected area to selection** button from the right **HUD** toolbar to select all the highlighted features from the model. Now, you can press **DELETE** key to remove these holes.
*   Press **ESC** to exit the tool.

SHARE TOPOLOGY TOOLS

The tools in the **SHARE TOPOLOGY** panel are used to check and modify interactions between various components when analysis is performed on the assembly; refer to Figure-40. Generally, when you perform a structural analysis on an assembly then effect of load is transferred between various components of the assembly but if you have poorly defined topology in assembly then you may get inaccurate results of analysis. Various tools of the panel are discussed next.

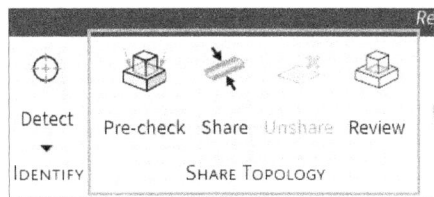

Figure-40. SHARE TOPOLOGY panel

Performing Pre-check

The **Pre-check** tool is used to check the problem areas of assembly model that can cause failure when performing analysis. The procedure to use this tool is given next.

• Click on the **Pre-check** tool from the **SHARE TOPOLOGY** panel in the **PREPARE** tab of the **Ribbon**. The **PRE-CHECK HUD** toolbar will be displayed and sections of the model that may fail topology connection test will be highlighted; refer to Figure-41.

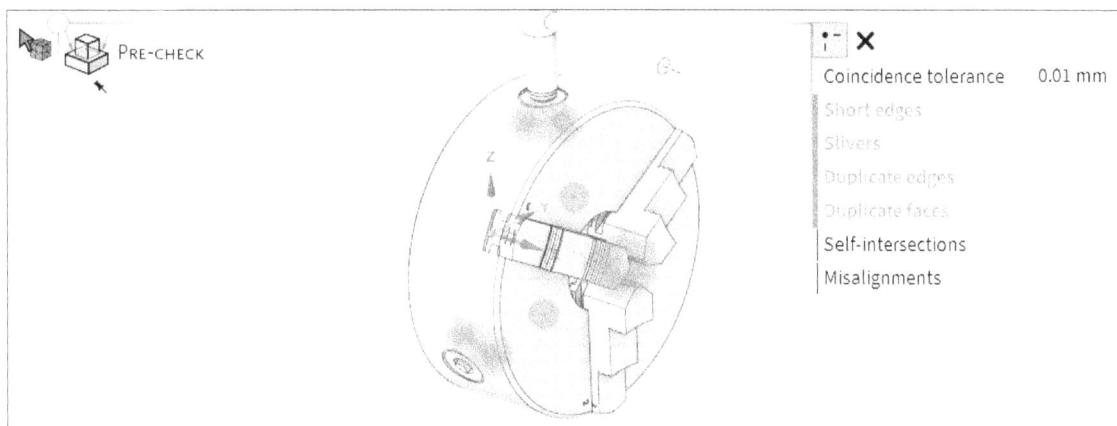

Figure-41. Pre-check results

• Select the **Slivers** toggle option from the right **HUD** toolbar to check for thin sections with high peaks generated when importing models.
• Select the **Misalignments** toggle option from the **HUD** toolbar to check for misalignments in the assembly components.
• Set the other parameters as discussed earlier and click on the **Convert highlighted items to selection** button from left **HUD** toolbar to select highlighted objects for performing operations like combining them.
• Press **ESC** to exit the tool.

Sharing Coincident Topology

The **Share** tool is used to make selected coincident faces and edges share the topology for seamless mesh generation when performing analysis. The procedure to use this tool is given next.

• Click on the **Share** tool from the **SHARE TOPOLOGY** panel in the **PREPARE** tab of the **Ribbon**. The **SHARE HUD** toolbar will be displayed and faces/edges that can be connected are highlighted when **Preview share topology** toggle button 🔍 is active; refer to Figure-42.

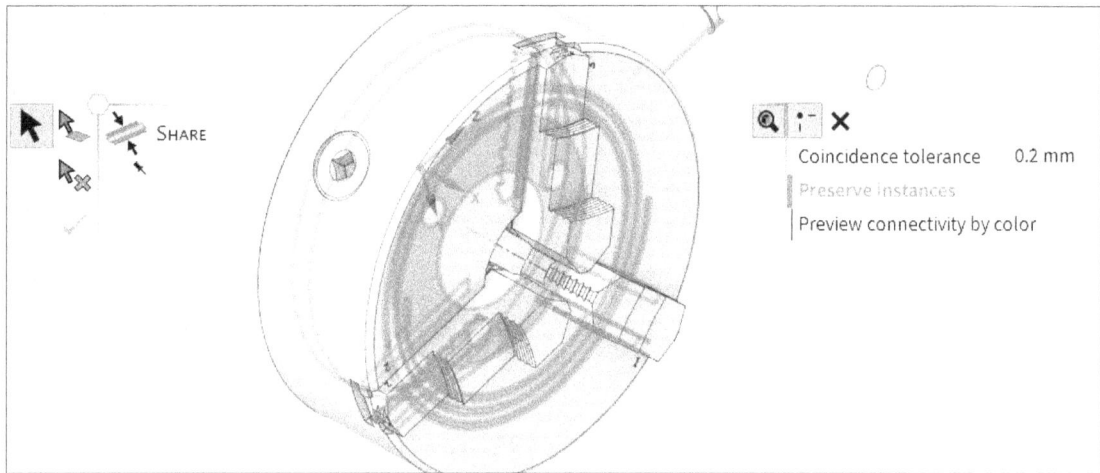

Figure-42. SHARE HUD toolbar

- Select the **Preserve instances** toggle option from right **HUD** toolbar if you want to keep all the instances connected as single body otherwise all the instances will be independent for selection.
- Select the **Preview connectivity by color** toggle option from the right **HUD** toolbar if you want to display connectivity of different regions in model by different colors; refer to Figure-43.

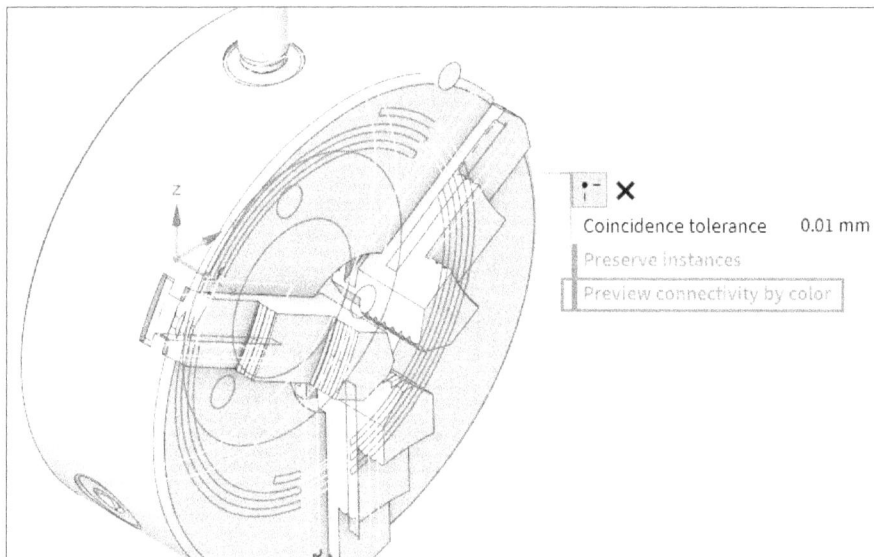

Figure-43. Preview connectivity by color option

- Set the other parameters as discussed earlier and click on the **OK** button from the **HUD** toolbar. The operation will be applied.
- Press **ESC** to exit the tool.

Unsharing Topology

The **Unshare** tool in the **SHARE TOPOLOGY** panel is used to reverse the operation performed by using **Share** tool. This tool makes all the bodies in model sharing topology to unshare. The procedure to use this tool is given next.

- Click on the **Unshare** tool from the **SHARE TOPOLOGY** panel in the **PREPARE** tab of the **Ribbon**. The **UNSHARE HUD** toolbar will be displayed and faces/edges with shared topology will be highlighted; refer to Figure-44.

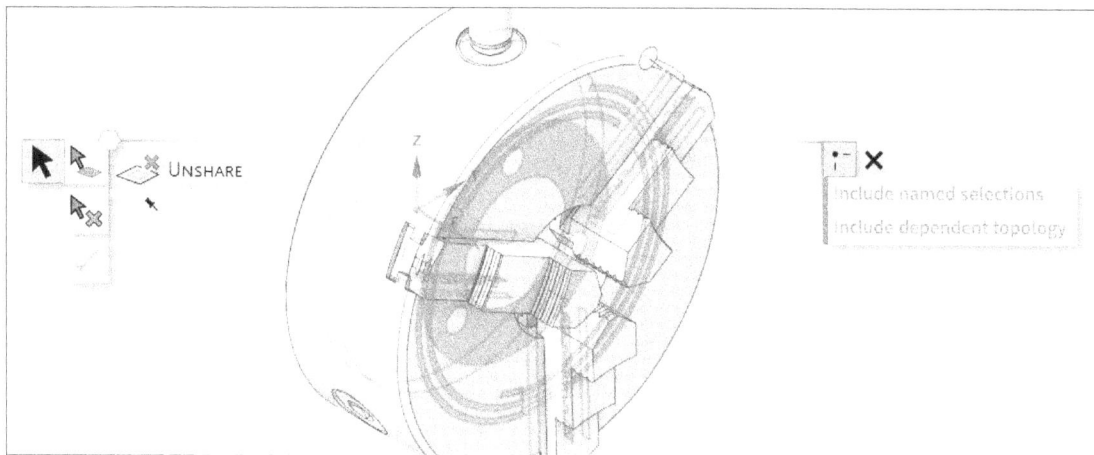

Figure-44. UNSHARE HUD toolbar

• Set the parameters as discussed earlier and click on the **OK** button to unshare faces and edges.

Reviewing Shared/Unshared Geometries

The **Review** tool is used to check shared faces and unshared faces of the model. The procedure to use this tool is given next.

• Click on the **Review** tool from the **SHARE TOPOLOGY** panel in the **PREPARE** tab of the **Ribbon**. The **SHARED TOPOLOGY REVIEW HUD** toolbar will be displayed; refer to Figure-45.

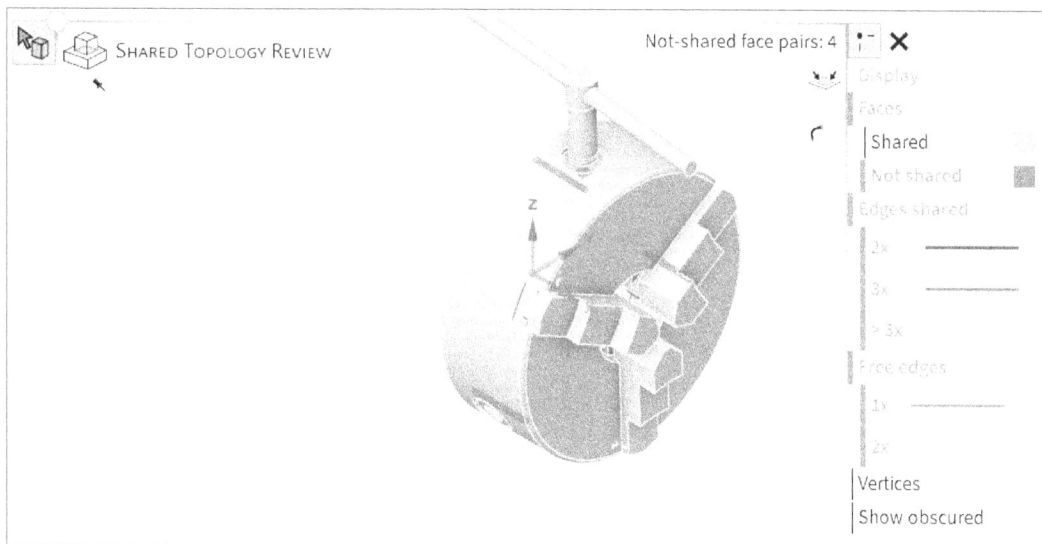

Figure-45. SHARED TOPOLOGY REVIEW HUD toolbar

• Select desired toggle options to check status of respective faces and edges.
• Press **ESC** to exit the tool.

GEOMETRY TRANSFER TOOLS

The tools of the **GEOMETRY TRANSFER** panel are displayed on clicking the **More** button at the right in the **PREPARE** tab of **Ribbon**; refer to Figure-46. Click on desired tool from the panel to transfer current Discovery model to respective Ansys application.

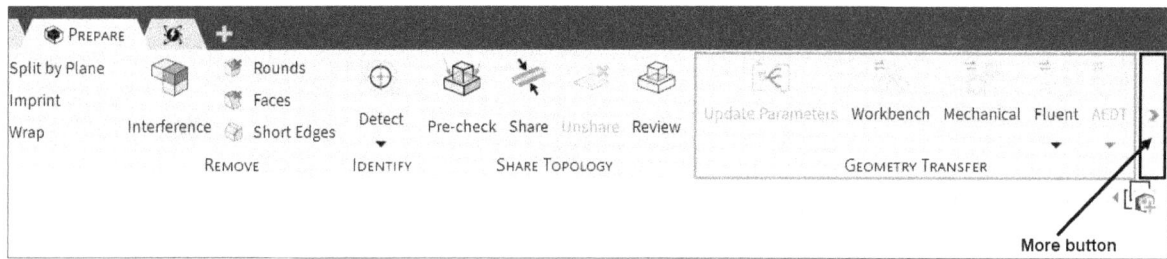

Figure-46. GEOMETRY TRANSFER panel

SELF-ASSESSMENT

Q1. In Ansys Discovery, which tab contains the tools for preparing a model for analysis?
A. ANALYZE
B. DESIGN
C. PREPARE
D. SIMULATE

Q2. Which of the following is NOT typically removed when preparing a model for analysis?
A. Chamfers
B. Aesthetic features
C. Load-bearing structures
D. Rounds on edges

Q3. The Extract tool is used to:
A. Create a new solid model
B. Automatically extract beam elements from a solid model
C. Assign beam types to curves
D. Create a 3D profile of beams

Q4. What must be selected before using the Extract tool in the BEAMS AND SHELLS panel?
A. Solid objects with uniform cross-sections
B. Only sketched curves
C. Only existing beam elements
D. Structural nodes

Q5. How can multiple line segments be selected when using the Assign tool?
A. By holding ALT
B. By holding CTRL
C. By double-clicking each segment
D. By using a selection box

Q6. The Display drop-down menu in the BEAMS AND SHELLS panel is used to:
A. Create a 3D model from line sketches
B. Hide 3D profiles of beams
C. Generate an analytical model
D. Perform meshing operations

Q7. The Orient tool is used to:
A. Change the orientation of beam profiles
B. Convert solid models into beam elements
C. Assign material properties to beams
D. Connect small gaps in wireframe models

Q8. What does the Connect tool do in the BEAMS AND SHELLS panel?
A. Connect small gaps in wireframe models
B. Assign beams to curves
C. Extend surfaces to other bodies
D. Assign loads to beam elements

Q9. Which option in the Connect tool allows connecting only free end points of beams?
A. Maximum distance
B. Select problem areas
C. Free ends only
D. Merge after connect

Q10. What is the purpose of the Midsurface tool?
A. To generate meshing elements
B. To create surfaces at the center of model walls for analysis
C. To generate a 3D profile of beams
D. To assign material properties to surfaces

Q11. Which selection method in the Midsurface tool is useful for large or complex models?
A. Use selected faces
B. Use range
C. Extend surfaces
D. Trim surfaces

Q12. The Extend tool is used to:
A. Extend surface edges to intersecting bodies
B. Assign beam types to curves
C. Convert beams into solid elements
D. Apply weld connections

Q13. In the Assign Bolts tool, what must be selected first?
A. Revolved coaxial faces
B. Solid bolt objects
C. Beam elements
D. Structural nodes

Q14. Which tool is used to apply welded connections at edges common to two or more faces?
A. Assign Beams
B. Assign Welds
C. Connect
D. Extract

Q15. The Volume Extract tool is primarily used in:
A. Structural Analysis
B. Computational Fluid Dynamics (CFD)
C. Thermal Analysis
D. Modal Analysis

Q16. What does the Enclosure tool do?
A. Creates an enclosure volume surrounding selected objects
B. Converts solid models into beam elements
C. Extracts midsurfaces for analysis
D. Connects beam elements

Q17. The Split by Plane tool is used to:
A. Assign loads to models
B. Divide a body into two or more sections along a plane
C. Connect small gaps in a wireframe model
D. Generate bolt connections

Q18. What is the primary purpose of imprinting in a model?
A. To apply material properties
B. To connect various sections for mesh generation
C. To delete unnecessary faces
D. To perform topology optimization

Q19. How can you select all points for imprinting operation?
A. Press ESC key
B. Click on the OK button in the HUD toolbar
C. Select the Wrap tool
D. Use the DELETE key

Q20. What does the Wrap tool do?
A. Wraps selected curves, faces, or solids around curved faces
B. Merges multiple bodies into one
C. Removes interference between bodies
D. Deletes unnecessary geometry

Q21. When should you select the "Imprint as edges" toggle option in the Wrap tool?
A. When creating embossed or engraved designs
B. When wrapping full bodies
C. When later performing splitting of the main body
D. When deleting the source geometry

Q22. What is the purpose of the Interference tool?
A. To check for overlapping bodies and remove interference
B. To delete short edges from a model
C. To wrap one body around another
D. To generate structured mesh

Q23. How can you ensure overlapping regions are deleted from the smaller body using the Interference tool?
A. Select "Subtract from smaller body" toggle option
B. Use the DELETE key
C. Click on the "Imprint as edges" toggle option
D. Enable "Preview connectivity by color"

Q24. What does the Rounds tool do?
A. Creates fillets and rounds on edges
B. Deletes selected rounds from the model
C. Detects and highlights short edges
D. Wraps faces onto a curved surface

Q25. What does the "Auto-shrink fill area" toggle option in the Rounds tool do?
A. Expands boundary of holes and voids
B. Reduces expanded boundaries of rounds applied to holes and voids
C. Removes fillets completely
D. Converts detected rounds into selection

Q26. Why are faces removed from a model during simplification?
A. To increase geometric complexity
B. To create a more detailed mesh
C. To fill voids/gaps and simplify meshing
D. To generate separate bodies

Q27. What is the function of the Short Edges tool?
A. Removes fillets and rounds
B. Highlights and removes edges shorter than a defined length
C. Merges small faces into a larger face
D. Deletes unnecessary points from the model

Q28. How can detected holes be saved for later operations?
A. By using the "Convert detected area to named selection" option
B. By pressing the DELETE key
C. By selecting the "Imprint as edges" option
D. By using the "Auto-shrink fill area" option

Q29. What is the purpose of the Pre-check tool?
A. To analyze and repair meshing errors
B. To check and highlight potential topology issues in an assembly
C. To automatically share topology between components
D. To delete short edges and voids

Q30. What does the "Slivers" toggle option in the Pre-check tool detect?
A. Misalignment of faces
B. Thin sections with high peaks generated during model import
C. Overlapping bodies
D. Gaps between faces

Q31. What is the main function of the Share tool?
A. To remove unnecessary topology from an assembly
B. To check interference between bodies
C. To make coincident faces and edges share topology for seamless meshing
D. To wrap faces onto curved bodies

Q32. What does the "Preview connectivity by color" toggle option do in the Share tool?
A. Highlights overlapping bodies
B. Displays connectivity of different regions using different colors
C. Deletes short edges
D. Converts detected areas into named selection

Q33. What does the Unshare tool do?
A. Merges multiple bodies into one
B. Removes mesh connectivity between shared bodies
C. Highlights topology errors in an assembly
D. Removes fillets and rounds

Q34. What is the function of the Review tool in Share Topology?
A. Deletes unshared faces
B. Highlights shared and unshared faces for inspection
C. Automatically merges overlapping bodies
D. Removes short edges

Q35. Where can you find tools for transferring geometry to other Ansys applications?
A. Under the SHARE TOPOLOGY panel
B. In the REMOVE panel
C. Under the GEOMETRY TRANSFER panel by clicking the More button in the PREPARE tab
D. Under the INTERFERENCE panel

Chapter 5

Ansys Discovery- Design Practical and Practice

Topics Covered

The major topics covered in this chapter are:

- *Practical 1*
- *Practical 2*
- *Practices*

INTRODUCTION

In previous chapters, you have learned to perform various design related operations. You have also worked on model repair and preparation tools. In this chapter, you will work on various practical examples and perform practices.

PRACTICAL 1

Create the model based on drawing given in Figure-1.

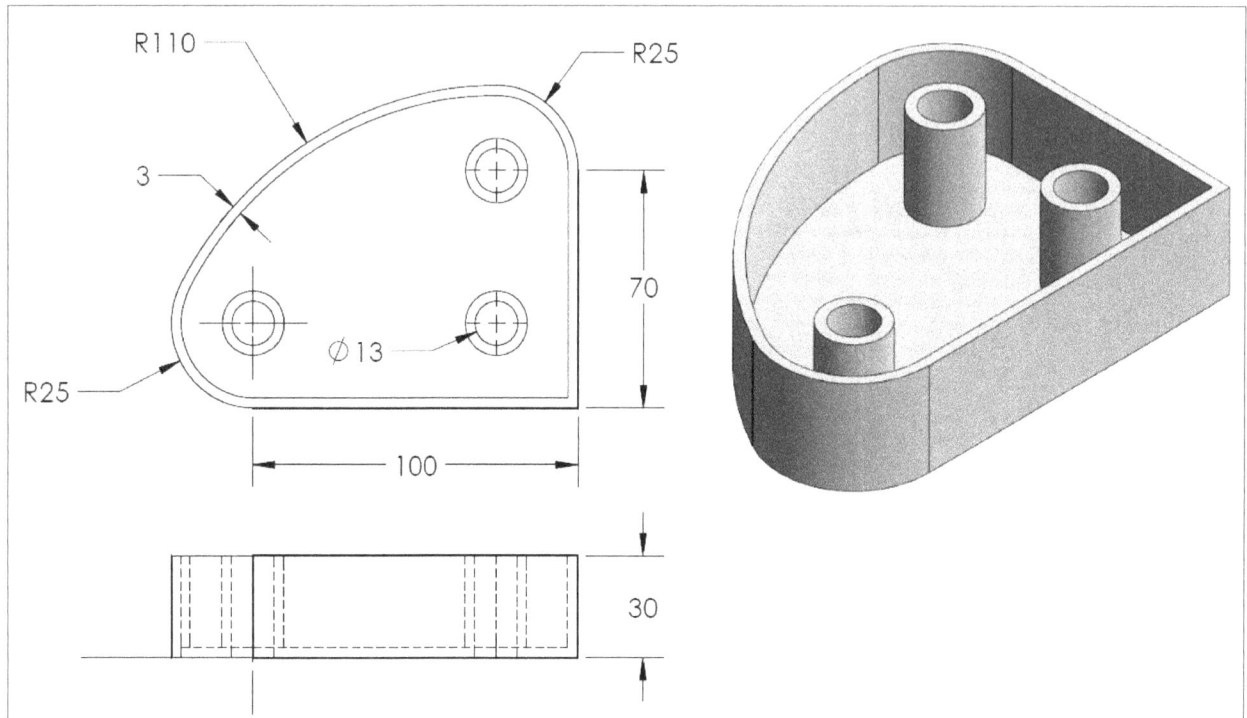

Figure-1. Practical 1 for Design

Steps:

*   Start Ansys Discovery if not started yet. The design environment will be active by default.
*   Click on the **Plan view** button from the **HUD** toolbar to make the sketching plane parallel to view.
*   Click on the **CIRCLE** tool from the **SKETCH** panel in the **DESIGN** tab of the **Ribbon**. The **CIRCLE HUD** toolbar will be displayed and you will be asked to specify center point of the circle.
*   Click at the origin and enter **50 mm** in the edit box to create the circle; refer to Figure-2. The circle will be created.

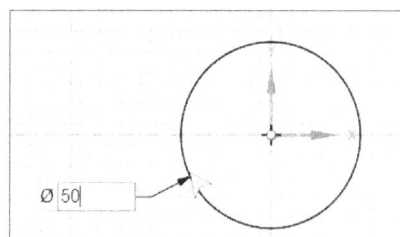

Figure-2. Specifying diameter of circle

- Click on the **LINE** tool from the **SKETCH** panel in the **DESIGN** tab of the **Ribbon**. You will be asked to specify start point of the line.
- Click at the bottom quadrant point to define start point of line and move the cursor to right. You will be asked to specify length of line; refer to Figure-3.

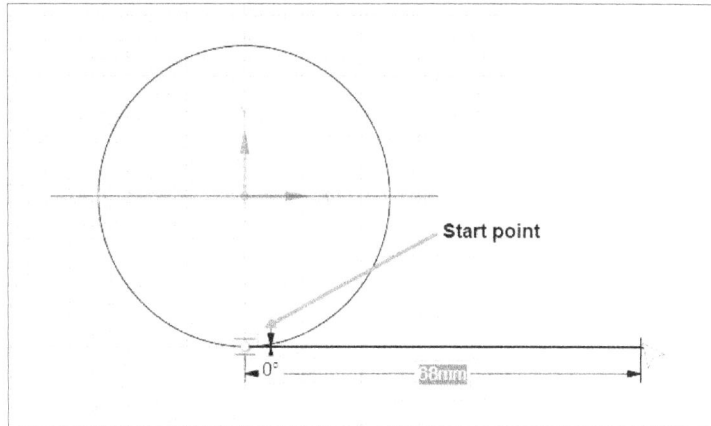

Figure-3. Start point specified

- Type **100 mm** in the input box and press **ENTER** to create the horizontal line. Move the cursor vertically upward and enter **70 mm** in the input box. The line will be created. Press **ESC** twice to exit the tool.
- Click on the **TANGENT ARC** tool from the **SKETCH** panel in the **DESIGN** tab of the **Ribbon**. You will be asked to select end point of curve for generating tangent arc.
- Click at the end point of the vertical line and move the cursor to left. Preview of arc will be displayed; refer to Figure-4.

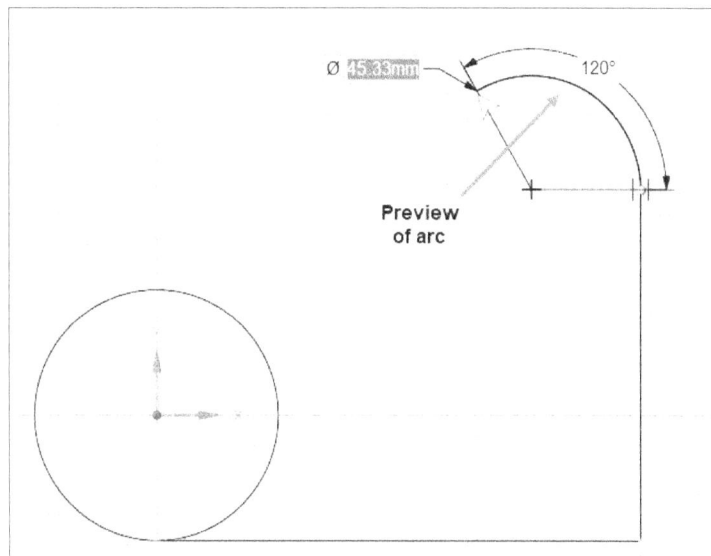

Figure-4. Preview of arc

- Click at the point where diameter is approximately **50 mm** and angle is approximately 120.
- Click on the end point of recently created arc and connect it to circle tangentially; refer to Figure-5.

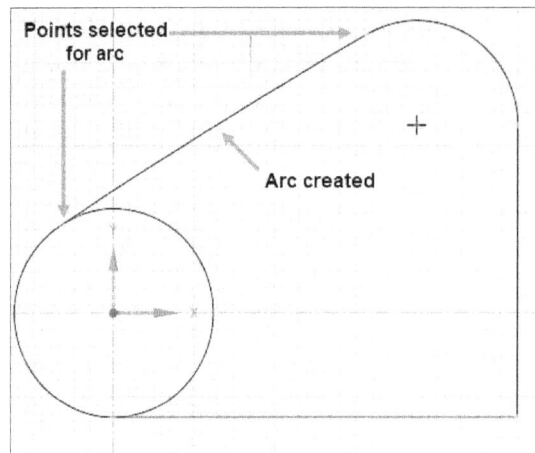

Figure-5. Arc created

- Click on the **TRIM AWAY** tool from the **SKETCH** panel in the **DESIGN** tab of the **Ribbon**. You will be asked to select sketch section to be trimmed.
- Select the section of circle as shown in Figure-6. The sketch will be displayed as shown in Figure-7.

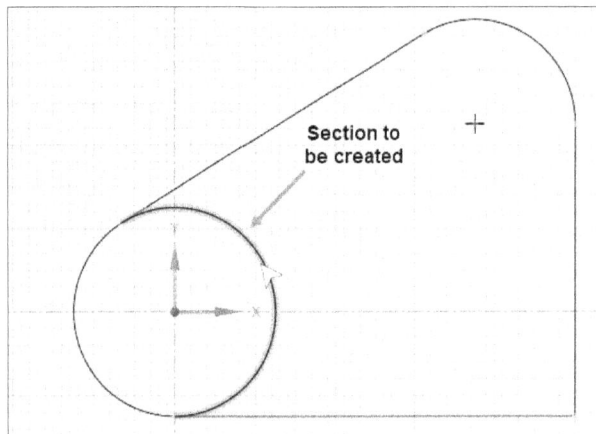

Figure-6. Section selected for trimming

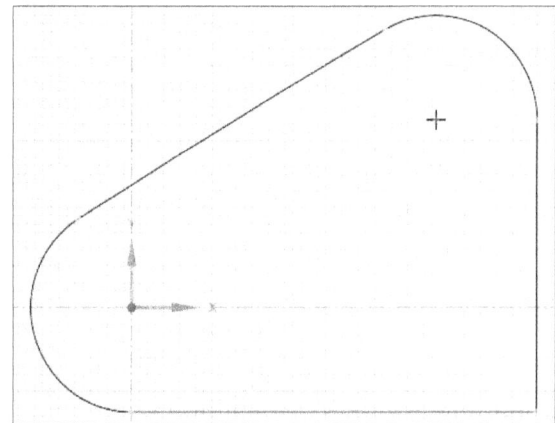

Figure-7. Sketch after trimming

- Click on the **CIRCLE** tool from the **SKETCH** panel in the **DESIGN** tab of the **Ribbon** and create circles of diameter **13** as shown in Figure-8.

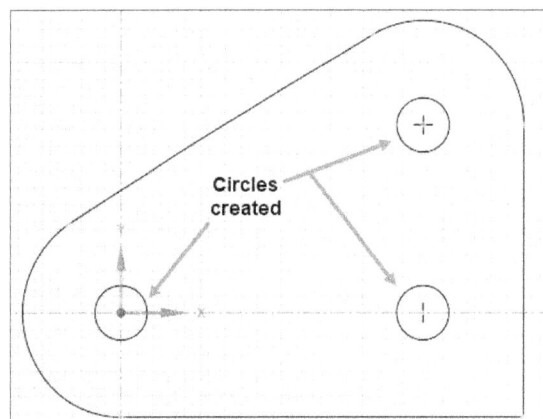

Figure-8. Circles created

- Click on the **Dimension** tool from the **CONSTRAINTS** panel in the **DESIGN** tab of the **Ribbon**. You will be asked to define dimensions.
- Apply the dimensions as shown in Figure-9.

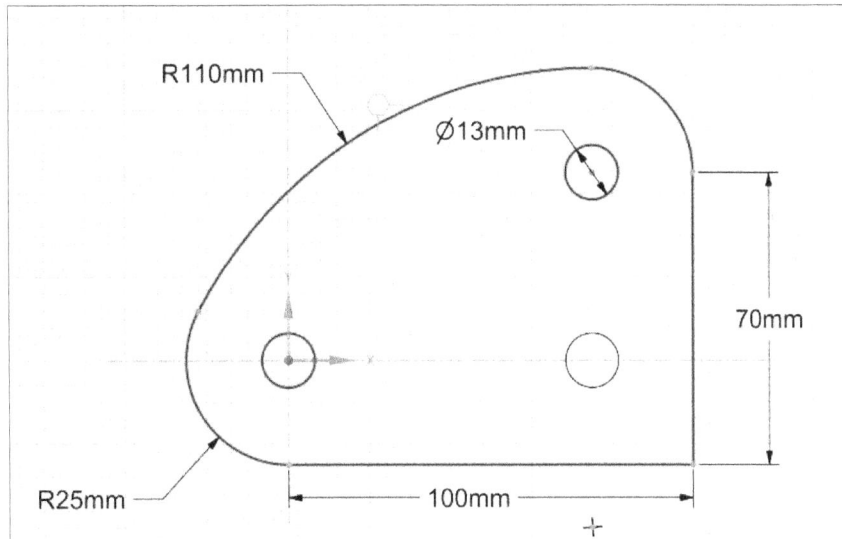

Figure-9. After applying dimensions

- After applying dimensions, click on the **Return to 3D** tool from the **HUD** toolbar. The **PULL HUD** toolbar will be displayed.
- Using Middle Mouse Button (MMB), reorient the sketch and double-click on the sketched surface. The surface will turn orange colored; refer to Figure-10.

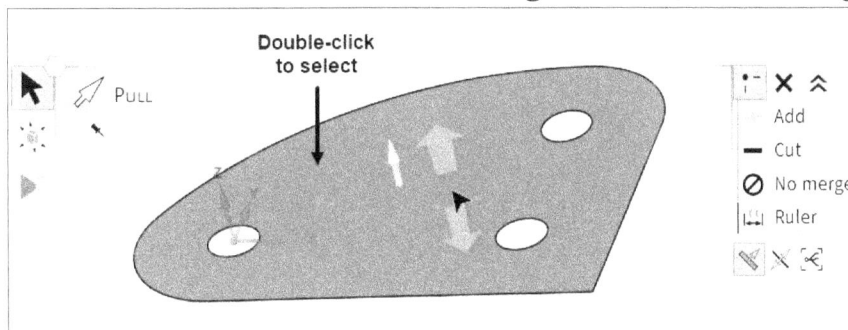

Figure-10. Selecting sketched surface

- Drag the surface upward and enter the value as **30** in the input box; refer to Figure-11. Press **ESC** twice to exit the tool.

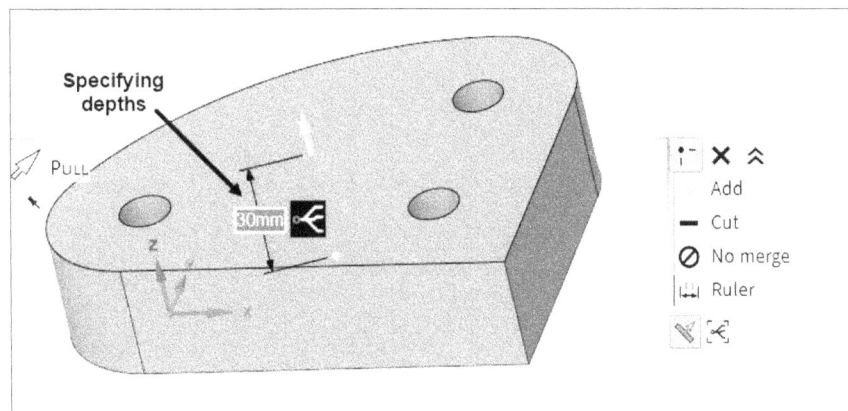

Figure-11. Specifying depth

- Click on the **Shell** tool from the **CREATE** panel in the **DESIGN** tab of the **Ribbon**. The **SHELL HUD** toolbar will be displayed with input box for specifying thickness of shell.
- Select the top face of model and enter thickness of shell as **3 mm** in the input box; refer to Figure-12.

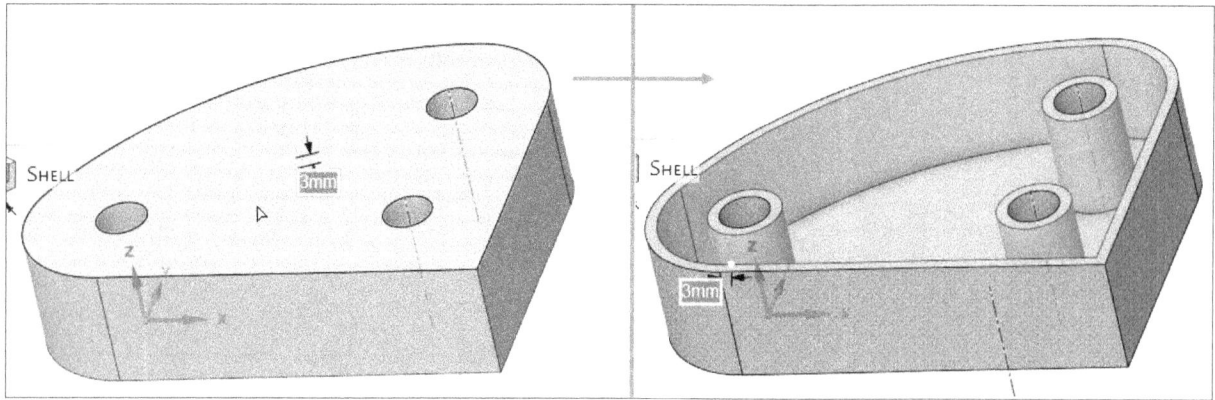

Figure-12. Applying shell operation

- Click on the **OK** button from the **HUD** toolbar and press **ESC** to exit the tool.
- Save the file at desired location in local drive as discussed earlier.

PRACTICAL 2

Create the model as shown in Figure-13. The drawing for the model is given in Figure-14.

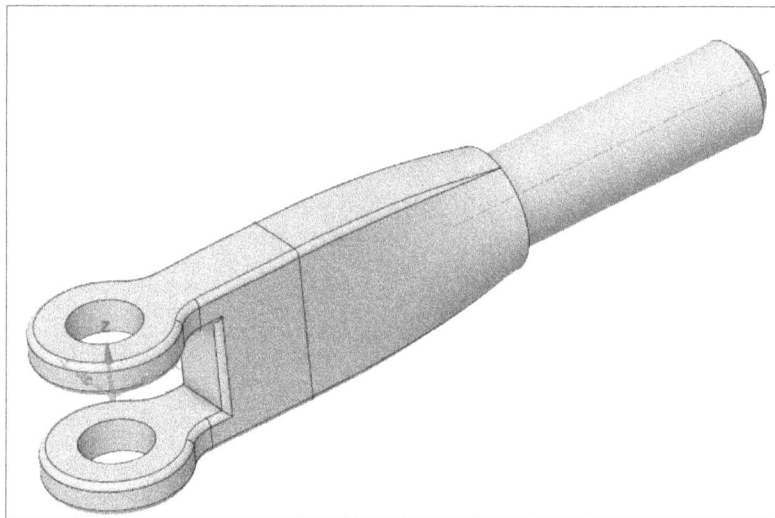

Figure-13. Practical 2 Design

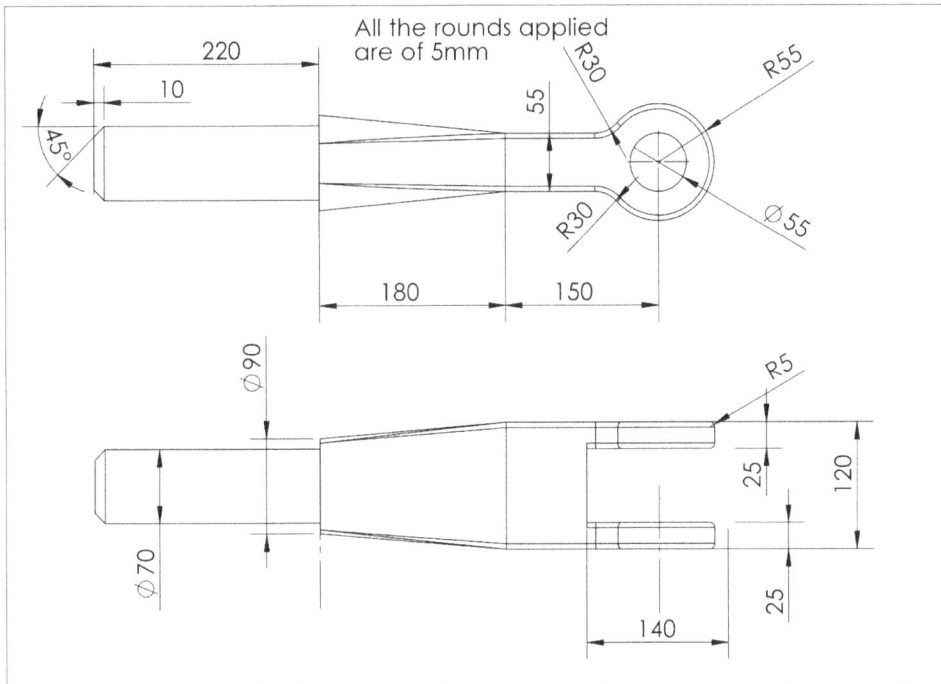

Figure-14. Practical 3 drawing views

Steps:

Creating First Extrude Feature

- Start Ansys Discovery if not started yet.
- Click on the **CIRCLE** tool from the **SKETCH** panel in the **DESIGN** tab of the **Ribbon**. The XY plane will be active as sketching plane and you will be asked to specify center point of the circle.
- Click on the **Plan view** tool from the **HUD** toolbar to make sketching plane parallel to screen.
- Click at the origin and create two circles of diameter 55mm and 100mm; refer to Figure-15.

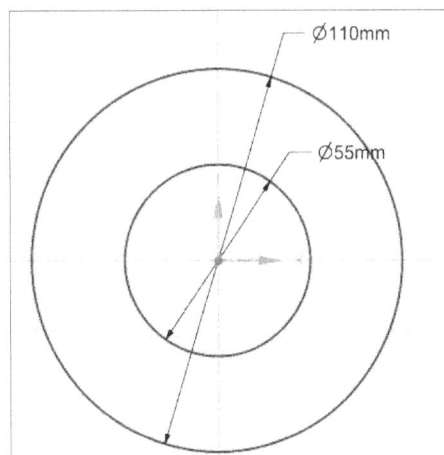

Figure-15. Circles created for practical

- Click on the **THREE-POINT RECTANGLE** tool from the **SKETCH** panel in the **DESIGN** tab of the **Ribbon**. You will be asked to specify corner point of rectangle.
- Create the rectangle as shown in Figure-16 and apply the dimensions as shown in Figure-17.

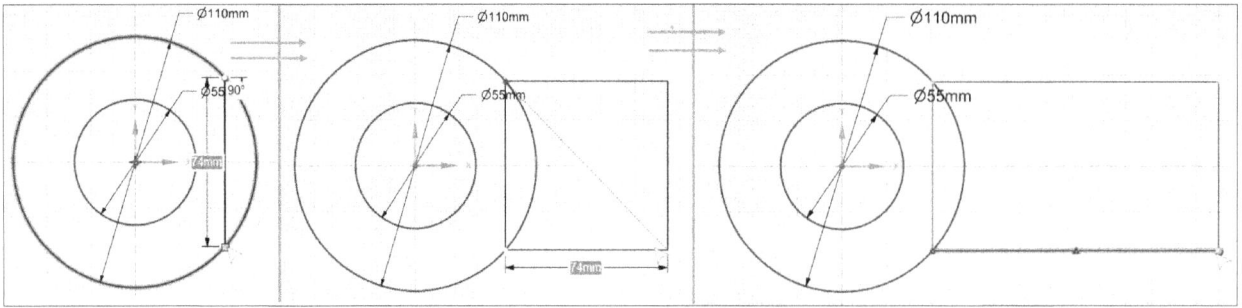

Figure-16. Rectangle created

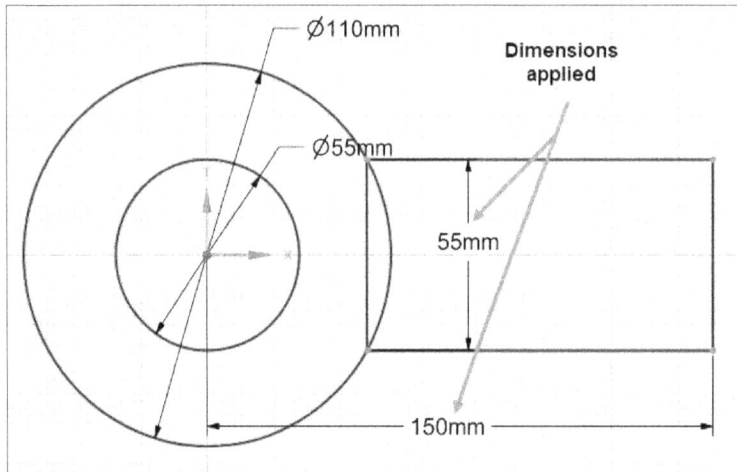

Figure-17. Applying dimensions to rectangle

- Click on the **TRIM AWAY** tool from the **SKETCH** panel in the **DESIGN** tab of the **Ribbon**. You will be asked to select section to be trimmed from the sketch.
- Select the sections of sketch as shown in Figure-18. Press **ESC** to exit the tool.

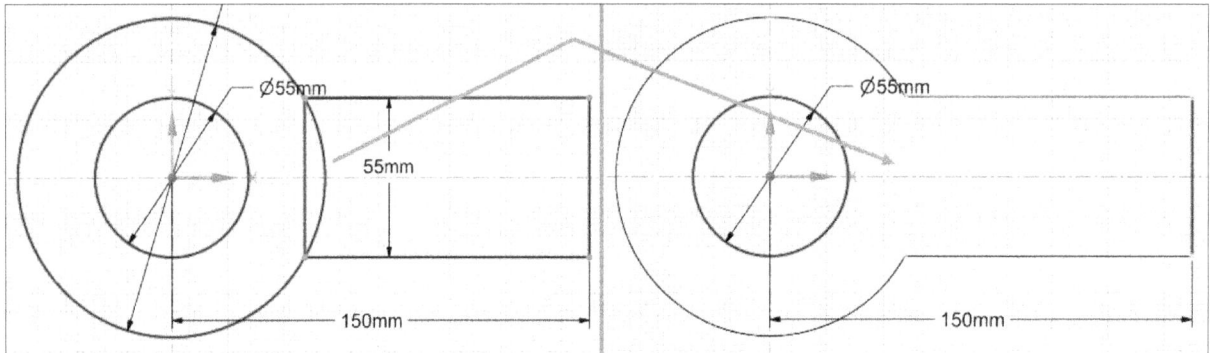

Figure-18. Trimming sketch

- Click on the **CREATE ROUNDED CORNER** tool from the **SKETCH** panel in the **DESIGN** tab of the **Ribbon** and select the curves of sketch as shown in Figure-19 to apply fillet.

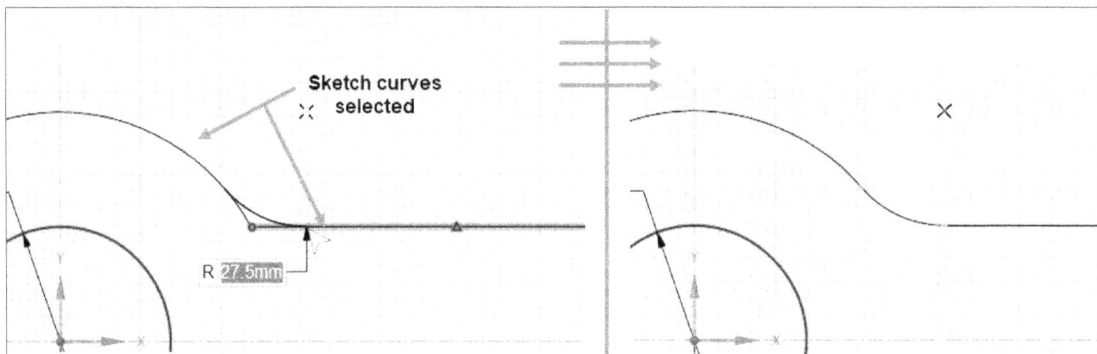

Figure-19. Selecting curves for round

- Similarly, apply round to other side in sketch; refer to Figure-20. Click on the **Dimension** tool and apply **30mm** radius to both fillets; refer to Figure-21.

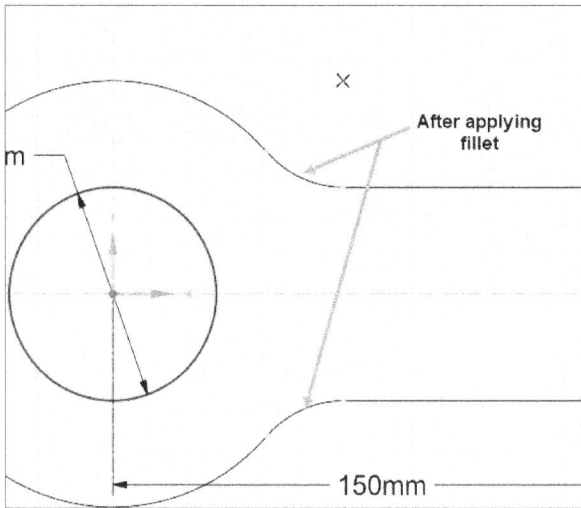

Figure-20. After applying fillet

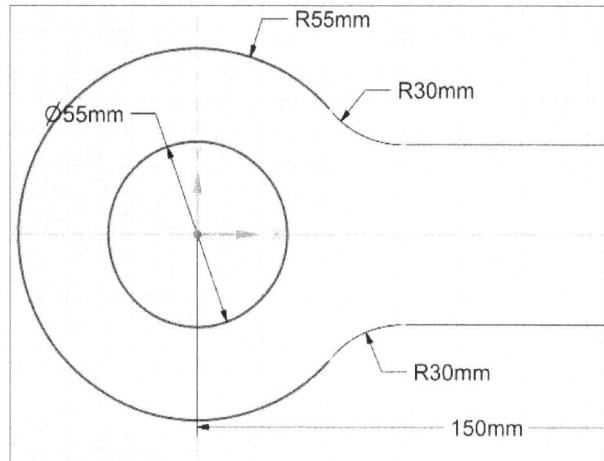

Figure-21. Applying radius dimension to fillets

- Click on the **Return to 3D** tool from the **HUD** toolbar. You will exit the sketching environment and the **PULL** tool become active.
- Select in the middle of sketch and drag using middle mouse button to reorient the model.
- Select and drag the sketch to create solid feature. An edit box to define depth of feature will be displayed; refer to Figure-22.

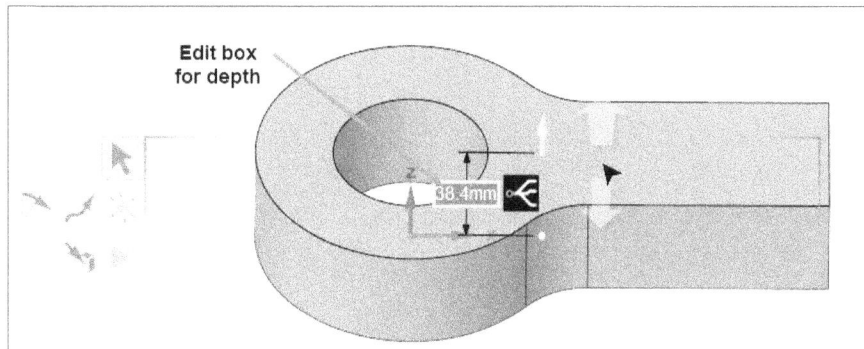

Figure-22. Edit box for depth

- Enter the value as **60mm** in the edit box.
- Using **MMB** reorient the model and select bottom face of the model. The face will become selected.
- Drag the face and enter **60mm** in the edit box to increase depth to other side as well; refer to Figure-23. Click in empty area and press **ESC** to exit the tool.

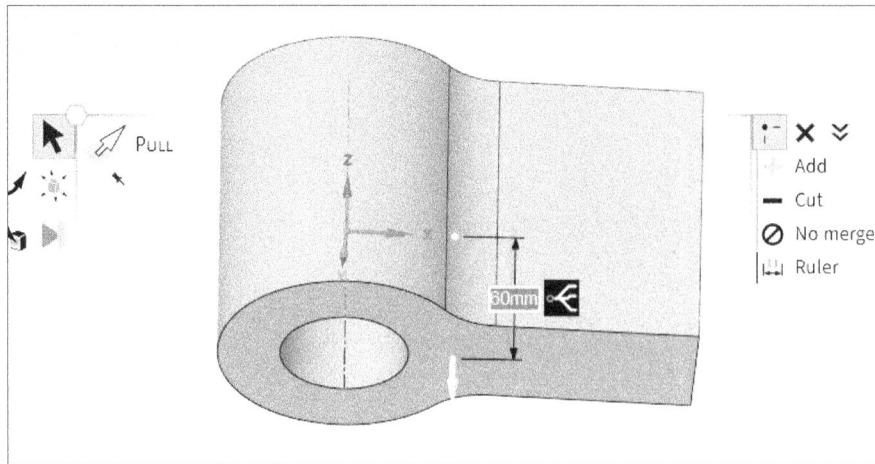

Figure-23. Extruded to other side

Creating Blend Feature

- For creating blend feature, two sketch sections are required. As per the drawing of this practical, blend feature joins rectangular face of current model with a circular shape at a distance of **180** mm. So, you need to create a plane at a distance of **180** mm from current model face to draw the circle. Click on the **PLANE** tool from the **CREATE** panel in the **DESIGN** tab of the **Ribbon**. The **PLANE HUD** toolbar will be displayed and you will be asked to select reference for creating plane.

- Select the face of model as shown in Figure-24. The plane will be created. Press **ESC** to exit the tool.

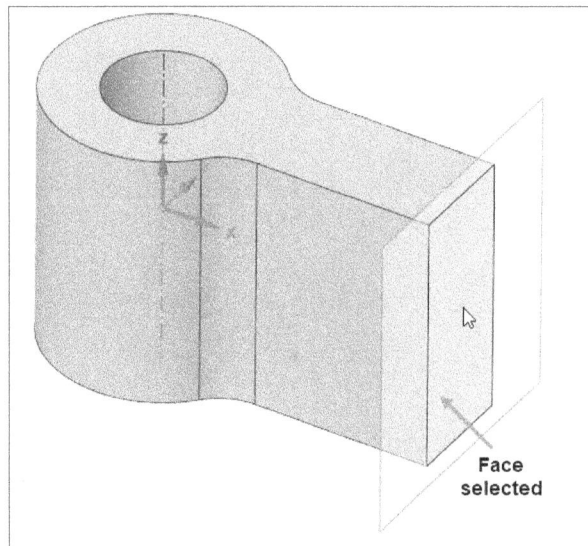

Figure-24. Face selected for creating plane

- Select the newly created plane and click on the **Move** tool from the **EDIT** panel in the **DESIGN** tab of the **Ribbon**. The drag handles will be displayed on the selected plane with **MOVE HUD** toolbar; refer to Figure-25.

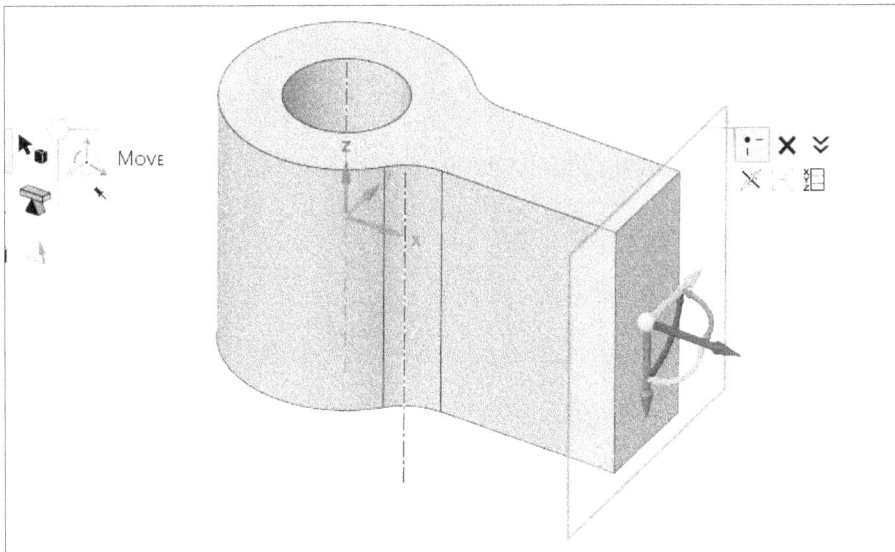

Figure-25. Handles displayed for moving plane

- Select the drag handle normal to selected face (Blue colored) and drag it outward by some distance. An edit box to define move distance will be displayed; refer to Figure-26.

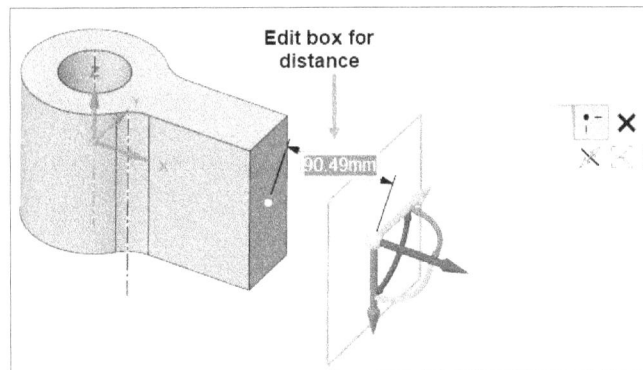

Figure-26. Edit box for distance

- Enter **180mm** value in the edit box to move the plane by respective value. Click in the empty area of screen and press **ESC** to exit the tool.
- Select the created plane and click on the **Circle** tool from the **SKETCH** panel in the **DESIGN** tab of the **Ribbon**. The sketching environment will become active.
- Click on the **Plan view** tool from the **HUD** toolbar and create circle of diameter **90mm** using origin as center of circle; refer to Figure-27.

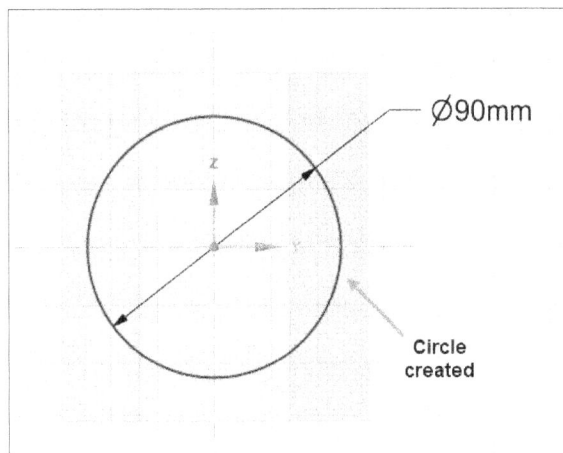

Figure-27. Circle created for blend

- After creating circle, click on the **Return to 3D** tool from the **HUD** toolbar to exit sketching environment and click on the **BLEND** tool from the **EDIT** panel in the **DESIGN** tab of the **Ribbon**. The **BLEND HUD** toolbar will be displayed.
- Select the recently created circle and rectangular face of the model while holding the **CTRL** key. Preview of blend feature will be displayed; refer to Figure-28.

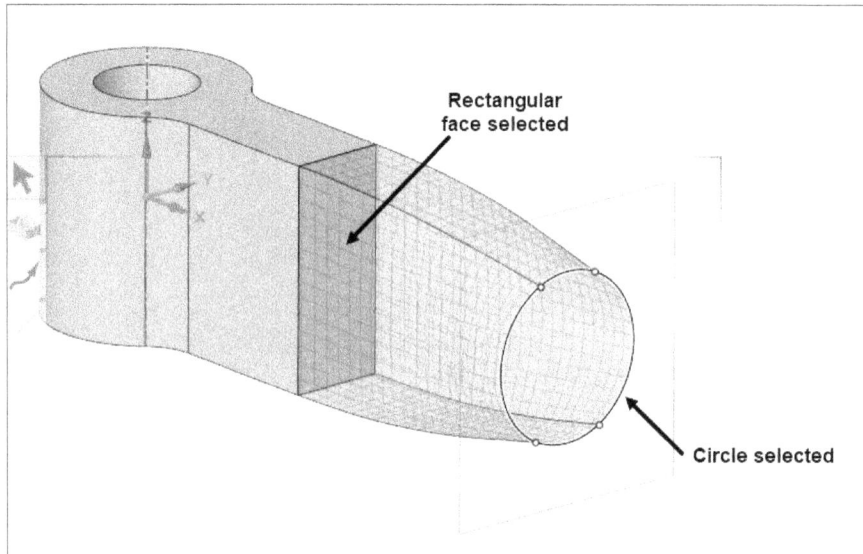

Figure-28. Preview of blend feature

- Click on the **OK** button from the **HUD** toolbar to create the feature. Press **ESC** to exit the tool.
- Select the circular flat face of the model; refer to Figure-29 and click on the **CIRCLE** tool from the **Ribbon**.

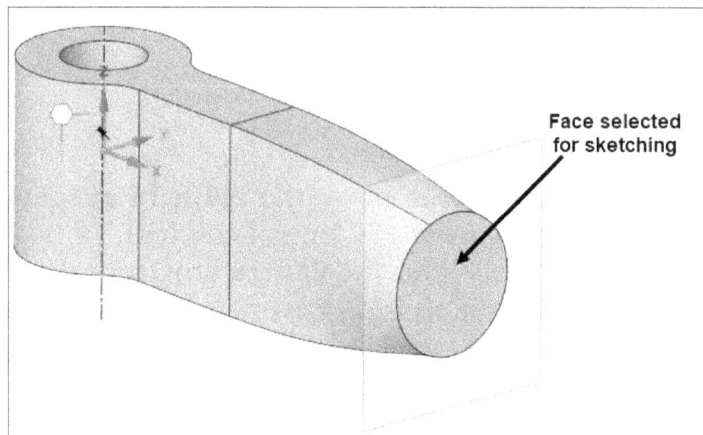

Figure-29. Face to be selected

- Create a circle of diameter **70mm** using origin as center of circle; refer to Figure-30.

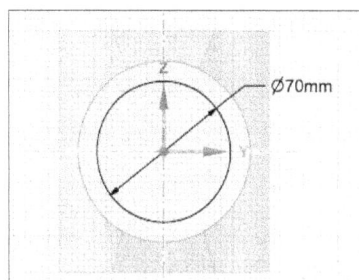

Figure-30. Creating circle at origin

- After creating circle, click on the **Return to 3D** tool from the **HUD** toolbar. The **PULL HUD** toolbar will be displayed.
- Click inside the recently created circle and drag outward by some distance. The edit box to define depth of extrude feature will be displayed.
- Enter the depth value as **220** in the edit box. Click in the empty area and press **ESC** to exit the tool.

Creating Extrude Cut Feature

- To create an extrude cut feature for removing material from front side, you need to create a plane at **25** depth from top face so that you can create extrude cut feature below that plane by specified value. Click on the **PLANE** tool from the **CREATE** panel in the **Ribbon**. The **PLANE HUD** toolbar will be displayed.
- Select the top face of model as shown in Figure-31. The plane will be created. Click in the empty area and press **ESC** to exit the tool.

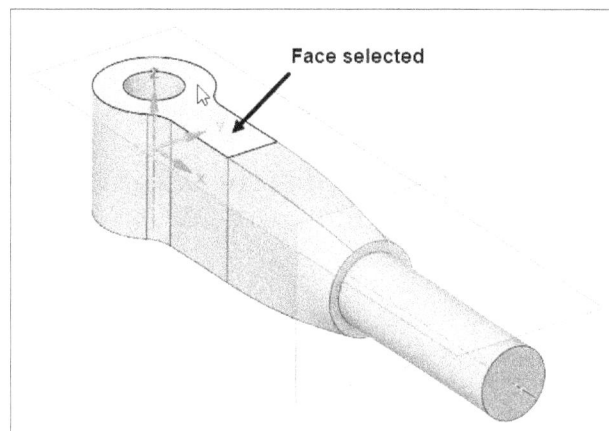

Figure-31. Face selected for plane

- Select the newly created plane and click on the **Move** tool from the **EDIT** panel in the **Ribbon**. The **MOVE HUD** toolbar will be displayed with move handles.
- Select the handle perpendicular to selected plane (blue handle) and drag it downward by some distance. An edit box to define distance will be displayed.
- Type the distance value as **25mm** in the edit box and press **ENTER**. The plane will move to specified distance. Click in the empty area and press **ESC** to exit the tool.
- Select the newly created plane and click on the **RECTANGLE** tool from the **Ribbon**. The sketching environment will become active.
- Make the sketch plane parallel to screen by selecting the **Plan view** tool from the **HUD** toolbar.
- Create a rectangle as shown in Figure-32. Use the **Dimension** tool to get sketch constrained.

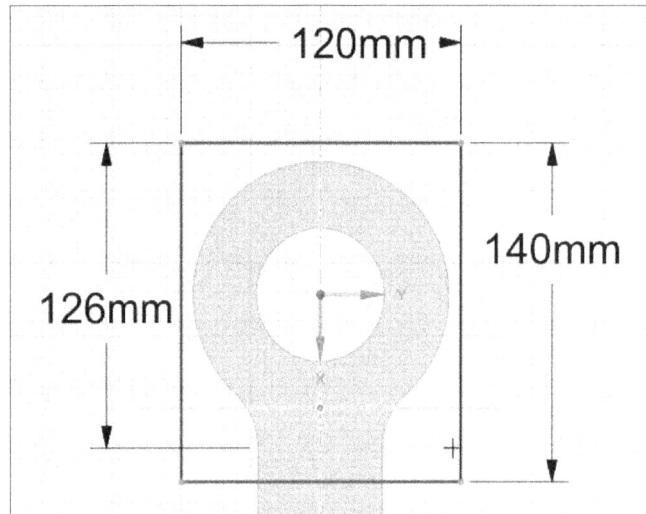

Figure-32. Sketch created for cut feature

- Click on the **Return to 3D** tool from the **HUD** toolbar. The **PULL HUD** toolbar will be displayed.
- Click on the **Cut** toggle button from the **HUD** toolbar and select the rectangle section to use it for extrude cut.
- Drag the sketch downward by some distance. The edit box to define depth of extrude cut will be displayed; refer to Figure-33.

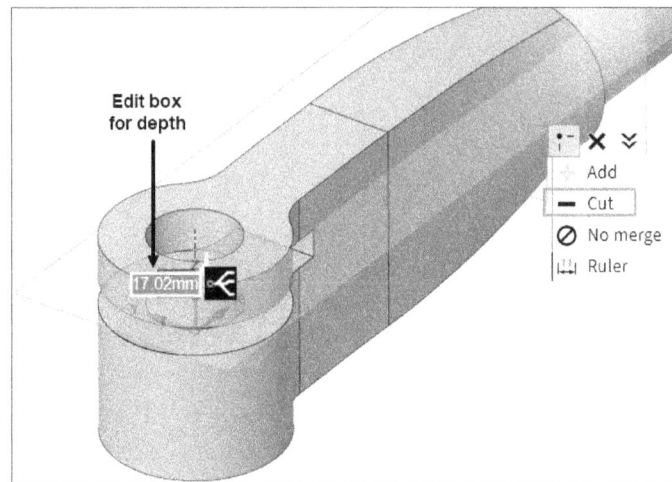

Figure-33. Edit box for extrude cut

- Enter **70mm** in the edit box to define depth of extrude cut. Preview of feature will be displayed; refer to Figure-34.

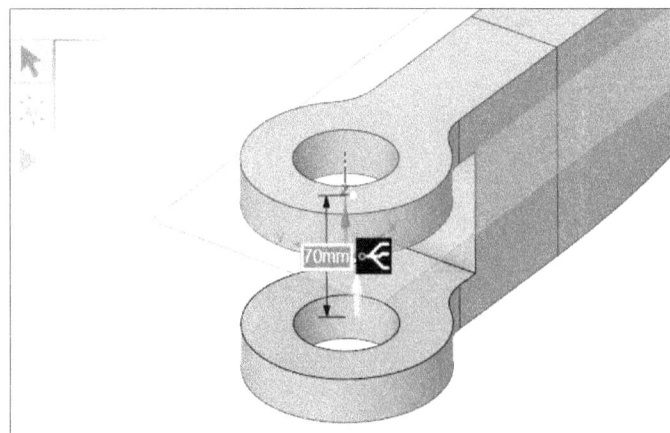

Figure-34. Preview of extrude cut

- Click in the empty area and press **ESC** to exit the tool.

Applying Fillet and Chamfer

- Select the sharp edges of the model as shown in Figure-35 and click on the **Pull** tool from the **Ribbon**. Drag handle to apply fillet will be displayed; refer to Figure-36.

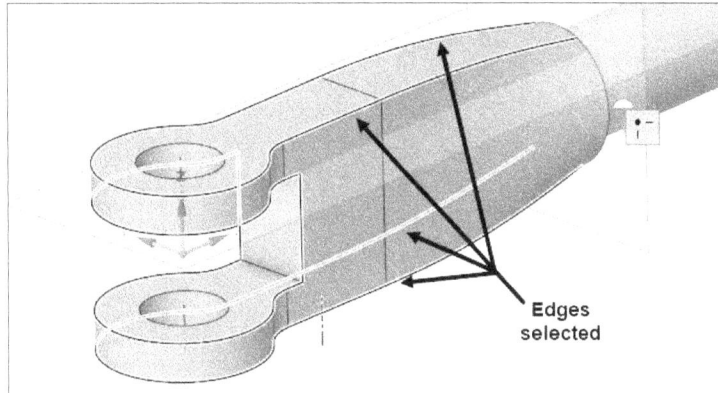

Figure-35. Edges selected for round

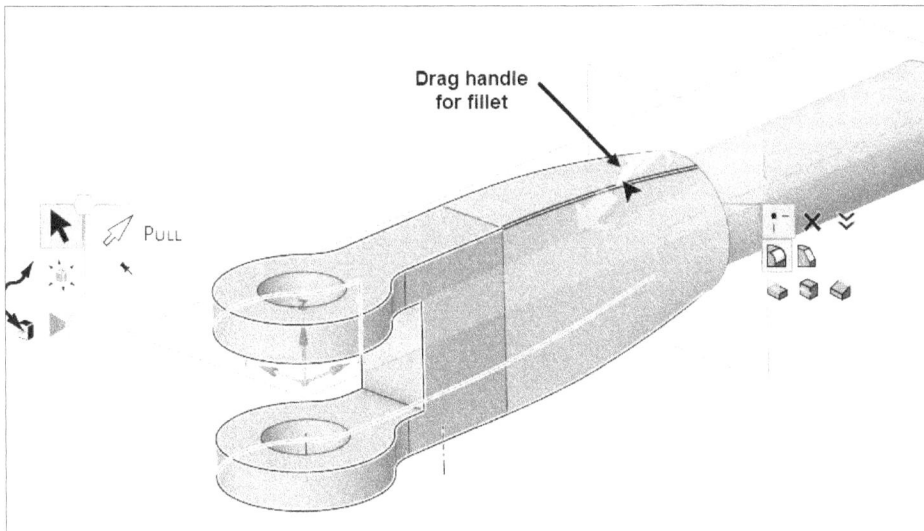

Figure-36. Drag handle for applying fillet

- Select the **History Tracking** toggle button from the **AUTOMATE** panel in the **DESIGN** tab of the **Ribbon**. The **History Tracking** window will be displayed which allows to modify the parameters of earlier created features; refer to Figure-37.

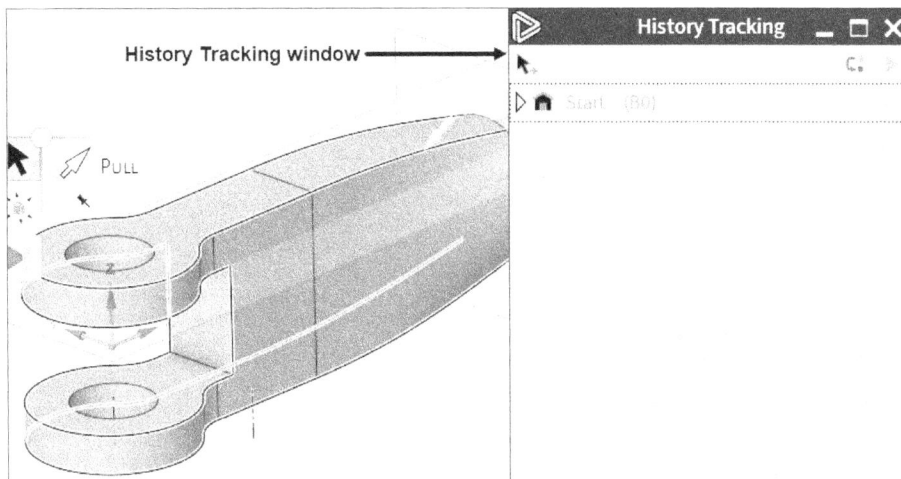

Figure-37. History Tracking window

- Drag the handle downward to apply round. An edit box to apply radius will be displayed; refer to Figure-38.

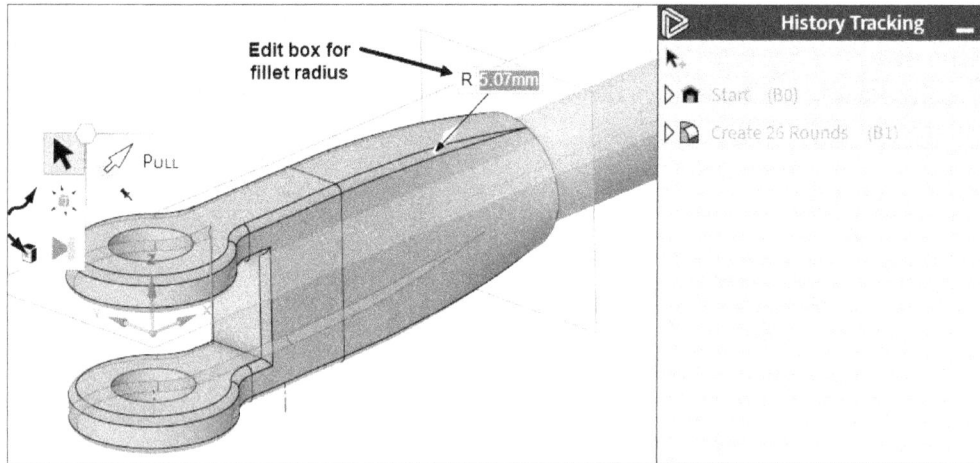

Figure-38. Edit box for fillet radius

- Type the value **5mm** in the edit box and press **ENTER** to apply fillet. Click in the empty area to exit selection.
- Select the circular edge of the model at the end of model and select the **Chamfer** button from the **HUD** toolbar; refer to Figure-39.

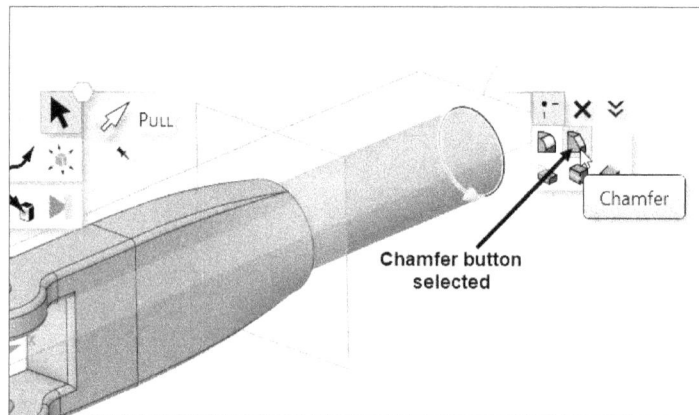

Figure-39. Chamfer button selected

- Drag the handle downward by some value. The chamfer will be applied.
- Expand the **Create Chamfer** node from the **History Tracking** window and enter the value as **10mm** in the **Distance** edit box; refer to Figure-40.

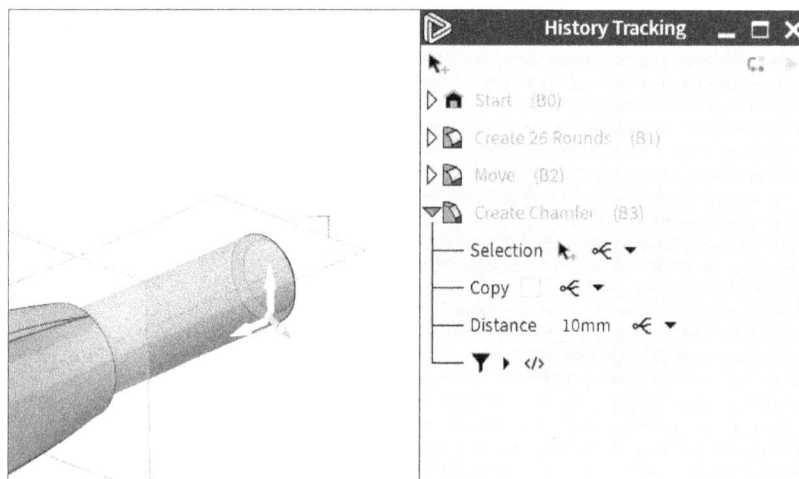

Figure-40. Specifying chamfer distance

- Click in the empty area and press **ESC** to exit the tool. Press **CTRL**+**S** to save the model as discussed earlier.

PRACTICE 1
Create the model by using the dimensions given in Figure-41.

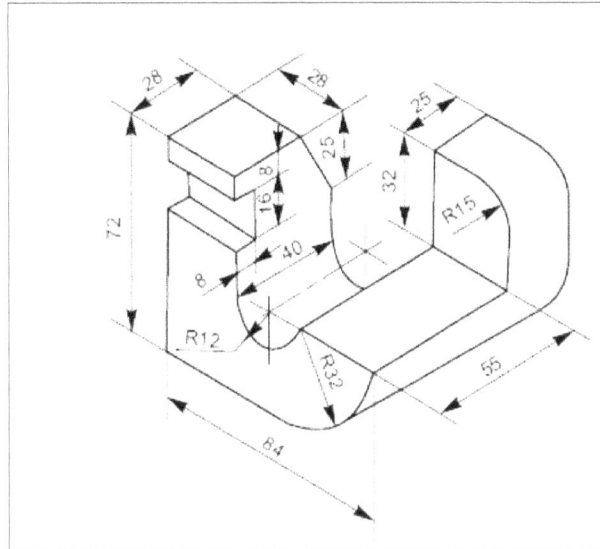

Figure-41. Practice 1

PRACTICE 2
Create the model by using the dimensions given in Figure-42.

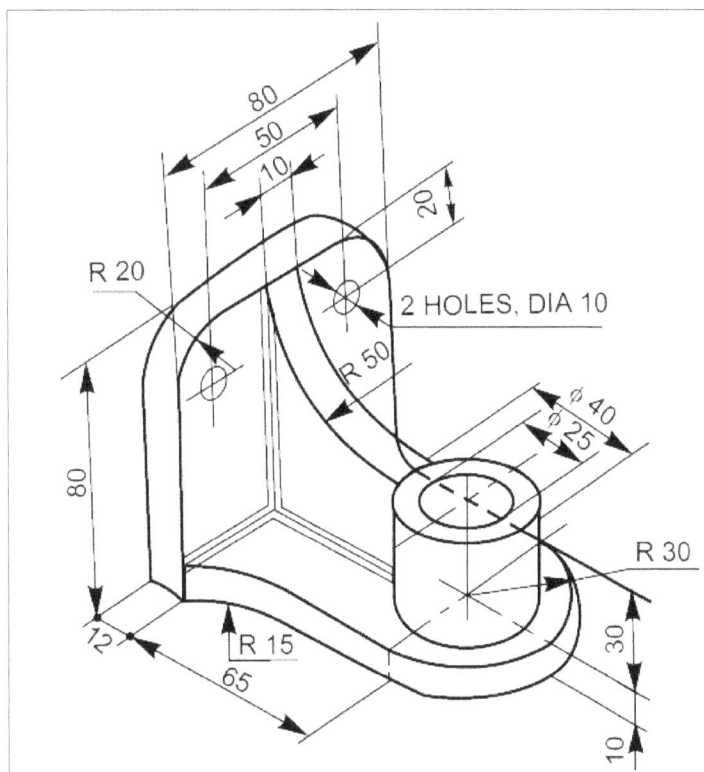

Figure-42. Practice 2

PRACTICE 3

Create the model by using the dimensions given in Figure-43.

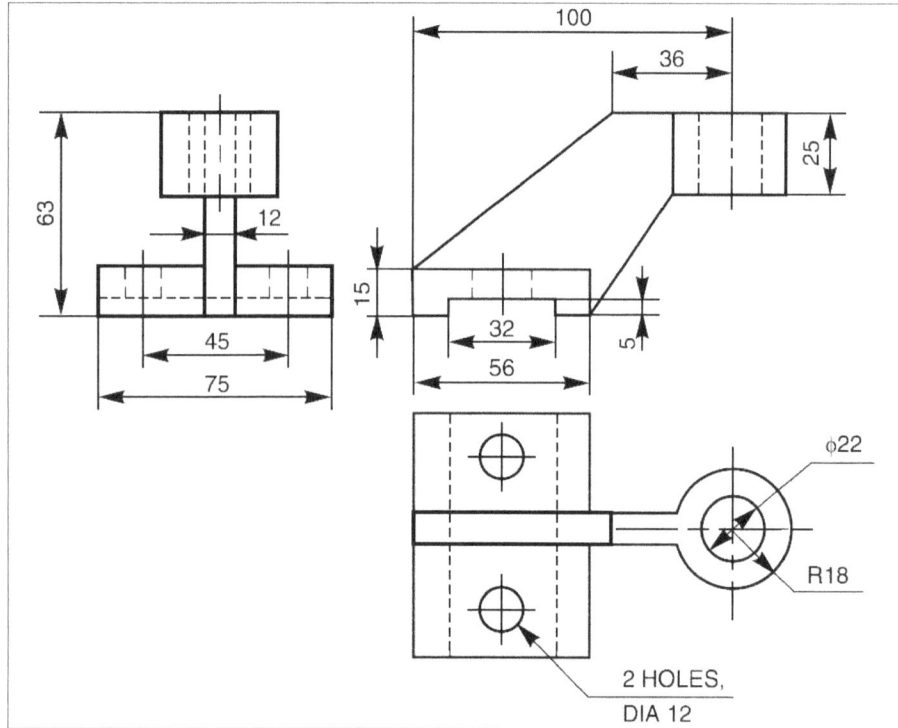

Figure-43. Practice 3

Chapter 6

Ansys Discovery-
Introduction to Analyses

Topics Covered

The major topics covered in this chapter are:

- *Stages of Analyses*
- *Quickscope*
- *Applying Material, Supports, and Force/Torque*
- *Applying Pressure, Displacement, Moment Load, Mass Load, Velocity, and Acceleration*
- *Applying Bearing Loads and Bolt Preloads*
- *Defining Contacts, Applying Joints, Spring Conditions,*
- *Setting Fluid Flow Domain, Fluid Flow Conditions, Solid Thermal Conditions*
- *Setting Electromagnetic conditions*
- *Setting Symmetry in Model*
- *Defining Simulation Options*

INTRODUCTION

In previous chapters, you have learned to create, modify, and prepare 3D models for analyses. You have learned to remove extra features from the model so that you can get solution of analysis faster. In this chapter, you will get familiar with user interface of Ansys Discovery and get an overview of analysis setup and solution process.

STAGES OF ANALYSES

In any parametric analysis and simulation software, there are three stages at which setting up an analysis and find solutions: Pre-processing, Processing, and Post-processing. These stages are discussed next.

Pre-Processing

Pre-processing is the stage at which model is prepared for running analysis. At this stage, you import/create the model, remove extra features, apply materials to model, apply loads and constraints, and define meshing parameters for the analytical model. Various steps involved in this processes can be given as:

* Importing the CAD geometry or creating it in Ansys Discovery. Note that you can use 2D geometry for elements like trusses and frames while complex objects are analyzed using 3D geometries.
* Assigning material properties of CAD geometry. Generally, properties of material include Modulus of elasticity, yielding strength, density, Poisson's ratio, and so on. When you are working on nonlinear analyses or Fatigue analyses then non linearity curves like S-N curve or E-N curve parameters are also required with general material properties.
* Defining boundary conditions which mainly includes defining constraints like fixed face, pin, and slide constraints. Note that constraints applied to model should represent the real environmental condition of the component undergoing analysis.
* Defining load conditions which may include force, torque, thermal loads like heat flux, electromagnetic force, and so on.
* Creating mesh from model with adequate quality and fineness.

Processing

Processing is the stage at which computer takes all the inputs provided in pre-processing and performs calculations based on equations related to selected analysis type. For example, when performing linear static analysis, system solves $K.u=F$ for each element created using meshing and then combines all the results to give you overall results of analysis. Generally, K which is stiffness matrix and F which is force matrix are known and u which is displacement matrix is solved. You will learn about various equations when working on different analyses during the course of this book.

Post-Processing

Post-processing is the stage at which result generated by system after processing is interpreted, analyzed, and visualized. Various steps and operations involved at this stage are given next.

* Generating Graphical Representations: The result values generated after solving analysis are used to create contour plots, vector plots, iso-surfaces, and so on.

- Understanding Results: After generating graphical results, you need to find out areas that need attention based on analysis. For example, if you have performed structural analyses then you need to find out the regions of model that have highest stresses and strains. You will also generate secondary results like Factor of Safety at this stage.
- Generating Reports: After generating and understanding the results, next step is to generate a report file on results that can be shared with other.

You will perform all these steps and generate results for different types of analyses throughout rest of the book. The tools to apply analysis related parameters are available in the **SIMULATION** tab of the **Ribbon**. To activate tools of this tab, you need to switch to **Explore** stage using **Stage Navigator**; refer to Figure-1. Various tools and options of this tab are discussed next.

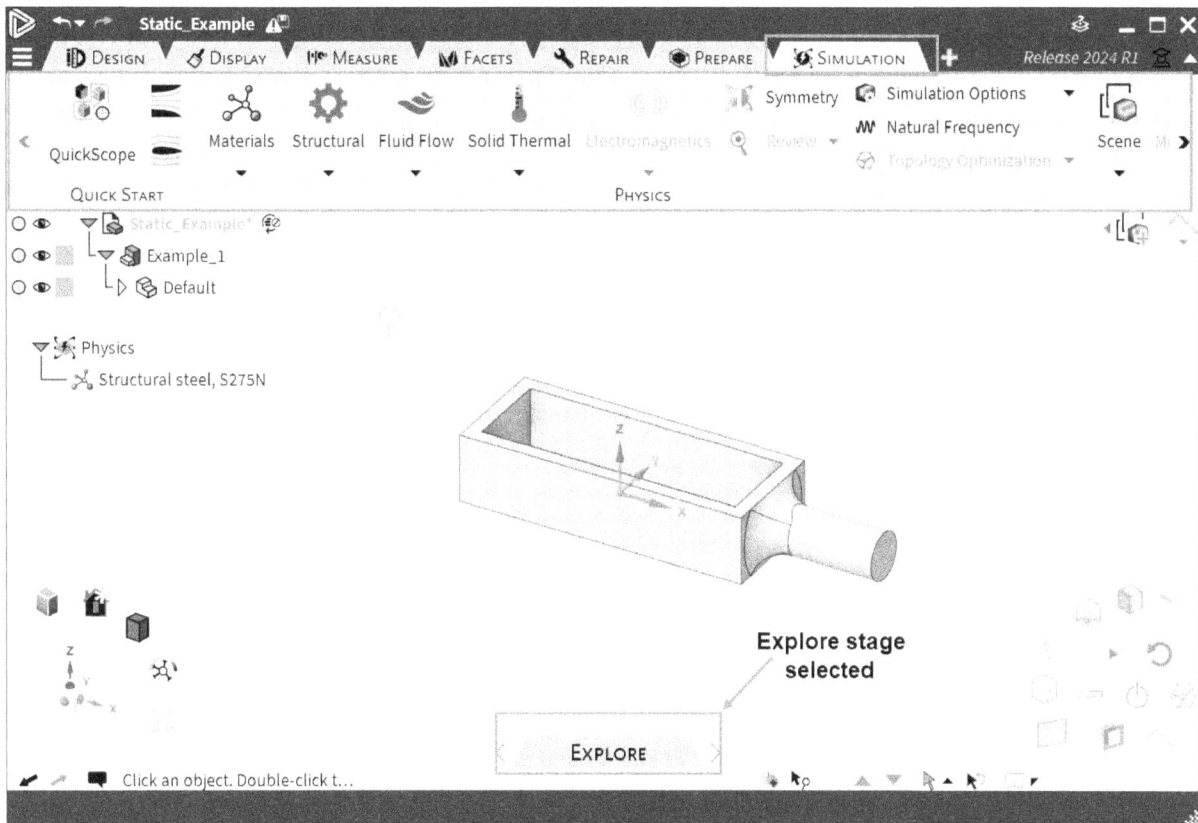

Figure-1. Explore stage selection

QUICKSCOPE

The **QuickScope** tool is used to define scope for performing analysis. In other words, you define what components and objects are to be included in the analysis. For example, if you have imported an assembly then you can define which components are to be included in the analysis. The procedure to use this tool is given next.

- Click on the **QuickScope** tool from the **QUICK START** panel in the **SIMULATION** tab of the **Ribbon**. The **QUICKSCOPE HUD** toolbar will be displayed and you will be asked to pick bodies to be included or excluded from the analysis; refer to Figure-2.

Figure-2. QUICKSCOPE HUD toolbar

- Select the **All** toggle option from the **Include** section of **HUD** toolbar to include all the components displayed in graphics area for analysis. Select the **None** toggle option if you do not want to include any component in the current analysis. Select the **Custom** toggle option if you want to manually select the components to be included in the analysis. Select the **Invert** toggle option if you want to flip the selections so that selected objects get de-selected and deselected objects will become selected.

- Select the **Invert appearance** toggle option from the **HUD** toolbar to flip the appearance of components in graphics area. Note that it will only flip display and does not affect selection.

- After setting desired parameters and selecting the bodies, press **ESC** to exit the tool. Note that there will be no change in displayed components in graphics area but when you check the **Design Tree** then you will find that only selected components are included in the simulation; refer to Figure-3.

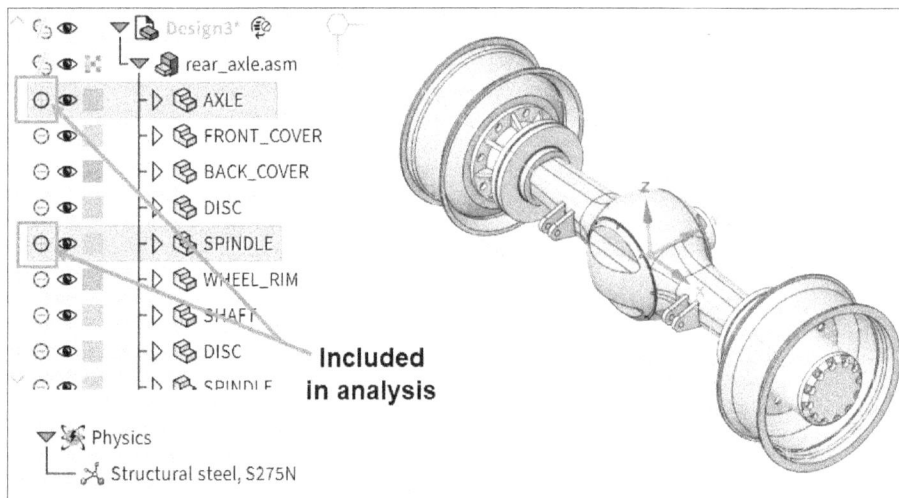

Figure-3. Design Tree for included bodies

- You can now hide rest of the components by clicking on the **Show/Hide** toggle button 👁/👁 from the **Design Tree**; refer to Figure-4.

Figure-4. Hiding bodies not needed

APPLYING MATERIAL

The options in the **Materials** drop-down are used to assign materials to different bodies in the graphics area. The procedure to apply materials is given next.

- Select desired body(s) from the **Design Tree** or graphics area to which you want to apply materials and click on the down button for the **Materials** drop-down in the **PHYSICS** panel of **SIMULATION** tab in the **Ribbon**. The list of materials will be displayed; refer to Figure-5.

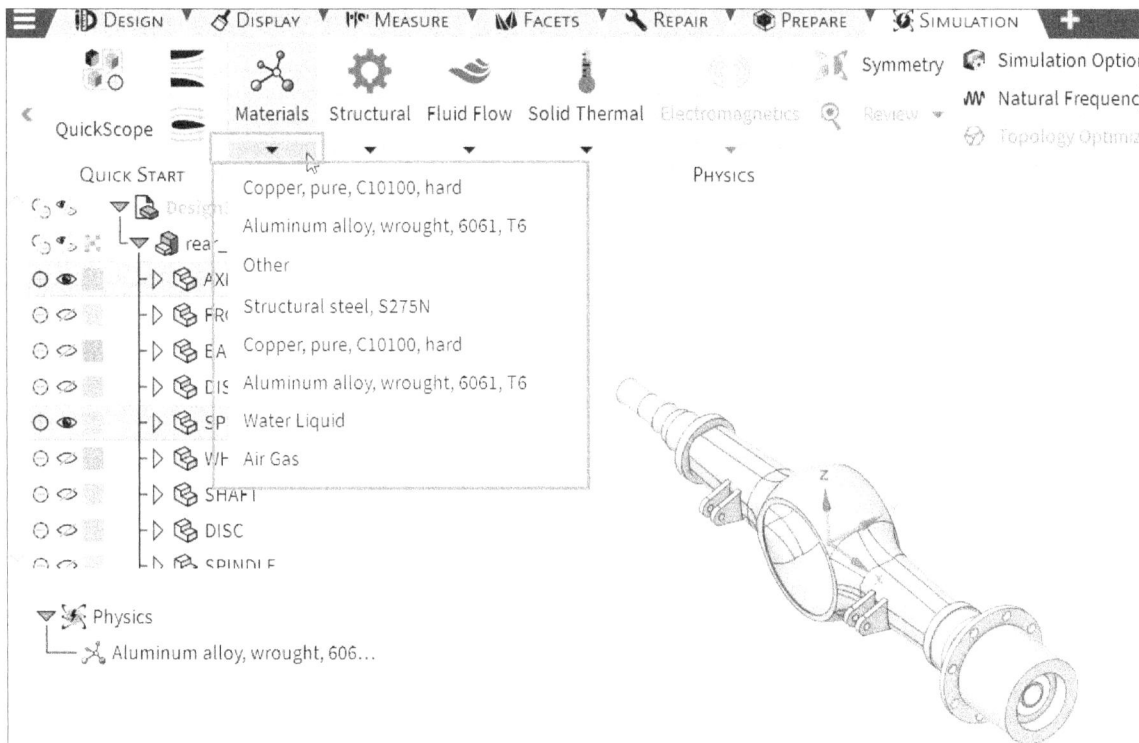

Figure-5. List of materials for selected bodies

- Select desired material from the list to apply it to selected bodies. Note that you can apply different materials to different bodies by individually selecting them from graphics area and then selecting related material from drop-down list.

- If you select the **Other** option from the **Materials** drop-down list then you will be asked to individually select the body to which other material is to be applied even if you have selected multiple bodies. Select the body to which you want to apply material that is not displayed in the drop-down list by default. The list of materials will be displayed in the **HUD** toolbar; refer to Figure-6. You can select desired material as discussed earlier.

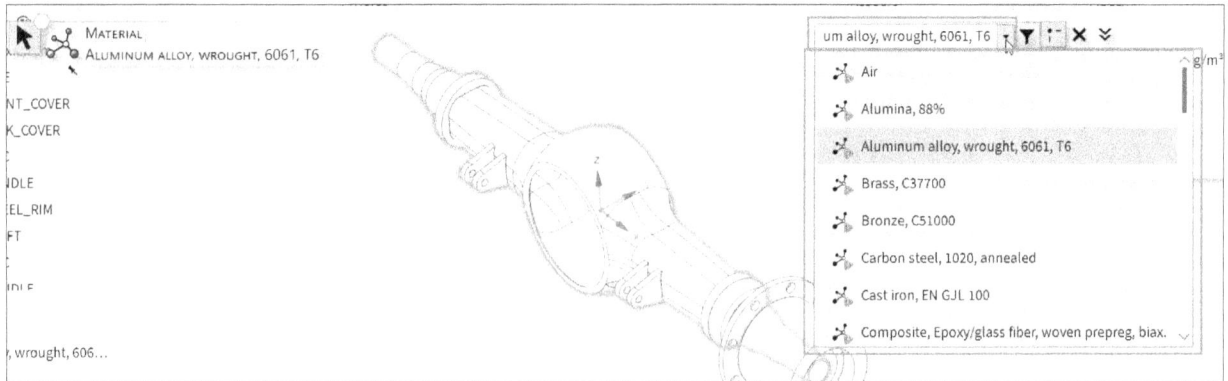

Figure-6. List of materials displayed in HUD toolbar

- After selecting material, you can tweak the value of basic parameters like Density, appearance color, environmental affects such as CO2 emission, energy consumed, etc. in the expanded **HUD** toolbar; refer to Figure-7.

Figure-7. Options to tweak material parameters

- After tweaking the values, you can save it to your local library by using the **Save to local library** option from the **HUD** toolbar.
- Press **ESC** twice to exit the tool. The materials will be applied to analytical model and will also be displayed in the **Solid Materials** node of **Physics Tree**; refer to Figure-8.

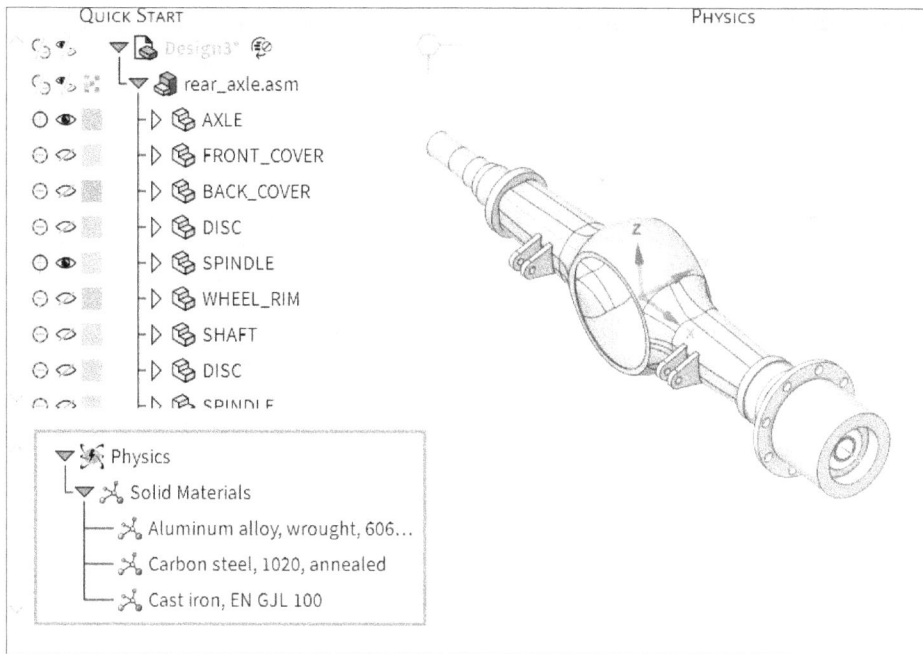

Figure-8. Materials added in the Physics Tree

APPLYING SUPPORTS

The **Support** tools are used to apply constraints to the analytical model. These constraints represent the real world condition of part in parametric form. For example, if you want to check the stress caused in a shaft under 150 N-m torque then it is important to restrict rotation of shaft when performing analysis otherwise there will be no stress induced in shaft because it will freely rotate due to applied torque; refer to Figure-9. In case of our example, when shaft will start to rotate under load of 150 N-m, the holes at the end of shaft which are connected with wheel will be at rest. This will be the time instance when shaft will be under maximum loading condition and as an engineering analyst we are supposed to check stress and strain of shaft at this precise time. There can be five different options to apply different types of support conditions for the analysis. The procedures to apply these support conditions are discussed next.

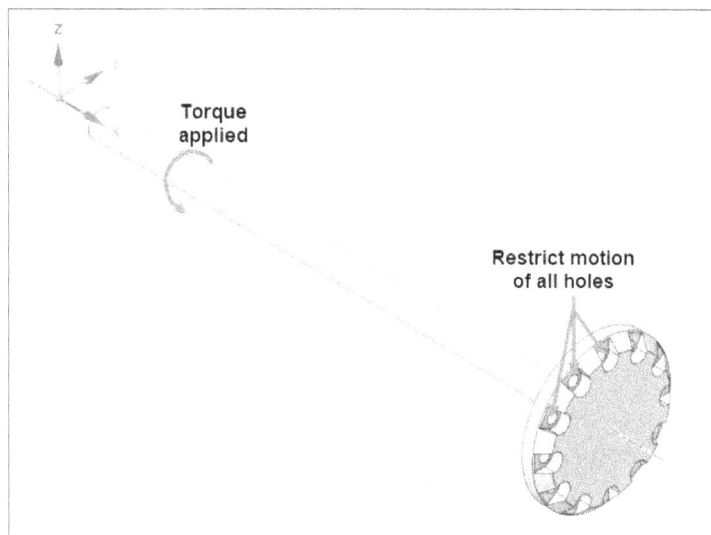

Figure-9. Shaft under torque

- Click on the **Support** tool from the **Structural** drop-down in the **PHYSICS** panel of **SIMULATION** tab in the **Ribbon**. The **STRUCTURAL HUD** toolbar will be displayed with **SUPPORT** options; refer to Figure-10.

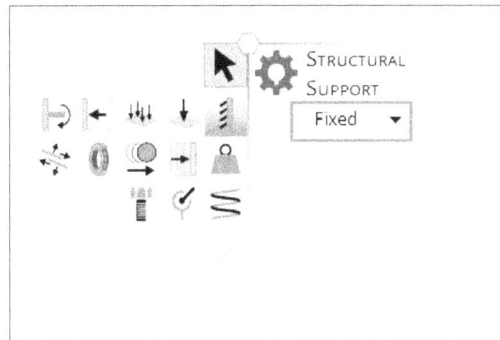

Figure-10. STRUCTURAL SUPPORT HUD Toolbar

Applying Fixed Support

By default, **Fixed** option is selected in the drop-down of **HUD** toolbar. This option is used when you want to restrict the motion of selected entity in all 6 Degrees of Freedom which means selected entity can neither translate nor rotate along X, Y, and Z axes. The procedure to use this option is given next.

- Select the **Fixed** option from drop-down in the **STRUCTURAL SUPPORT HUD** toolbar. You will be asked to select face(s) to be fixed.
- Select desired faces from the model to be fixed; refer to Figure-11 and click on the **OK** button from the **HUD** toolbar.

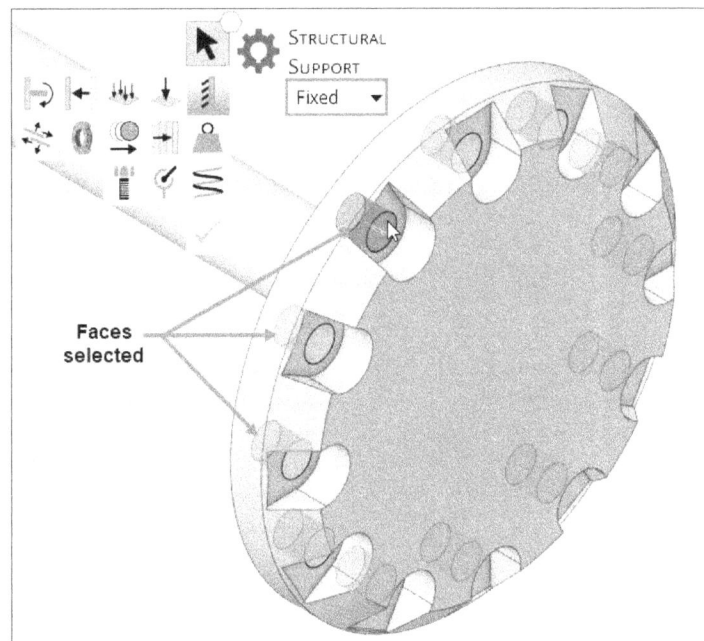

Faces selected

Figure-11. Fixing faces

Applying Sliding Support

The **Sliding** support is applied when you want to allow translational movements along selected face and rotational movement about only axis which is normal to selected face. Note that in case of cylindrical face selection, the axis for rotation will be center axis of cylindrical face. Select the **Sliding** option from the drop-down in the **STRUCTURAL SUPPORT HUD** toolbar. The procedure to apply this support is similar to Fix support.

Applying Hinged Support

The **Hinged** support is applied when you want to allow only rotational motion about axis of selected cylindrical face. Note that you can apply this support to cylindrical faces only. The procedure to use this tool is given next.

- Select the **Hinged** option from the drop-down in the **STRUCTURAL SUPPORT HUD** toolbar. You will be asked to select cylindrical face(s).
- Select desired face to which you want to apply hinged support; refer to Figure-12.

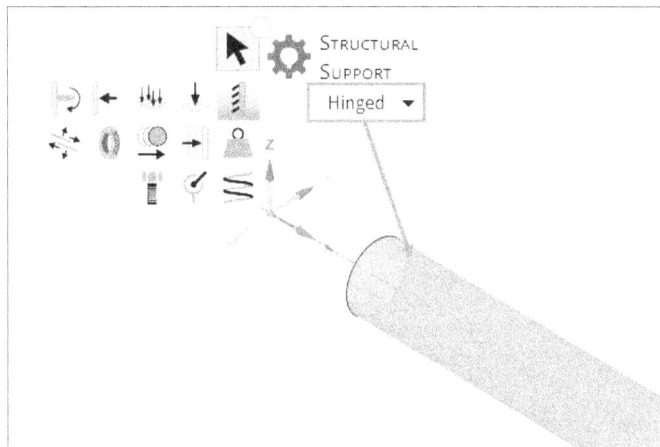

Figure-12. Applying hinged support

- Click on the **OK** button to apply the changes in the model.

Applying Ball Support

The **Ball** option is used where you want to allow the selected faces to rotate in all three directions but they cannot translate (move linearly) in any direction. This support is generally applied to represent ball joint type of connections. The procedure to use this tool is given next.

- Select the **Ball** option from the drop-down in the **STRUCTURAL SUPPORT HUD** toolbar. You will be asked to select spherical face(s) to apply the support.
- Select desired ball shaped faces to apply the support; refer to Figure-13 and click on the **OK** button. The support will be applied.

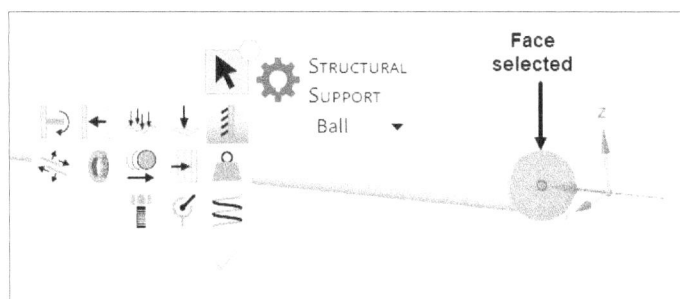

Figure-13. Face selected

Applying Displaced Support

The **Displaced** option is used to manually define the locking/unlocking of various degrees of freedom for selected face. This option is generally used when you want to specify non-standard constraints in the model. For example, you can allow 15 mm displacement for selected face along X axis and 5 degree rotation about Y axis while keeping rest degrees of freedom locked by using this option. The procedure to use this tool is given next.

* Select the **Displaced** option from the drop-down in the **STRUCTURAL SUPPORT HUD** toolbar. You will be asked to select face(s) to apply the support.
* Select desired face(s) from the model. The options to apply support will be displayed; refer to Figure-14.

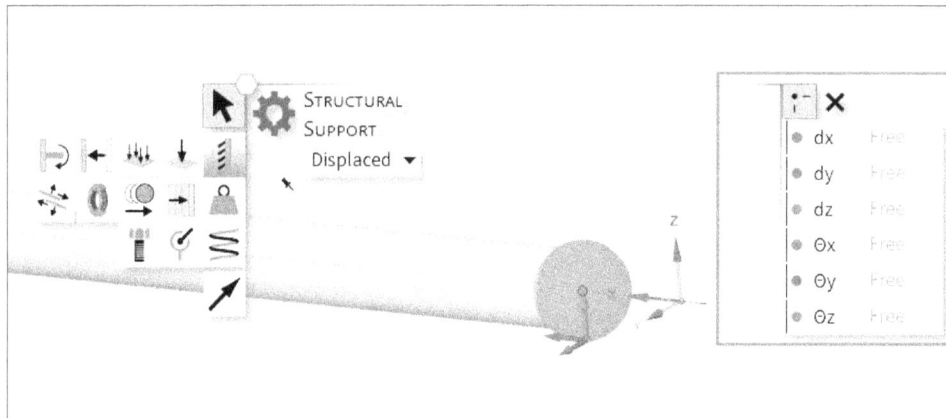

Figure-14. Displaced support options

* Select desired toggle buttons from the right **HUD** toolbar to activate respective degrees of freedom. Select the **dx**, **dy**, and **dz** toggle buttons to allow translation in respective direction. Similarly, select the **θx**, **θy**, and **θz** toggle buttons to allow respective rotations. After selecting the toggle buttons, specify the values in the edit boxes to allow respective movement. If you specify **0** in edit box then that movement will be restricted.
* After specifying desired parameters, press **ENTER** to apply displacement support. Press **ESC** to exit the tool

APPLYING FORCE/TORQUE

The **Force** tool is used to apply force and torque to selected face(s). The procedure to use this tool is given next.

* Click on the **Force** tool from the **Structural** drop-down in the **PHYSICS** panel of **SIMULATION** tab in the **Ribbon**. The options related to force will be displayed; refer to Figure-15.

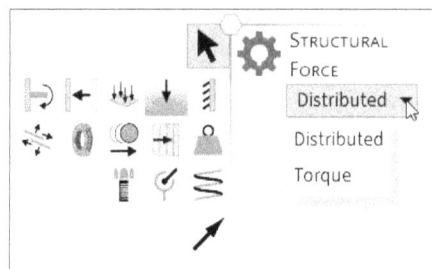

Figure-15. Force options in HUD toolbar

Applying Distributed Force

- Select the **Distributed** option from the drop-down if you want to apply distributed force on selected faces.
- After selecting the option, select the face(s) on which you want to apply the force. The options in the **HUD** toolbar will be displayed as shown in Figure-16.

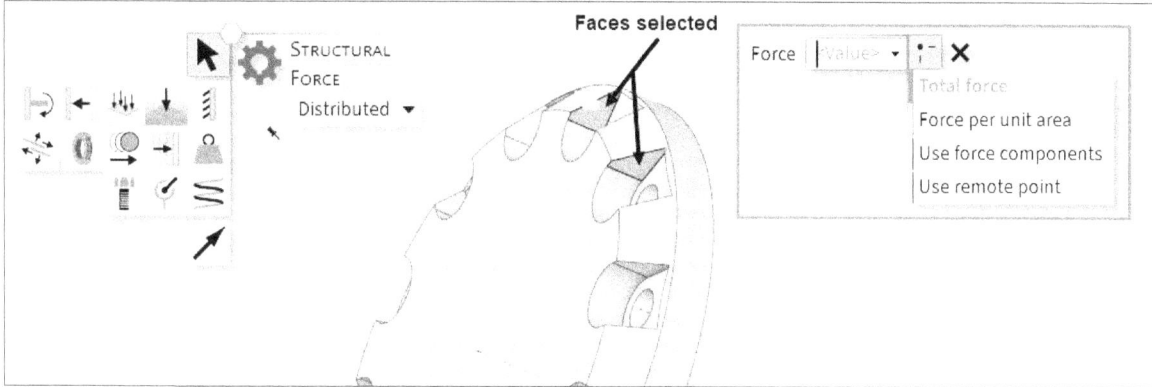

Figure-16. Force options

- Select the **Total force** option if you want to specify total force value for all selected faces. Select the **Force per unit area** option if you want to specify value of force per unit area of selected face generally specified in MPa or lb/ft². By default, force is applied normal/perpendicular to selected face(s). If you want to define X, Y, and Z components of force individually then select the **Use force components** toggle option from the right **HUD** toolbar. The edit boxes to define each component of force will be displayed; refer to Figure-17. Specify desired values in the edit boxes to define net value of force.

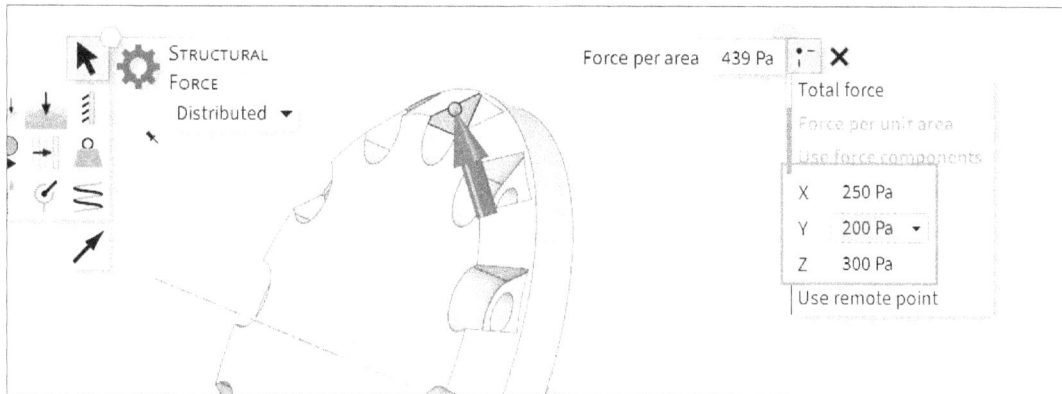

Figure-17. Edit boxes to define force components

- Select the **Use remote point** toggle option from the right **HUD** toolbar if you want to apply remote force on the face(s). You will be asked to specify origin location of force in X, Y, and Z edit boxes below the toggle option; refer to Figure-18.

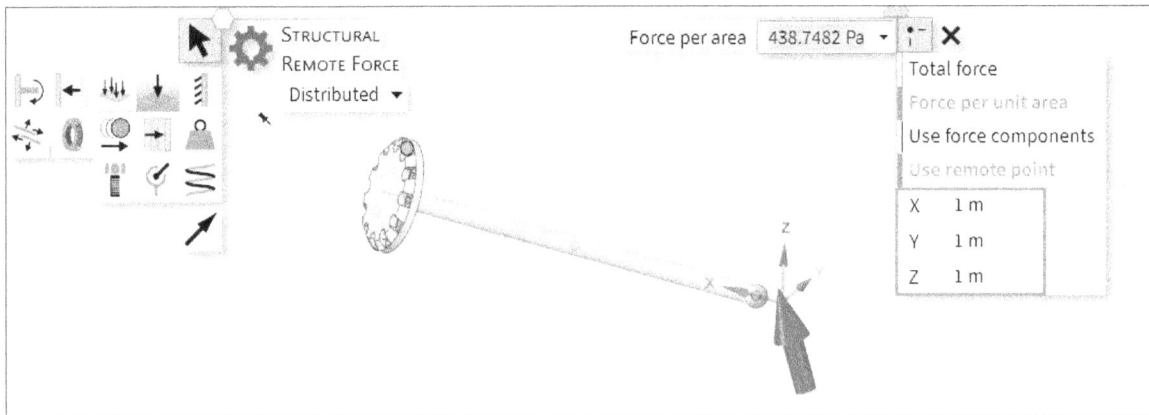

Figure-18. Remote force parameters

• Set desired parameters in the **HUD** toolbar and press **ENTER** to apply the load.

Applying Torque

Torque is a type of force that tends to rotate the object about an axis at specified distance. The SI unit of torque is (N-m)Newton-meter and it can be defined in lb-ft, kg-m, and other related units that represent force and length. The procedure to apply the torque is given next.

When talking about cars, there is a lot of confusion between torque and horsepower of car. Torque is the force that tends to rotate objects so in case of car, it is the force generated at connecting rod to rotate crank shaft. Horsepower or bhp is the power generated by engine to do any work. You can use engine to generate electricity, pull car, run air conditioner, lift objects, and so on. In case of car, when you need to carry heavy load at slow speed or quick acceleration then car should have high torque but when you want the car to achieve high total speed or superfast cooling inside the large cabin of car then you should select car with high bhp.

• Select the **Torque** option from the drop-down in the **STRUCTURAL FORCE HUD** toolbar. The options will be displayed as shown in Figure-19.
• Select the **Use force components** toggle option to individually define components of torque along X, Y, and Z edit boxes.
• Specify desired values in the edit boxes and press **ENTER** to apply torque.

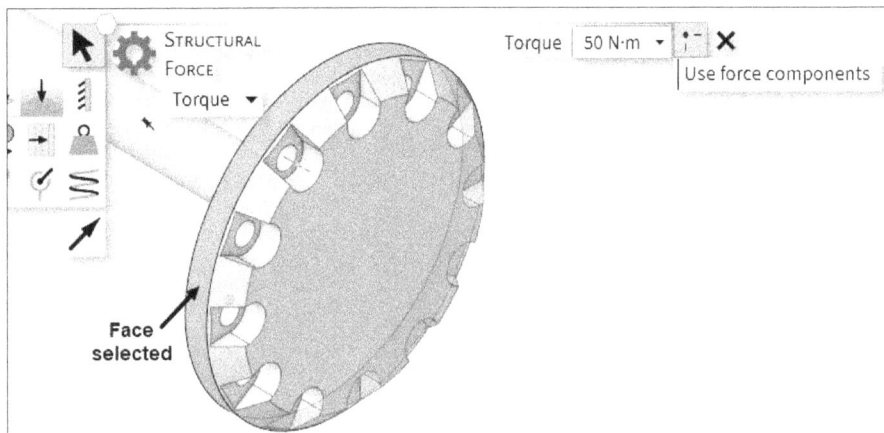

Figure-19. Torque options

• Press **ESC** to exit the tool.

APPLYING PRESSURE

The **Pressure** tool is used to apply pressure load on selected face(s). Note that pressure is the force per unit area and you can apply this load using the **Force** tool as well. The procedure to use this tool is given next.

- Click on the **Pressure** tool from the **Structural** drop-down in the **PHYSICS** panel of **SIMULATION** tab in the **Ribbon**. The **STRUCTURAL PRESSURE HUD** toolbar will be displayed; refer to Figure-20.

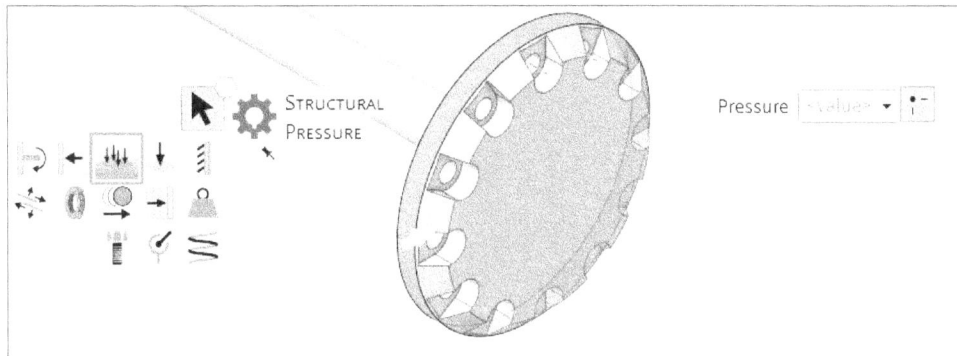

Figure-20. Pressure tool

- Select the face on which you want to apply pressure and enter desired value in the input box of **HUD** toolbar. The pressure will be applied. Press **ESC** to exit the tool.

APPLYING DISPLACEMENT

The **Displacement** tool is used to apply specified linear and rotational displacement to selected faces. In other words, the face selected for applying displacement load will move by specified displacement value in defined direction. This load is usually applied when you want to check the deformation of a workpiece shaping processes like forging, stamping, hammer press, etc. The procedure to use this tool is given next.

- Click on the **Displacement** tool from the **Structural** drop-down in the **SIMULATION** tab of the **Ribbon** or select the **Displacement** button from the **STRUCTURAL HUD** toolbar if active. The **STRUCTURAL DISPLACEMENT HUD** toolbar will be displayed; refer to Figure-21.

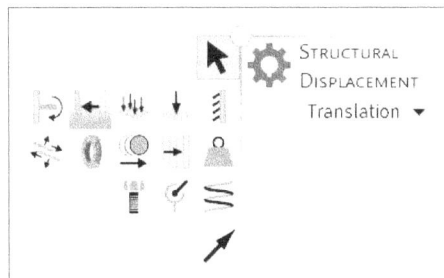

Figure-21. STRUCTURAL DISPLACEMENT HUD Toolbar

- Select the **Translation** option from the drop-down in **HUD** toolbar to apply linear displacement. Select the **Rotation** option from the drop-down in **HUD** toolbar to apply rotational displacement. Select the **Combined** option from the drop-down in the **HUD** toolbar to apply combination of rotational and translational displacement.

- After selecting desired option from the drop-down, select the face(s) on which you want to apply the loads. The options to define values of translation and rotation displacement will be displayed; refer to Figure-22.

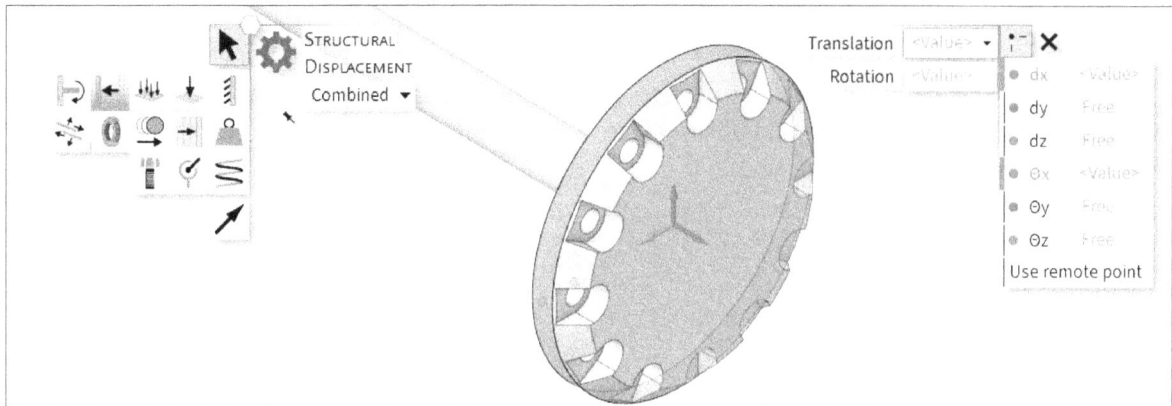

Figure-22. Translation and rotation options

- Set the parameters as discussed earlier and press **ENTER** to apply the load.

APPLYING MOMENT LOAD

The **Moment** tool is used to apply loads that tends to bend/twist the object about an axis/point. The procedure to use this tool is given next.

- Click on the **Moment** tool from the **Structural** drop-down in the **SIMULATION** tab of the **Ribbon** or select the **Moment** button from the **STRUCTURAL HUD** toolbar if active. The **STRUCTURAL MOMENT HUD** toolbar will be displayed.
- Select the face on which you want to apply moment load. The options to define components of bending loads will be displayed; refer to Figure-23.

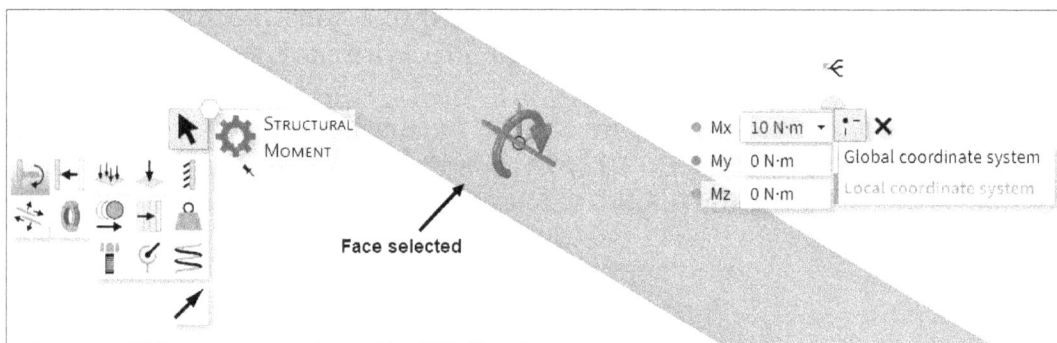

Figure-23. Face selected for bending load

- Select desired toggle option from the right **HUD** toolbar and specify related value to define bending moment in respective edit box.
- After specifying the value(s), press **ENTER** to apply the load. Press **ESC** to exit the tool.

APPLYING MASS LOAD

The Mass load is applied when you do not want to include parts in analysis but you want to consider effects of their mass on analytical model. The procedure to use this tool is given next.

- Click on the **Mass** tool from the **Structural** drop-down in the **SIMULATION** tab of the **Ribbon** or select the **Mass** button from the **STRUCTURAL HUD** toolbar if active. The **STRUCTURAL MASS HUD** toolbar will be displayed.
- Select the face on which you want to apply mass load. The **Mass** input box will be displayed in **HUD** toolbar and you will be asked to specify value of mass load.
- Type desired value in the input box and press **ENTER**. Specified amount of mass will be added to selected face. Note that you can use a remote point to apply this load by selecting **Use remote point** option as discussed earlier. Press **ESC** to exit the tool.

APPLYING VELOCITY

The **Velocity** tool is used to apply kinematic load "velocity" to selected body. This type of load is generally applied to test collision or flow of objects. The procedure to use this tool is given next.

- Click on the **Velocity** tool from the **Structural** drop-down in the **SIMULATION** tab of the **Ribbon** or select the **Velocity** button from the **STRUCTURAL HUD** toolbar if active. The **STRUCTURAL Velocity HUD** toolbar will be displayed.
- Select the body to which you want to apply velocity load. The **Velocity** input box will be displayed in **HUD** toolbar and you will be asked to specify the value of velocity.
- Enter desired value in the input box as discussed earlier. The load will be applied. Press **ESC** to exit the tool.

APPLYING ACCELERATION

The **Acceleration** tool is used to apply kinematic load "acceleration" to selected body. This type of load is generally applied to test collision or flow of objects. The procedure to use this tool is given next.

- Click on the **Acceleration** tool from the **Structural** drop-down in the **SIMULATION** tab of the **Ribbon** or select the **Acceleration** button from the **STRUCTURAL HUD** toolbar if active. The **STRUCTURAL Acceleration HUD** toolbar will be displayed.
- Select the body to which you want to apply acceleration load. The **Acceleration** input box will be displayed in **HUD** toolbar and you will be asked to specify the value of acceleration.
- Enter desired value in the input box as discussed earlier. The load will be applied. Press **ESC** to exit the tool.

APPLYING BEARING LOADS

The Bearing Loads are applied to represent actual radial and axial loads that are generated in the housing of bearing. In Ansys Discovery, axial load is named as thrust force. The procedure to apply bearing loads is given next.

- Click on the **Bearing Load** tool from the **Structural** drop-down in the **SIMULATION** tab of the **Ribbon** or select the **Bearing Load** button from the **STRUCTURAL HUD** toolbar if active. The **STRUCTURAL BEARING LOAD HUD** toolbar will be displayed.
- Select the cylindrical face to which you want to apply bearing load. The **Radial force** and **Thrust force** input boxes will be displayed in **HUD** toolbar and you will be asked to specify the values of loads; refer to Figure-24.

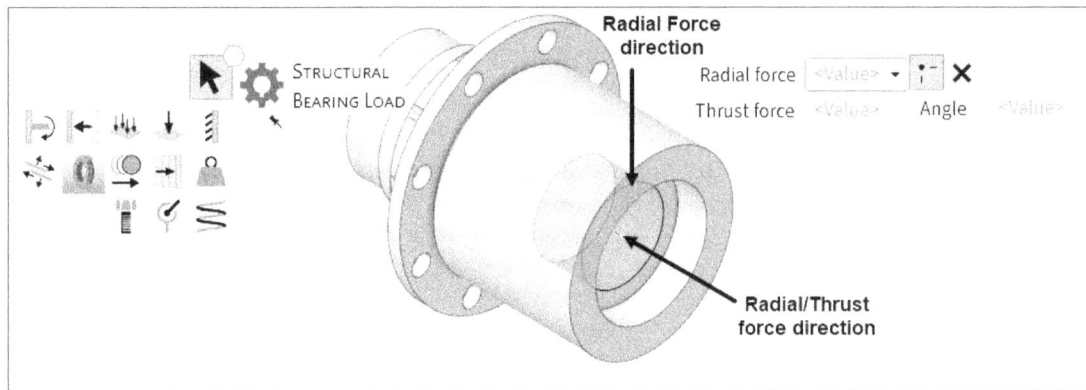

Figure-24. BEARING LOAD options

- Enter desired values in the input boxes to define radial and axial forces on selected face(s). The load will be applied. Press **ESC** to exit the tool.

BOLT PRELOAD

The **Bolt Preload** tool is used to apply pre-tension load along the axis of bolted faces. The procedure to use this tool is given next.

- Click on the **Bolt Preload** tool from the **Structural** drop-down in the **SIMULATION** tab of the **Ribbon** or select the **Bolt Preload** button from the **STRUCTURAL HUD** toolbar if active. The **STRUCTURAL BOLT PRELOAD HUD** toolbar will be displayed.
- Select the central axis of bolt to which you want to apply preload. The **Clamp force** input box will be displayed in **HUD** toolbar and you will be asked to specify the value of force; refer to Figure-25.

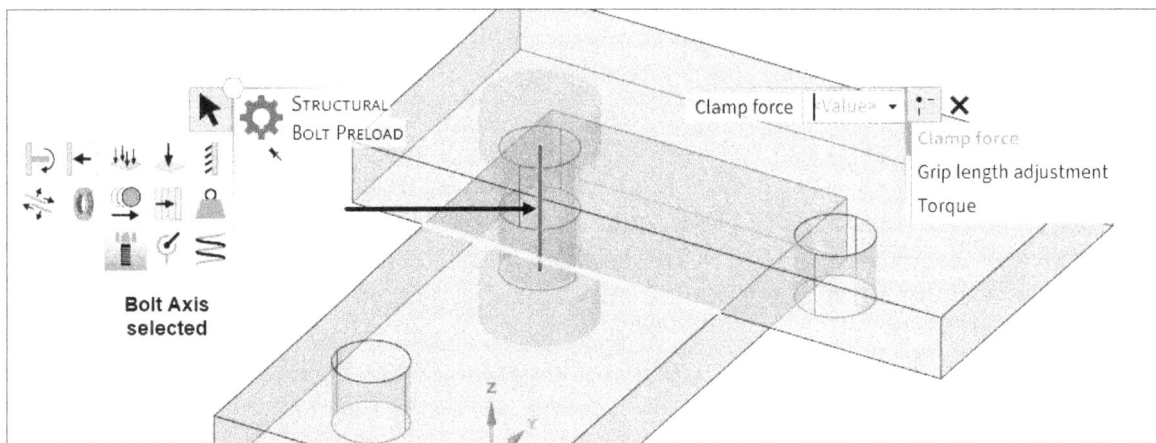

Figure-25. Clamp force input box

- Select the **Clamp force** toggle option from the **HUD** toolbar if you want to define force at which two faces connected by bolt are clamped.
- Select the **Grip length adjustment** toggle option from the **HUD** toolbar if you want to define the amount of shortening that happens in grip length of bolt under load.
- Select the **Torque** toggle option from the **HUD** toolbar if you want to define the tightening torque value. On selecting this toggle option, the **Torque coefficient** and **Specify bolt diameter** options will be displayed in **HUD** toolbar; refer to Figure-26. The torque coefficient value define the amount of clamp force that will be transferred to bolted faces when a certain amount of tightening torque is applied to bolt. For example, a 0.2 torque coefficient defines that when 100 Nm

torque is applied to tighten bolt then 100x0.2 = 20N clamping force is applied to bolted faces. Select the **Specify bolt diameter** toggle option if you want to define shank diameter of the bolt. The specified value will be used to transmit load in radial direction to cylindrical faces of bolt representing slight expansion of diameter of bolt under load.

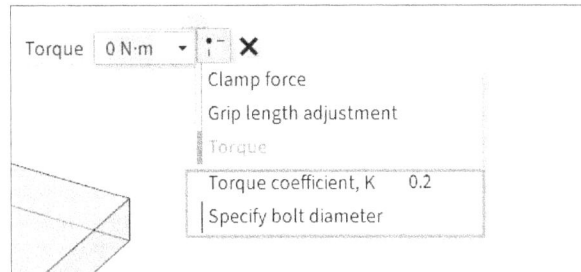

Figure-26. Bolt Torque options

- Enter desired load value in the Input box(es) to apply load and press **ESC** to exit the tool.

DEFINING CONTACTS

The **Contact** tool is used to define how elements of two components are connected to each other at common layer when performing meshing. At the meshing stage, solids are converted to small finite number of elements at which analysis calculations can be performed easily. You will learn about meshing later. The procedure to apply contacts is given next.

- Click on the **Contact** tool from the **Structural** drop-down in the **SIMULATION** tab of the **Ribbon** or select the **Contact** button from the **STRUCTURAL HUD** toolbar if active. The **STRUCTURAL CONTACT HUD** toolbar will be displayed and you will be asked to select the face(s) to apply contact conditions.
- Select the face(s) to which you want to apply the contact. The **HUD** toolbar will be displayed as shown in Figure-27.

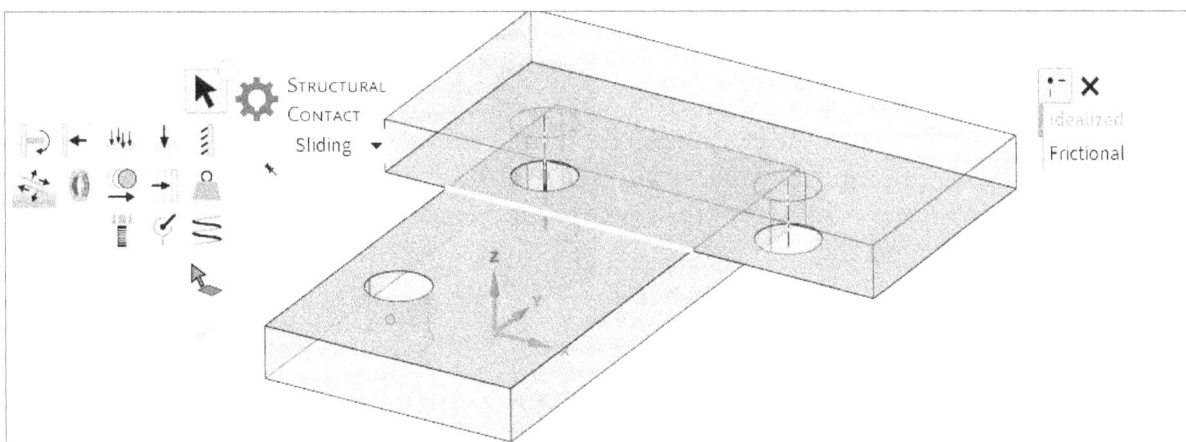

Figure-27. Applying contact

Bonded Contact

- Select **Bonded** option from the drop-down in left **HUD** toolbar if you want the selected faces to be glued together like permanent weld. In this case, all the loads applied on one face will proportionally affect the other face. In simple words, if there is no relative motion between two faces which is generally the case in mechanical assemblies, you apply **Bonded** option. On selecting **Bonded** option, the **Bonding**

service limits toggle option will be displayed in the right **HUD** toolbar. Select this toggle option if you define the strength limits after which bonded contact can break and cause contact failure; refer to Figure-28. This is generally the limitation of fastening method used to bond two components physically.

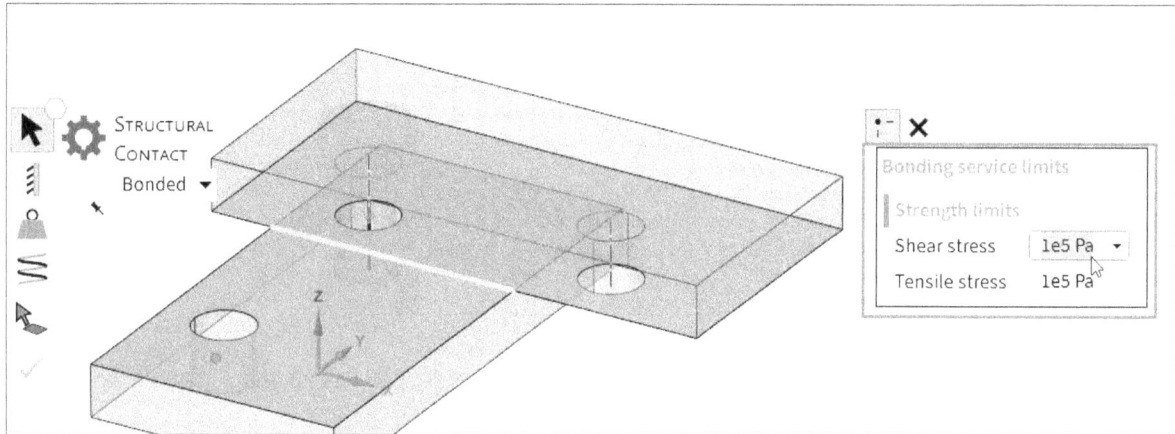

Figure-28. Bonding service limits option

Sliding Contact

- Select the **Sliding** option from the drop-down to allow sliding of one face over the other when under load. This option is used when you want to check deformation of workpiece in sheet metal press or other similar cases. Note that the two bodies at contact faces cannot penetrate into each other, they can only slide over each other. On selecting this option, **Idealized** and **Frictional** toggle options are displayed in the right **HUD** toolbar. Select the **Idealized** toggle option if you want to allow friction free sliding between selected faces without separation which is not generally possible in real world. You can use this option for finding ideal condition results. Select the **Frictional** toggle option if you want to define the friction coefficient between two sliding faces. On selecting this toggle option, an edit box to define friction coefficient will be displayed.

Prevented Contact

- Select the **Prevented** option from the drop-down in left **HUD** toolbar to keep two selected faces separated. In this case, stresses will not be transferred between two selected faces. Although, it is not possible in real world but selecting this option in Ansys sometimes makes work of design engineer faster and reliable. Select this option when you are not interested in interaction of certain components that are near to each other which can affect the analysis results.

- After setting desired parameters, click on the **OK** button from the **HUD** toolbar or press **ENTER** to apply the contacts.

APPLYING JOINTS

The **Joint** tool is used to apply connection between two faces which can be fixed or can provide specific motion. For example, you can apply hinge joint which will allow rotational motion about common axis only. The procedure to use this tool is given next.

- Click on the **Joint** tool from the **Structural** drop-down in the **SIMULATION** tab of the **Ribbon** or select the **Joint** button from the **STRUCTURAL HUD** toolbar if active. The **STRUCTURAL JOINT HUD** toolbar will be displayed.

Fixed Joint

- Select the **Fixed** option from the drop-down in the **STRUCTURAL JOINT HUD** toolbar to make two selected faces fixed with respect to each other. Some may confuse between Fixed support and Fixed joint. Note that applying this joint will allow the global movement of joint set but they cannot move with respect to each other. Take the example of steering wheel connected with steering rod. The steering wheel of car has fixed joint with the steering rod and the assembly of steering wheel with steering rod can rotate with respect to column for turning the wheel.

- After selecting the **Fixed** option, select the first face to which you want to apply the fixed joint. If there is a connected face then it will be highlighted automatically; refer to Figure-29. Click on the **Select secondary surface** toggle button from the left **HUD** toolbar and select the other connected face; refer to Figure-30. After performing desired selection, click on the **OK** button from the **HUD** toolbar. Press **ESC** to exit the tool.

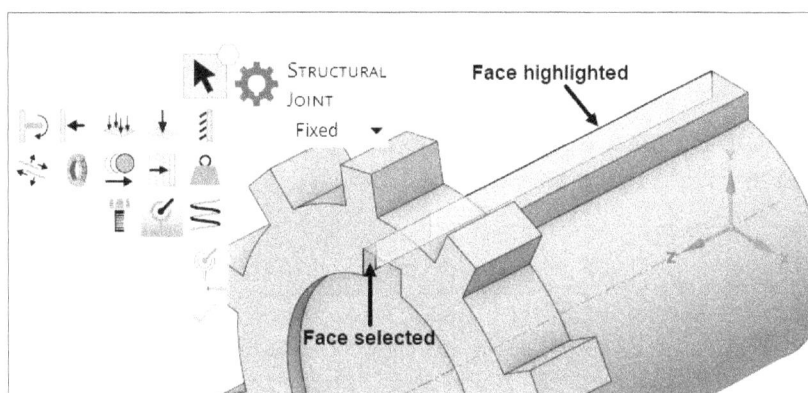

Figure-29. Face highlighted for fixed joint

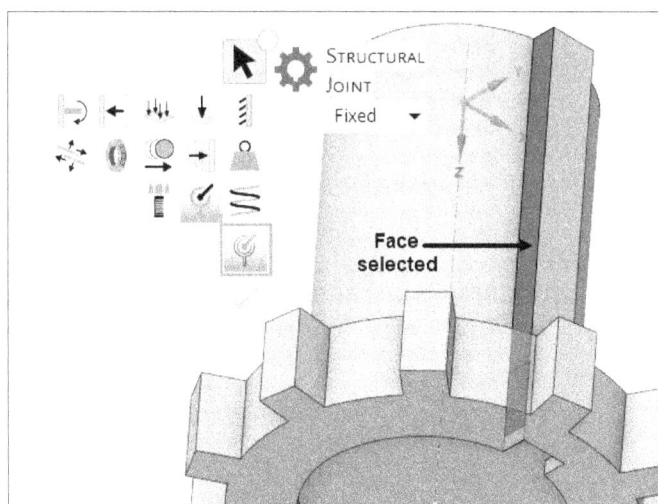

Figure-30. Secondary face selected

Hinged Joint

- Select the **Hinged** option from the drop-down in the **STRUCTURAL JOINT HUD** toolbar to allow rotation of one face about common axis with respect to other selected face. Like joints in doors at home; refer to Figure-31.

Figure-31. Door hinge

- After selecting the **Hinged** option from the drop-down, select first set of faces to be used in hinge joint; refer to Figure-32.

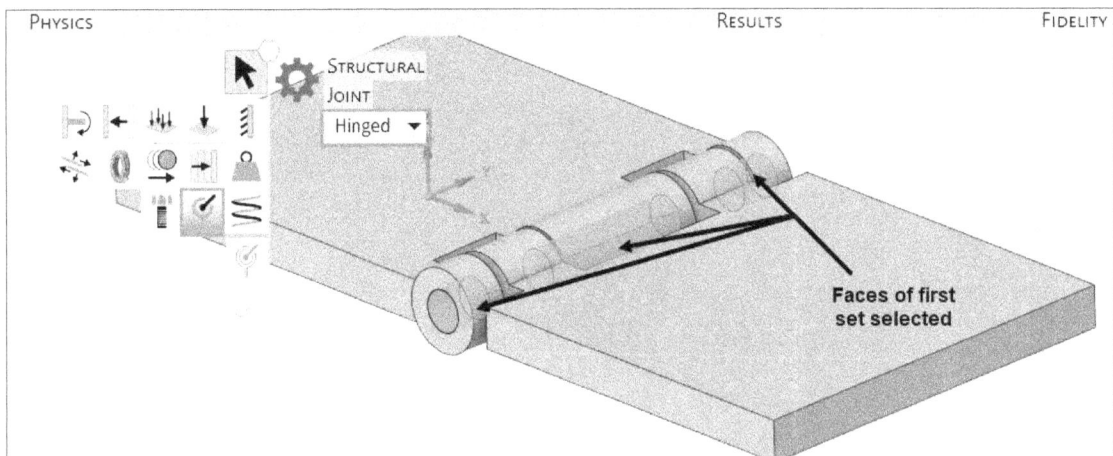

Figure-32. Faces of first set for hinge

- Click on the **Select secondary surface** button from the **HUD** toolbar and select the set of faces from other part; refer to Figure-33.

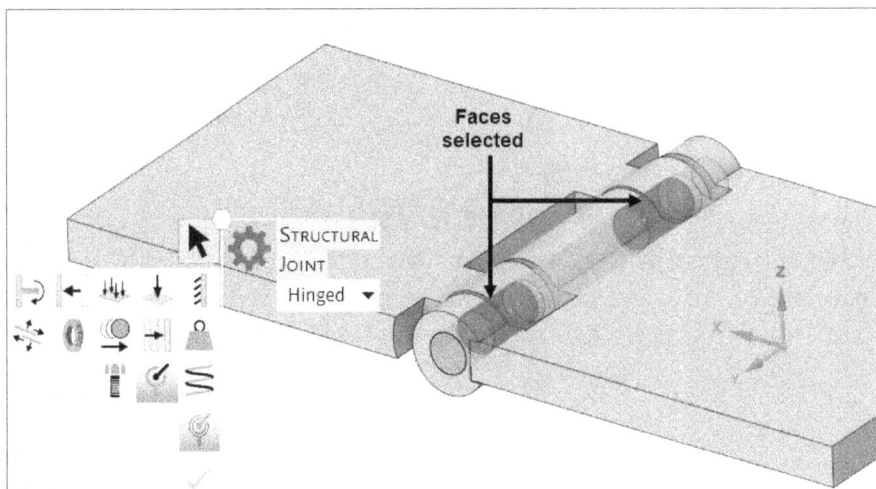

Figure-33. Faces of secondary set

• Click on the **OK** button from the **HUD** toolbar to apply the joint.

Applying Spherical Joint

• Select the **Spherical** option from the drop-down in the **STRUCTURAL JOINT HUD** toolbar to connect two bodies at a single spherical node like in a ball-socket joint; refer to Figure-34.

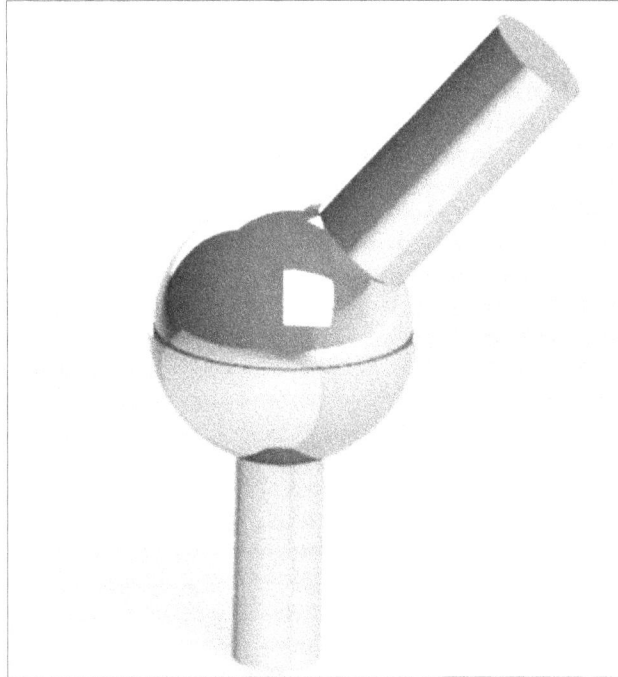

Figure-34. Spherical joint example

• After selecting the option from the drop-down, select faces of the model for primary and secondary selection sets of joint; refer to Figure-35.

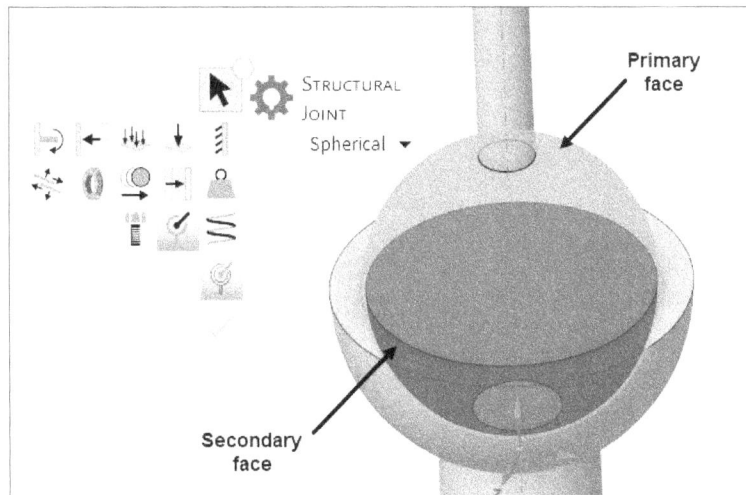

Figure-35. Faces selected for spherical joint

• Click on the **OK** button from the **HUD** toolbar to apply the joint. Press **ESC** to exit the tool.

APPLYING SPRING CONDITION

The **Spring** tool is used to represent spring condition of specified type and stiffness between two selected faces. If you have springs modeled in your design and you are not concerned about deformation of the spring under load then you can replace it with virtual representation of the spring. The procedure to use this tool is given next.

- Click on the **Spring** tool from the **Structural** drop-down in the **SIMULATION** tab of the **Ribbon** or select the **Spring** button from the **STRUCTURAL HUD** toolbar if active. The **STRUCTURAL Spring HUD** toolbar will be displayed and you will be asked to select the faces between which spring condition will be applied.
- Select the primary face from the model and then click on the **Select secondary surface** button from the **HUD** toolbar. You will be asked to select face/point at other end of the spring which will act as ground.
- Select desired face/point from the graphics area. The options to define spring properties will be displayed in the **HUD** toolbar; refer to Figure-36.

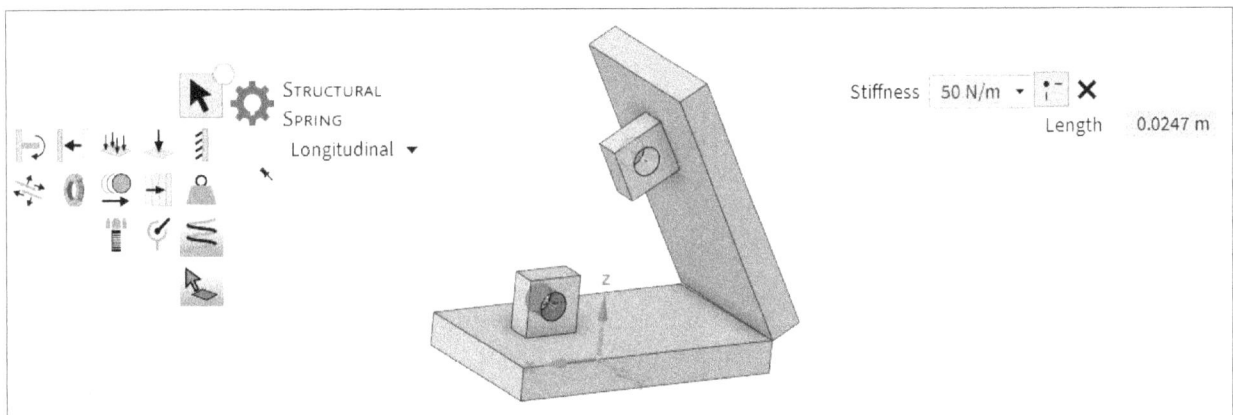

Figure-36. STRUCTURAL SPRING HUD Toolbar

- Select the **Longitudinal** option from the drop-down in the left **HUD** toolbar to create spring that resists compression between two selected faces. Select the **Torsional** option from the drop-down if you want to resist twisting between selected faces. On releasing the load, springs revert to their original shape and position.
- After selecting desired option from the drop-down, specify the stiffness value of spring in the **Stiffness** edit box. The length of spring will be displayed in the **Length** edit box of the right **HUD** toolbar.
- After setting desired parameters, press **ENTER** to create feature and press **ESC** to exit the tool.

SETTING FLUID FLOW DOMAIN

In any CFD (Computational Fluid Dynamics) tool, there are two different types of CFD analyses can be performed - External Fluid Flow Analysis (like Aerodynamics test) and Internal Fluid Flow Analysis (like fluid flow analysis of a pipe). Defining this analysis type is the first step for performing any CFD analysis. The tools to define fluid flow analysis type are available in the **QUICK START** panel of **SIMULATION** tab in the **Ribbon**; refer to Figure-37. The procedures to define the domains are discussed next.

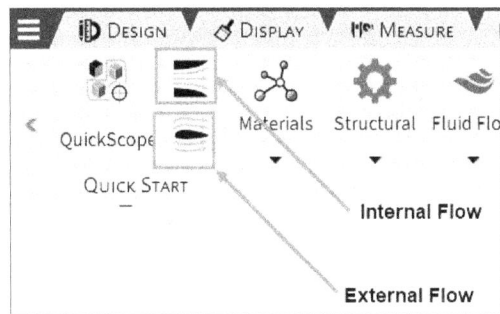

Figure-37. Tools to define Flow type

Defining Internal Flow

The **Internal Flow** tool in **QUICK START** panel of **Ribbon** is used to initiate internal flow analysis study in Ansys Discovery. The procedure to use this tool is given next.

- Click on the **Internal Flow** tool from the **QUICK START** panel in the **SIMULATION** tab of the **Ribbon**. The **INTERNAL FLOW HUD** toolbar will be displayed and you will be asked to select face for defining fluid inlet; refer to Figure-38.

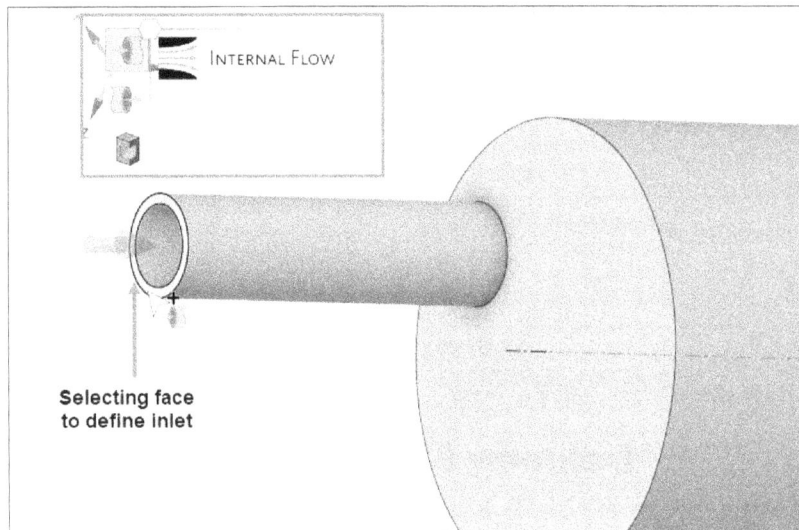

Figure-38. INTERNAL FLOW HUD toolbar

- Select desired face from the model to be used as entry point for the fluid. You will be asked to select face for defining outlet.
- Select desired face from model to be used as exit point of the fluid; refer to Figure-39. The software will automatically generate fluid volume and apply the default boundary conditions; refer to Figure-40.

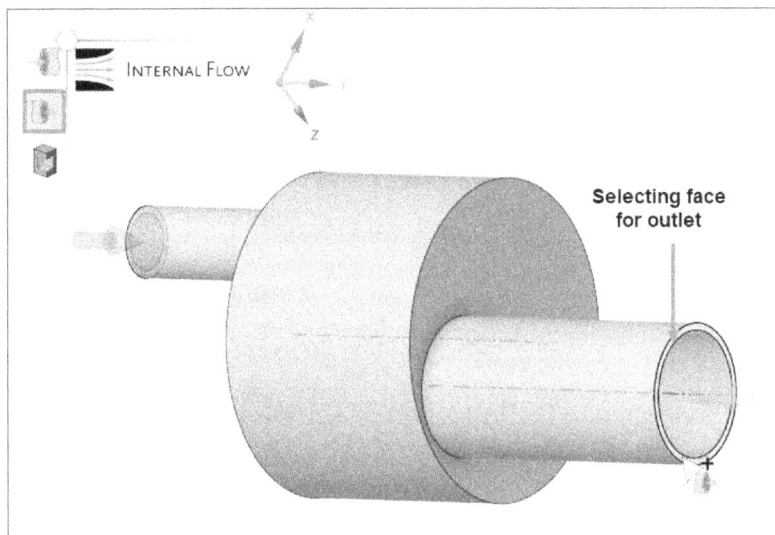

Figure-39. Selecting face for outlet

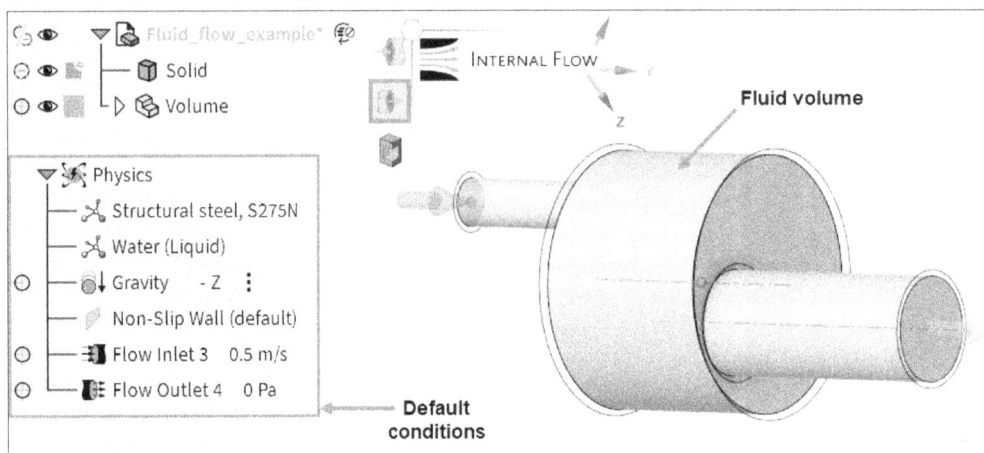

Figure-40. Default conditions applied to model

Defining External Flow

The **External Flow** tool is used to generate fluid dynamics setup where effect of objects in path of fluid flow is studied. The procedure to use this tool is given next.

- Click on the **External Flow** tool from the **QUICK START** panel in the **SIMULATION** tab of the **Ribbon**. The **EXTERNAL FLOW HUD** toolbar will be displayed; refer to Figure-41 and you will be asked to define direction of fluid inlet.
- Select desired arrow from the graphics area to define direction. You will be asked to select ground plane for enclosure.
- Select desired face of the model based on which gravity direction will be calculated and setup will be generated; refer to Figure-42.

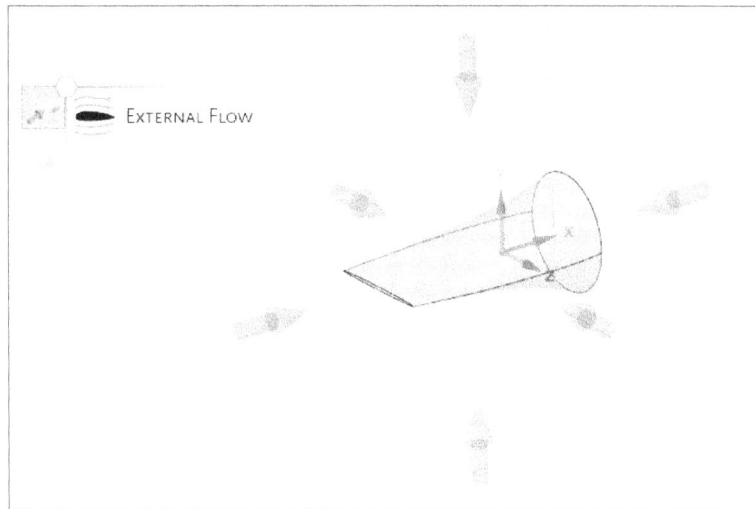

Figure-41. EXTERNAL FLOW HUD toolbar

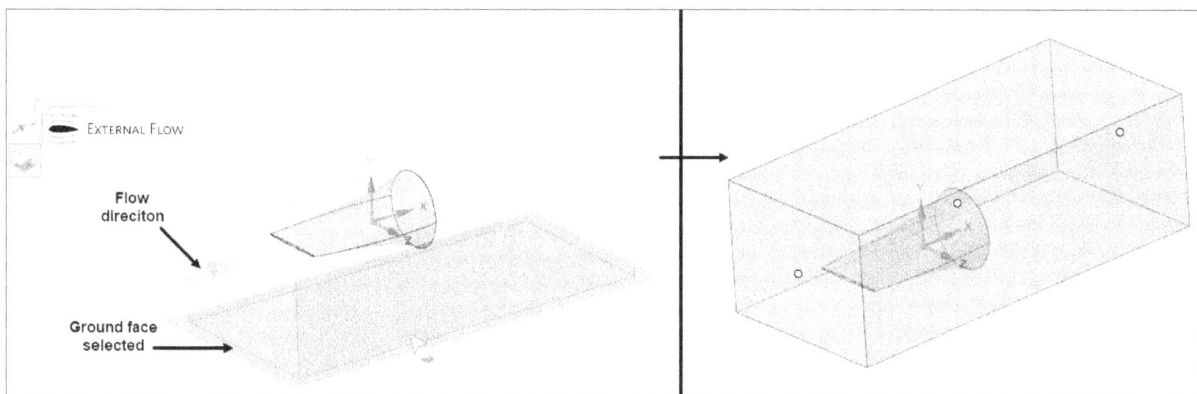

Figure-42. Setting external flow

FLUID FLOW CONDITIONS

The tools of **Fluid Flow** drop-down are used to define boundary conditions related to flow of fluids in analysis domain; refer to Figure-43. These tools are displayed when you are working on a model on which fluid dynamics analysis is to be performed and you have not applied boundary conditions from **Structural** drop-down. Note that if you delete all the boundary conditions applied from **Structural** drop-down in the model then also the tools of **Fluid Flow** drop-down will become active. Various tools of this drop-down are discussed next.

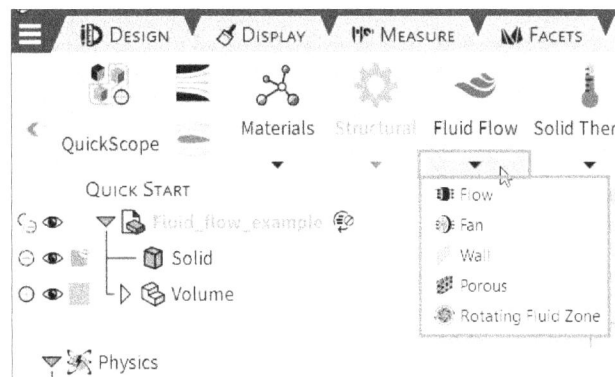

Figure-43. Fluid Flow drop-down

Creating Flow Openings

The **Flow** tool is used to define additional inlet and outlet conditions for the model. The procedure to use this tool is given next.

- Click on the **Flow** tool from the **Fluid Flow** drop-down in the **PHYSICS** panel of **SIMULATION** tab in the **Ribbon**. The **FLUID FLOW FLOW HUD** toolbar will be displayed; refer to Figure-44.

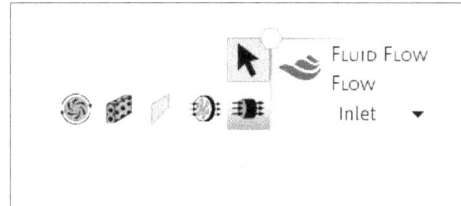

Figure-44. FLUID FLOW FLOW HUD toolbar

- Select the **Inlet** option from the drop-down if you want to define entry of fluid and select the **Outlet** option from the drop-down in **HUD** toolbar if you want to define exit parameters for fluid. The options in right **HUD** toolbar will be displayed as shown in Figure-45.

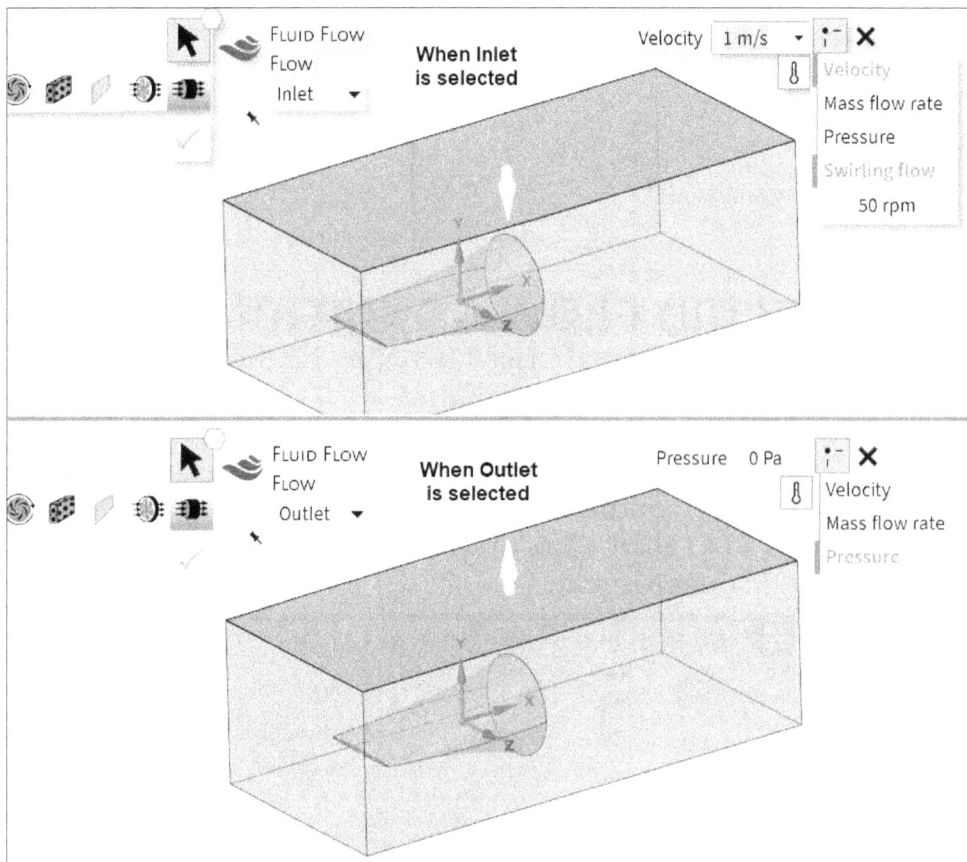

Figure-45. Fluid flow options

- Select the **Velocity** toggle option if you want to specify speed of fluid flow in length per unit time format. When **Velocity** toggle option is selected for Inlet boundary condition then **Swirling flow** toggle option is displayed in the right **HUD** toolbar. Select the **Swirling flow** toggle option if you want to include twisting of fluid in the flow. On selecting this toggle option, the edit box to define swirling rate of fluid will be displayed. Specify desired value in RPM (rounds per minute).

- Select the **Mass flow rate** toggle option from the right **HUD** toolbar if you want to define the rate of fluid flow in mass per unit time value like g/s (gram/second), Kg/s, and lb/s
- Select the **Pressure** toggle option from the right **HUD** toolbar if you want to define pressure value at inlet/outlet point. Generally, atmospheric pressure value 1 atm is applied at the outlet of model to define open environment condition for outlet.
- Select the **Define temperature** toggle button from the right **HUD** toolbar; refer to Figure-46 to specify initial temperature of fluid at inlet/outlet.

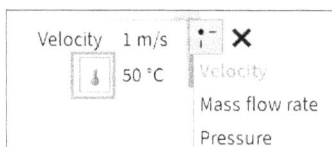

Figure-46. Define temperature toggle button

- Click in the empty area of graphics window to create the boundary condition.

Creating Fan Boundary Condition

The **Fan** tool is used to provide forced flow of fluid in to or out of the fluid domain. In other words, you can create an intake fan or an exhaust fan to provide fluid flow. The procedure to use this tool is given next.

- Click on the **Fan** tool from the **Fluid Flow** drop-down in the **SIMULATION** tab of **Ribbon**. The **FLUID FLOW FAN HUD** toolbar will be displayed.
- Select the face of model to which you want to apply force flow of fluid. The options to define parameters for fan will be displayed in the right **HUD** toolbar; refer to Figure-47.

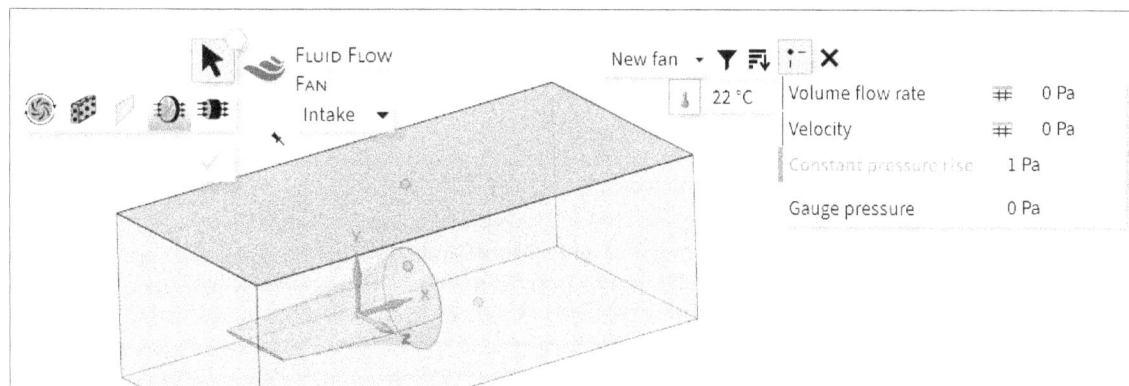

Figure-47. Options for fan

- Specify desired value in the **Volume flow rate** edit box to define the amount of fluid in volume which will be entering the domain via fan. Click on the **Volume flow dependent values** button for the edit box to specify values of volume flow rate with respect to pressure; refer to Figure-48. Enter desired values of volume flow rate and related pressure in the empty fields of table. New fields will automatically keep generating.

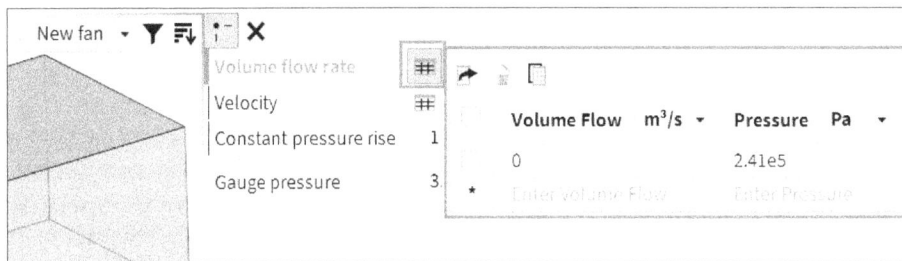

Figure-48. Volume flow dependent values table

- Specify desired value in the **Velocity** edit box to define the speed of fluid flow via fan. You can use the **Velocity dependent values** button in the same way as discussed for **Volume flow dependent values** button.
- Specify desired value in the **Constant pressure rise** edit box to define how much pressure value of fluid will be increased when fluid passes through the fan. For intake fan, it is the total pressure and for exhaust fan, it is the static pressure.
- Specify desired value in the **Gauge pressure** edit box to define the pressure value of fluid after exiting the fan in case of intake fan and pressure value before the fan in case of exhaust fan type.
- Set the other parameters as discussed earlier and click on the **OK** button from **HUD** toolbar to apply the boundary condition.

Defining Walls

The **Wall** tool is used to define various fluid flow wall conditions for the analysis and act as separator between different types of fluids in same domain. You can use wall conditions to restrict fluid flow, apply twisting in flow, generate linear motion in fluid, and so on. The procedure to use this tool is given next.

- Click on the **Wall** tool from the **Fluid Flow** drop-down in the **PHYSICS** panel of **SIMULATION** tab in the **Ribbon**. The **FLUID FLOW WALL HUD** toolbar will be displayed.
- Select the face that you want to be defined as wall. The options in the **HUD** toolbar will be displayed as shown in Figure-49.

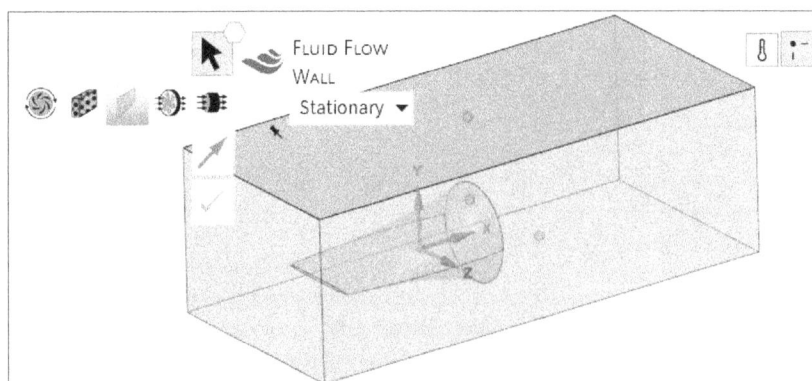

Figure-49. Options for Fluid flow wall

- Select the **Stationary** option from the drop-down at the left in the **HUD** toolbar if you want the wall to be fixed without allowing flow of fluid past it and there is no motion in the wall. Also, selecting this option makes velocity of fluid zero and fluid adheres to the assigned wall/face. Select the **Free slip** option from the drop-down if you want the fluid to flow without friction on assigned wall/face. This option is generally used when effect of shear force in boundary faces of enclosure body is to be ignored. For example when you are not concerned about the effect generated by contact between fluid and plastic pipes through which fluid is flowing. Select the **Rotating** option from the drop-down if you want the assigned wall/face to rotate at specified rotational speed. After selecting this option, specify desired rotation value in the edit box of right **HUD** toolbar. This option is generally used when you want to represent a rotating pump or fan in the fluid domain to alter flow of fluid. Select the **Translating** option from the drop-down if you want to simulate linearly moving boundary faces of fluid domain. On selecting this option, the edit boxes to define motion along X and Y axes will be displayed; refer to Figure-50. Specify desired values in the X and Y edit boxes to define motion of selected face. The translating wall boundary condition is generally applied in analysis setups like fluid flow change due to piston movement.

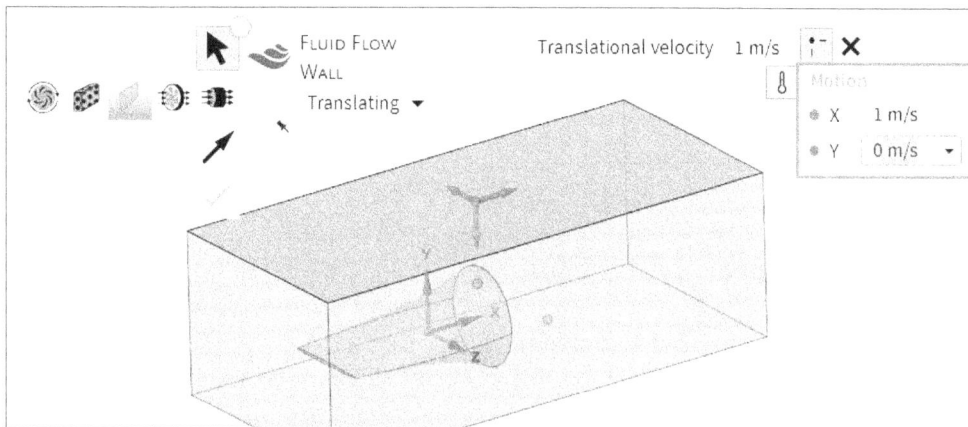

Figure-50. Motion X and Y edit boxes

Defining Thermal Conditions

- Select the **Define Thermal Conditions** toggle button ⬛ from the right **HUD** toolbar to define thermal conditions for the wall faces. On selecting this toggle button, the options to define thermal condition are displayed; refer to Figure-51.

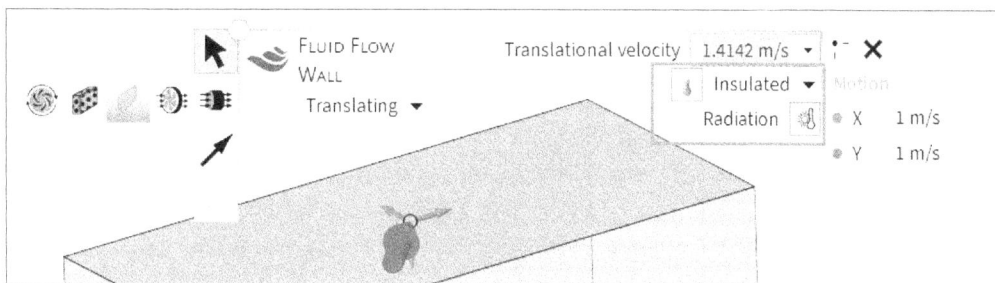

Figure-51. Thermal options in HUD toolbar

- Select the **Insulated** option from the drop-down to create an adiabatic wall that does not let heat transfer or escape through it. This condition is generally used to simulate ideal conditions in heat exchange or similar applications so that you can compare practical results of system with ideal cases.
- Select the **Convection** option from the drop-down to define heat convection (transfer) rate and external temperature at which heat flows from/to the wall. On selecting this option, **External temperature** and **Convection coefficient** edit boxes will be displayed in the right HUD toolbar; refer to Figure-52. Specify desired value in the **External temperature** edit box to define the temperature outside the walls. Specify desired value in the **Convection coefficient** edit box to define the rate at which heat will move from current wall to nearby areas due to convection.

Formula used for heat convection is Q_{heat} = Convection coefficient x Surface Area of wall x (wall surface temperature - external temperature)

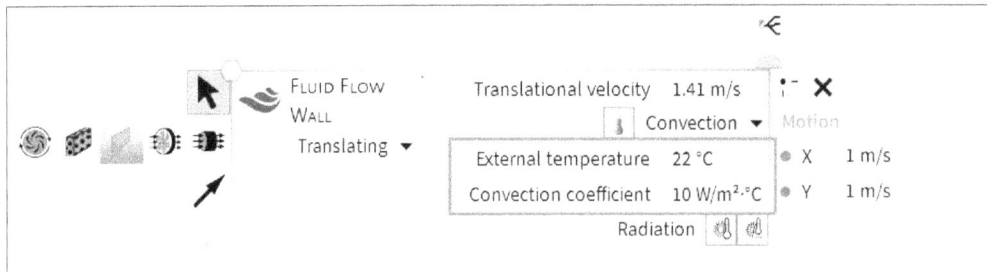

Figure-52. Convection options

- Select the **Heat flux** option from the drop-down if you want to define the heat energy that can pass through per unit area of wall for given amount of time. On selecting this option, the edit box to define value of heat flux will be displayed below the drop-down.
- Select the **Heat flow** option from the drop-down if you want to define the total amount of heat that flows from wall to nearby objects.
- Select the **Temperature** option from the drop-down if you want to define the temperature of wall.
- Select the **Radiation** option from the drop-down if you want to define the heat transfer occurring via radiation mechanism. Note that radiation is the only heat transfer mechanism which do not need any medium and can occur in space. On selecting this option, the **External emissivity** edit box will be displayed. Specify desired value in edit box to define ability of surface to emit heat as compared to perfect black body. The value ranges from 0 to 1. Generally, painted surfaces have emissivity of 0.8 to 0.95 and shining metals have emissivity of 0.1 to 0.4. For radiation, $Q_{heat} = \epsilon \times \sigma \times A \times (T_{wall}^4 - T_{ex}^4)$. Note that temperature values are at 4th order in this formula so when temperature is high then heat transfer via radiation dominates all the other methods of heat transfer.
- Select the **Radiation** toggle button from the right **HUD** toolbar to define the internal emissivity of radiation. Internal emissivity is used to define the heat flow occurring inside the faces of wall due to radiation when fluid boundary and outer boundary faces of wall are in equilibrium. On selecting this toggle button, the **Internal emissivity** edit box will be displayed for defining the value.
- After setting desired parameters, click on the **OK** button from the **HUD** toolbar. The wall boundary conditions will be applied.

Defining Porous Fluid Flow Conditions

The **Porous** tool is used to simulate filter type materials which allow passage of fluid with some restrictions. The procedure to use this tool is given next.

- Click on the **Porous** tool from the **Fluid Flow** drop-down in the **PHYSICS** panel of **SIMULATION** tab in the **Ribbon**. The **FLUID FLOW POROUS HUD** toolbar will be displayed and you will be asked to select the body.
- Select the body and then face that you want to be defined as porous medium. The options in the **HUD** toolbar will be displayed as shown in Figure-53.

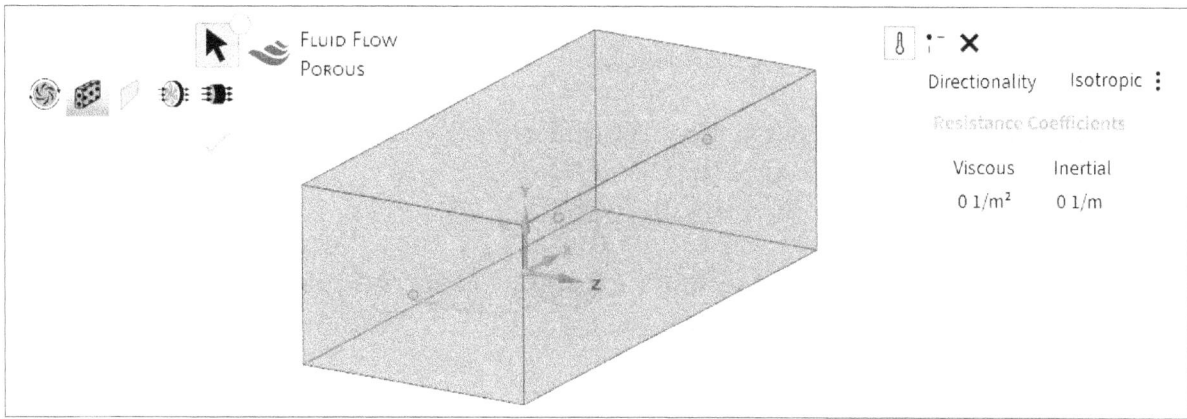

Figure-53. FLUID FLOW POROUS HUD toolbar

- Select the **Isotropic** option from the **Directionality** drop-down in the right **HUD** toolbar to apply same flow resistance coefficient to all the faces of selected body. The options will be displayed as shown in Figure-53. Specify viscous resistance coefficient value in the **Viscous** edit box. This value is generally used to define the amount of resistance created by fluid when an object passes through it. Unit of this parameter is $1/m^2$ in SI unit system. Specify desired value in the **Inertial** edit box to define the resistance against movement of object due to inertia of fluid. Note that this coefficient represents effect of flow of fluid against/along the motion of object.

- Select the **Bidirectional** option from **Directionality** drop-down in the right **HUD** toolbar to define different values of Viscous and Inertial resistance coefficients when going along the fluid flow and across the fluid flow. The options will be displayed as shown in Figure-54. Specify desired values in the edit boxes to define respective resistance coefficients.

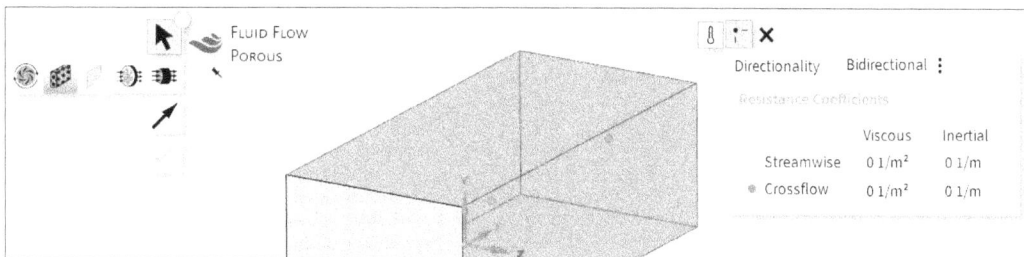

Figure-54. Bidirectional option

- Select the **Orthotropic** option from the **Directionality** drop-down to define different values of resistance coefficients along X, Y, and Z axis of fluid domain; refer to Figure-55. Specify the parameters as discussed earlier and click on the **OK** button from the **HUD** toolbar to apply porous boundary condition to body/face.

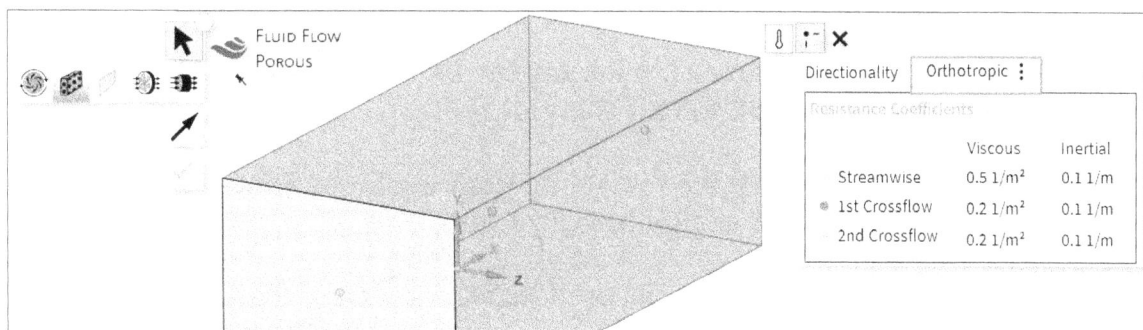

Figure-55. Orthotropic option

Note: The formula used for calculating momentum loss caused in the flow of fluid due to porous material can be given as:

Total Momentum Loss = Loss by viscous resistance + Loss by inertial resistance
Loss by Viscous resistance = C_v . (L/D) . (Q^2/g)
Loss by Inertial resistance = C_i . (L/D) . (Q^2/g)

Here, C_v is Viscous resistance coefficient and C_i is inertial resistance coefficient.
L is length of porous material
D is hydraulic diameter
Q is Volumetric flow rate
g is gravitational acceleration

Defining Rotating Fluid Zone

The **Rotating Fluid Zone** tool is used to create a cylindrical zone in the model where fluid will rotate at specified rotational speed. The procedure to use this tool is given next.

- Click on the **Rotating Fluid Zone** tool from the **Fluid Flow** drop-down in the **PHYSICS** panel of **SIMULATION** tab in the **Ribbon**. The **FLUID FLOW ROTATING FLUID ZONE HUD** toolbar will be displayed and you will be asked to select a cylindrical body.
- Select the cylindrical body created to represent rotating fluid boundary. The options in the **HUD** toolbar will be displayed as shown in Figure-56.

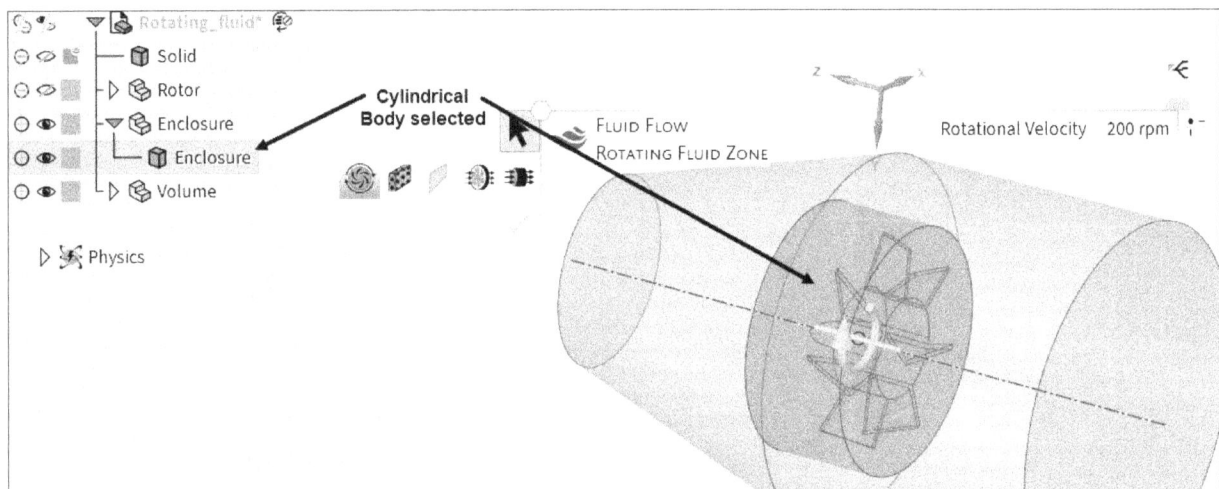

Figure-56. ROTATING FLUID ZONE HUD toolbar

- Specify desired value in the **Rotational Velocity** edit box to define the rotational speed of the zone.
- After setting desired parameters, click on the **OK** button from the **HUD** toolbar to apply rotation and press **ESC** to exit the tool.

Note: There are few conditions applied on using Rotating Fluid Zone in Ansys discovery or rather in any CFD software, which are given next.

- The rotating fluid zone must be located on a body that features cylindrical or conical outer surfaces.
- It is not possible to designate both a rotating fluid zone and a porous zone on the same body.

- A rotating fluid zone cannot have a fluid-fluid interface with another rotating fluid zone.
- Only wall conditions of "Rotating with the fluid zone" or "Stationary" are permitted, with the default behavior for walls in a rotating fluid zone being "Rotating with the fluid zone." User-defined moving walls are not allowed in this context.
- Rotating fluid zones are not applicable in simulations involving compressible flow or in time-dependent fluid flow scenarios. In some cases involving solid-body rotation, discrepancies may arise between the solutions in the Explore and Refine modes.
- Results for rotating fluid zones are displayed in the absolute reference frame.
- For simulations involving closed domains, you may encounter issues with convergence. In the Refine mode, it might be necessary to lower the target residual and increase the maximum number of iterations.
- In fluid-solid heat transfer simulations, the influence of the rotation of a solid thermal region on the fluid will only be considered if a face of that region is in contact with the rotating fluid zone.
- Flow conditions cannot be established on the surfaces representing the rotating device or the outer cylindrical/conical surfaces of a rotating fluid zone.
- It is not possible to create non-coaxial bodies within a single rotating fluid zone.
- If a face on the solid thermal region is adjacent to faces on both a rotating and a stationary fluid zone, it is essential to imprint the face to facilitate heat transfer between the solid thermal region and both fluid zones.
- Resolving rotating geometries, such as thin blades, in Explore mode may prove challenging and could result in suboptimal outcomes. To achieve better results, it is advisable to use a higher fidelity setting.
- In the Explore mode, all flow conditions associated with a rotating fluid zone are defined relative to the absolute frame. For instance, if a rotating fluid zone contains a swirl inlet, the rotational velocity should be specified in the absolute frame rather than relative to the rotating fluid zone.
- In Refine, all flow conditions in a rotating fluid zone are also defined relative to the absolute frame, with the exception of a mass flow outlet, where flow direction aligns with the upstream flow.

SOLID THERMAL CONDITIONS

The tools in the **Solid Thermal** drop-down are used to define thermal conditions for solid faces/walls; refer to Figure-57.

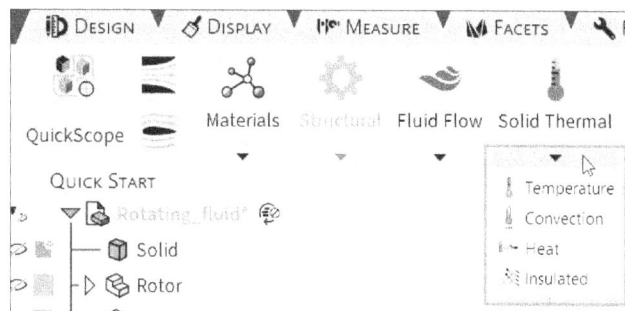

Figure-57. Solid Thermal drop-down

Setting Temperature

The **Temperature** tool is used to assign specified temperature to selected face(s). The procedure to use this tool is given next.

- Click on the **Temperature** tool from the **Solid Thermal** drop-down in the **PHYSICS** panel of **SIMULATION** tab in the **Ribbon**. The **THERMAL TEMPERATURE HUD** toolbar will be displayed.
- Select the solid face to which you want to apply temperature. The edit box to specify temperature will be displayed in the **HUD** toolbar; refer to Figure-58.

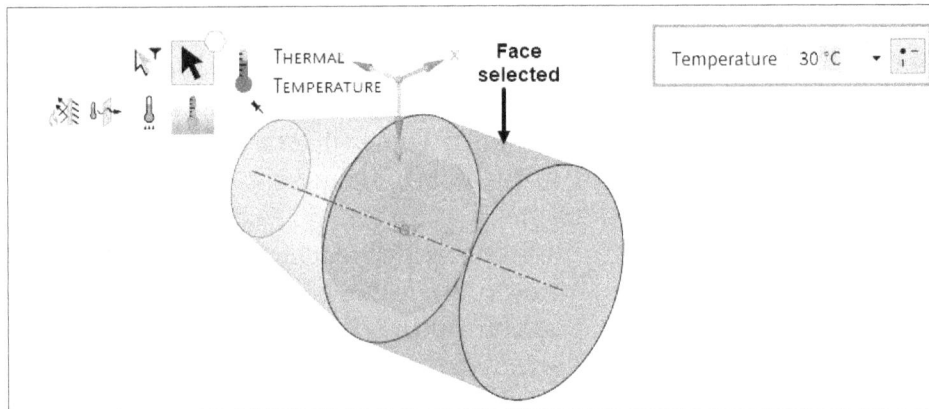

Figure-58. Temperature edit box

- Type desired value in the edit box and press **ENTER**. The temperature will be applied.

Setting Convection Conditions

The **Convection** tool is used to define heat convection rate at selected face(s) by which heat will transfer from/to current face depending on environmental temperature. The procedure to use this tool is given next.

- Click on the **Convection** tool from the **Solid Thermal** drop-down in the **PHYSICS** panel of **SIMULATION** tab in the **Ribbon**. The **THERMAL CONVECTION HUD** toolbar will be displayed.
- Select the solid face to which you want to apply convection boundary condition. The edit boxes to specify convection coefficient and convection temperature will be displayed in the **HUD** toolbar; refer to Figure-59.

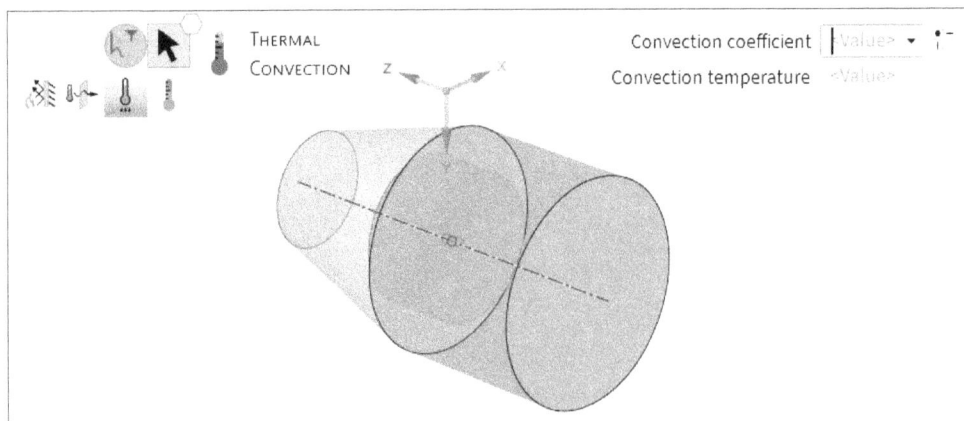

Figure-59. THERMAL CONVECTION HUD toolbar

- Specify desired value in the **Convection coefficient** edit box to define the heat transfer rate from face to surrounding. The **SI** unit for convection coefficient is **W/m².K**
- Specify desired value in the **Convection temperature** edit box to define temperature of surrounding fluids.
- After setting desired parameters, press **ENTER** to create the boundary condition.

Applying Heat Boundary Conditions

The **Heat** tool is used to define the rate of heat generation at selected faces. The procedure to use this tool is given next.

- Click on the **Heat** tool from the **Solid Thermal** drop-down in the **PHYSICS** panel of **SIMULATION** tab in the **Ribbon**. The **THERMAL HEAT HUD** toolbar will be displayed.
- Select the solid face to which you want to apply heat boundary condition. The options to define heat rate will be displayed in the **HUD** toolbar; refer to Figure-60.

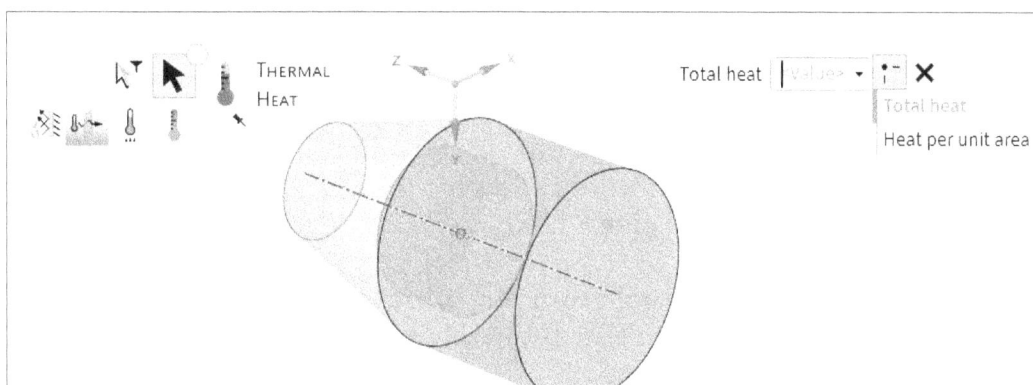

Figure-60. Options to define heat rate

- Select the **Total heat** toggle option from the right **HUD** toolbar if you want to define the total heat rate of full selected face(s).
- Select the **Heat per unit area** toggle option if you want to define the heat rate per unit area of the selected face.
- Specify desired value in the input box and press **ENTER**. The heat boundary condition will be applied. Press **ESC** twice to exit the tool.

Applying Insulated Boundary Condition

The **Insulated** tool is used to make selected face(s) thermally insulated. It means heat cannot pass through selected faces. The procedure to use this tool is given next.

- Click on the **Insulated** tool from the **Solid Thermal** drop-down in the **PHYSICS** panel of **SIMULATION** tab in the **Ribbon**. The **THERMAL INSULATED HUD** toolbar will be displayed.
- Select the solid face to which you want to apply insulated boundary condition and click on the **OK** button. The condition will be applied.

ELECTROMAGNETIC BOUNDARY CONDITIONS

The options of **Electromagnetics** drop-down are used to apply boundary conditions to represent electromagnetic loads in the model. Common example of this type of load is radio frequencies transmitted through a thin antenna in mobile phones. The options of this drop-down are active when you have selected **REFINE** stage from **Stage Selector**; refer to Figure-61. The tools of this drop-down are discussed next.

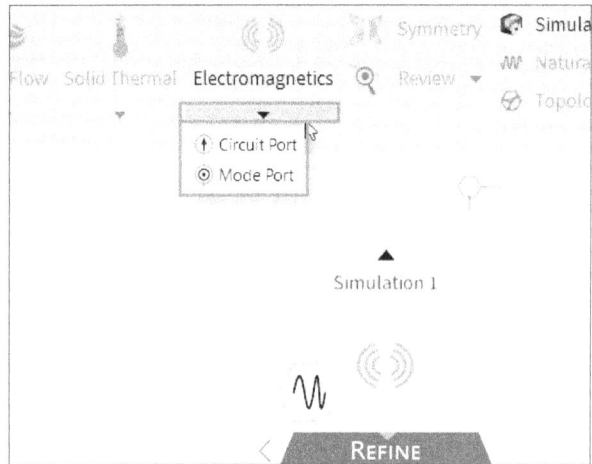

Figure-61. Electromagnetics drop-down

Applying Circuit Port

The **Circuit Port** tool is used to represent overall effect of a circuit board connected in the model. When we are considering electromagnetic effect of various components of circuit like resistors, transistors, coils, etc., most of the time we are interested in overall effect of the whole arrangement of components combined in the design. Using **Circuit Port** tool, we can define the parameters like impedance, voltage, phase, and so on at the input/output terminals of a port to represent net circuit parameter. The procedure to use this tool is given next.

- Click on the **Circuit Port** tool from the **Electromagnetics** drop-down in the **PHYSICS** panel of **SIMULATION** tab in the **Ribbon**. The **ELECTROMAGNETICS CIRCUIT PORT HUD** toolbar will be displayed.
- Select an edge of the model to define input edge of electrical signal. The options to define parameters at input will be displayed and an arrow will be displayed showing possible location for secondary edge; refer to Figure-62.

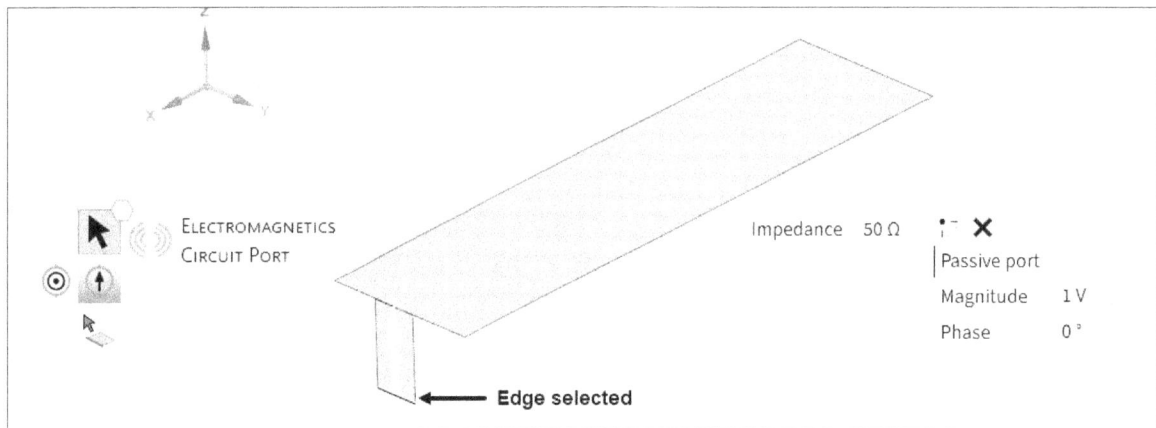

Figure-62. ELECTROMAGNETICS CIRCUIT PORT HUD toolbar

- Specify desired parameters in the input boxes of **HUD** toolbar to define impedance, voltage value, and phase angle for the input/output signal.
- Select the **Passive port** toggle option if you want to use selected geometry as receiver which does not generate power or provide signal amplification by itself.
- Click on the **Select secondary edge** button from the toolbar if you want to define secondary edge for calculating impedance. Otherwise, press **ENTER** to use default secondary edge after specifying the value. The boundary condition will be applied

and electromagnetic region will be created with circuit port feature in **PHYSICS Tree**; refer to Figure-63.

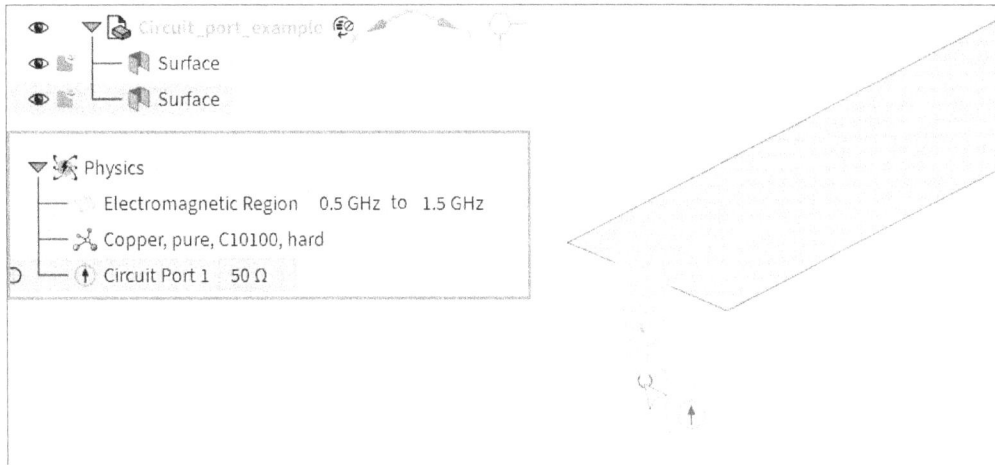

Figure-63. Features created in PHYSICS Tree

• Press **ESC** to exit the tool.

Applying Mode Port

The **Mode Port** tool is used to define port/point of entry/exit of electromagnetic waves. While circuit ports are defined for electric signals, mode ports also called wave ports are defined for electromagnetic waves for analyzing antennas, waveguides, and resonators. The procedure to use this tool is similar to **Circuit Port** tool discussed earlier.

APPLYING SYMMETRY CONDITION

The **Symmetry** tool is used to apply symmetric condition to selected faces. If your part can be split into two halves which are mirror copy of each other with respect to center plane then you can use symmetric condition to simplify your analysis design. The procedure to use this tool is given next.

• Click on the **Symmetry** tool from the **PHYSICS** panel in the **SIMULATION** tab of the **Ribbon**. The **SYMMETRY HUD** toolbar will be displayed.
• Select the face of half part which lies at mirror plane; refer to Figure-64 and click on the **OK** button to apply symmetry. Note that applying this condition will represent our model as shown in Figure-65 for results. Press **ESC** to exit the tool.

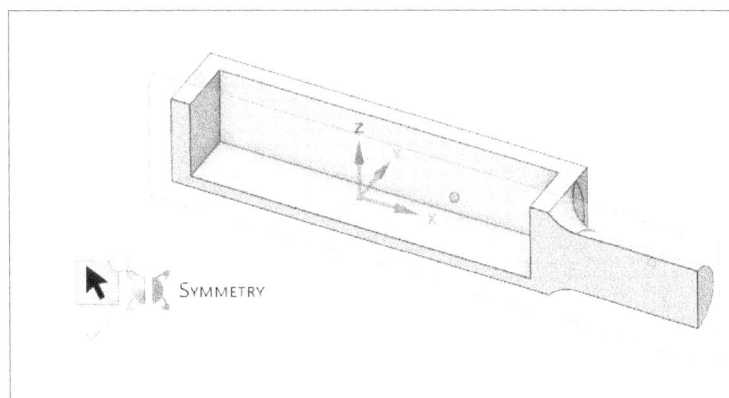

Figure-64. Face selected for symmetry

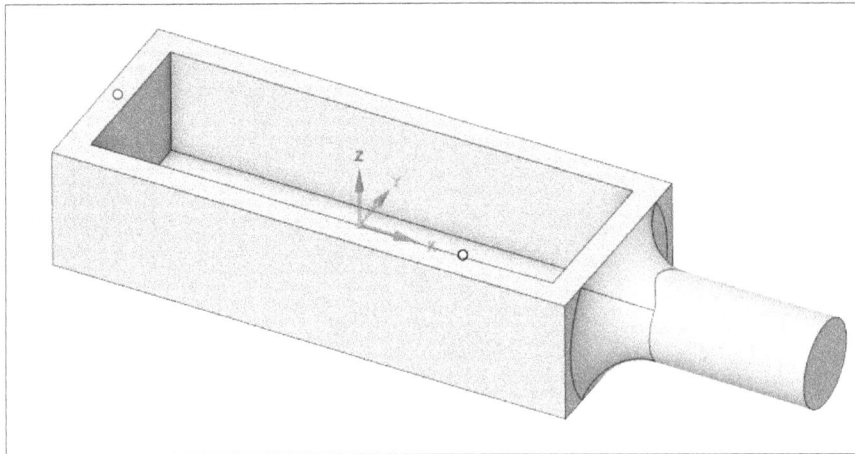

Figure-65. Full model

REVIEW TOOLS

The tools in **Review** drop-down of **Ribbon**; refer to Figure-66 are used to check and modify connections between various solid components of assembly or connection between fluids and solid components of assembly for performing analysis. These tools are discussed next.

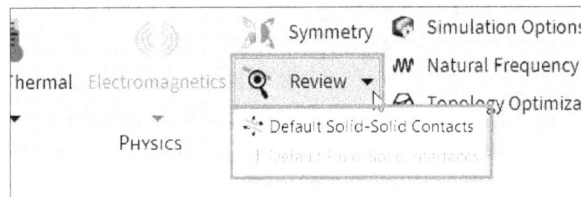

Figure-66. Review tools

Reviewing Default Solid-Solid Contacts

The **Default Solid-Solid Contacts** tool is used to check contacts between solids in the model and apply default settings automatically. The procedure to use this tool is given next.

- Click on the **Default Solid-Solid Contacts** tool from the **Review** drop-down in the **PHYSICS** panel of **SIMULATION** tab in the **Ribbon**. The **CONTACT REVIEW HUD** toolbar will be displayed; refer to Figure-67.

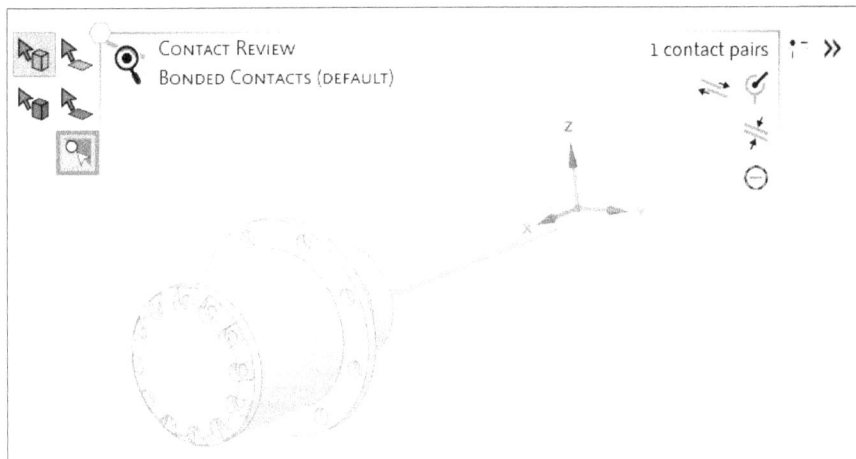

Figure-67. CONTACT REVIEW HUD toolbar

- In the left **HUD** toolbar, there are various selection tools to define which contact pairs are to be reviewed. Select the **Primary Body** selection button from the

HUD toolbar to select the body whose contacts are to be reviewed. On selecting body, all the faces of this body which have contacts with other objects will be highlighted for review. Select the **Primary Face** selection button 🐾 from the **HUD** toolbar if you want to review faces connected with the face for reviewing contacts. On selecting a primary body, you may get multiple connected bodies highlighted for reviewing contacts. In such cases, if you want to select a specific secondary body for evaluating common contacts then select the **Secondary Body** selection button from the toolbar and select desired body from the graphics area; refer to Figure-68. Similarly, you can use the **Secondary Face** selection button to select contacts common to selected faces of the two bodies. Select the **Select Markers** selection button 🔍 to individually select contacts from the graphics area for review.

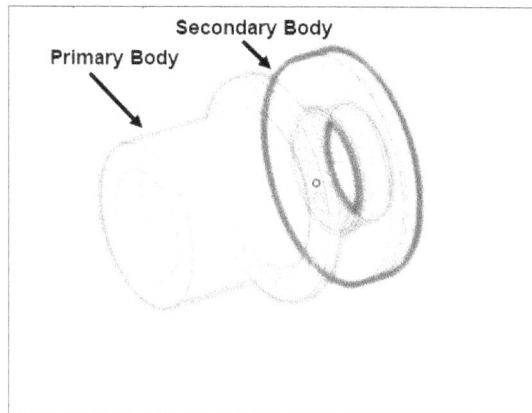

Figure-68. Selecting bodies for contacts

- Select the **Convert to sliding** button from the right **HUD** toolbar to convert highlighted contacts to sliding type contact; refer to Figure-69. Note that on hovering the cursor on this button, you will be asked to select the **Idealized** or **Frictional** option to define what type of sliding contact do you want to generate.

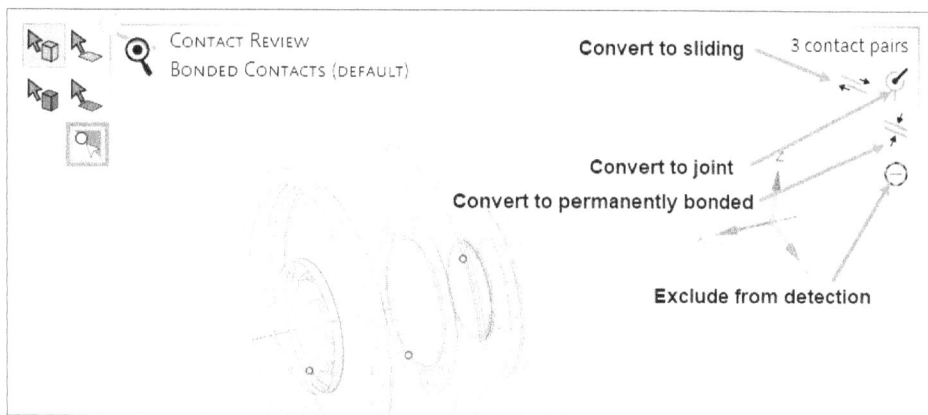

Figure-69. Options of right HUD toolbar

- Select the **Convert to joint** button from toolbar to convert highlighted contacts into joints. System will create fixed joints between the faces.
- Select the **Convert to permanently bonded** button from toolbar to convert highlighted contacts into permanent bonded contacts which will act as if they are sections of same body.
- Select the **Exclude from detection** button from the toolbar to exclude highlighted contacts from automatic detection by this tool.
- Expand the right **HUD** toolbar to check more options; refer to Figure-70.

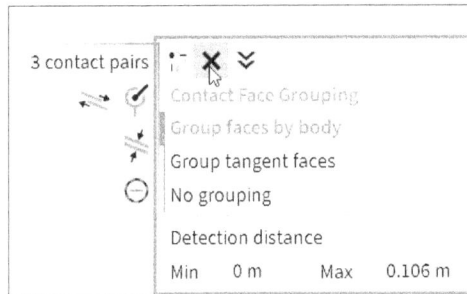

Figure-70. Expanded HUD toolbar

- Select the **Group faces by body** toggle option from the **Contact Face Grouping** section of the **HUD** toolbar to automatically group faces based on parent body. Select the **Group tangent faces** toggle option from the toolbar if you want to group all the tangent faces in one selection. Select the **No grouping** toggle option if you want to select all the face pairs individually. These options are used to define how pairs of faces in contact are displayed and are available for modification. For example, if you have selected **Group tangent faces** option then you will not be able to modify contact properties of individual faces in the tangent group but you will be able to modify contact properties of multiple faces in group at once.
- Specify desired values in the **Min** and **Max** edit boxes of the **Detection distance** section of **HUD** toolbar to set limits within which two nearby faces will be said to be in contact.
- After creating desired contact changes, press **ESC** to exit the tool.

Reviewing Default Fluid Solid Interface

The **Default Fluid-Solid Interface** tool is used to check and change default boundary condition applied at the interface of fluid and solid wall. This tool is active when you have a Fluid dynamics problem in graphics area with solid body in contact to fluid. The procedure to use this tool is given next.

- Click on the **Default Fluid-Solid Interface** tool from the **Review** drop-down in the **PHYSICS** panel of **SIMULATION** tab in the **Ribbon**. The **INTERFACE REVIEW HUD** toolbar will be displayed; refer to Figure-71.

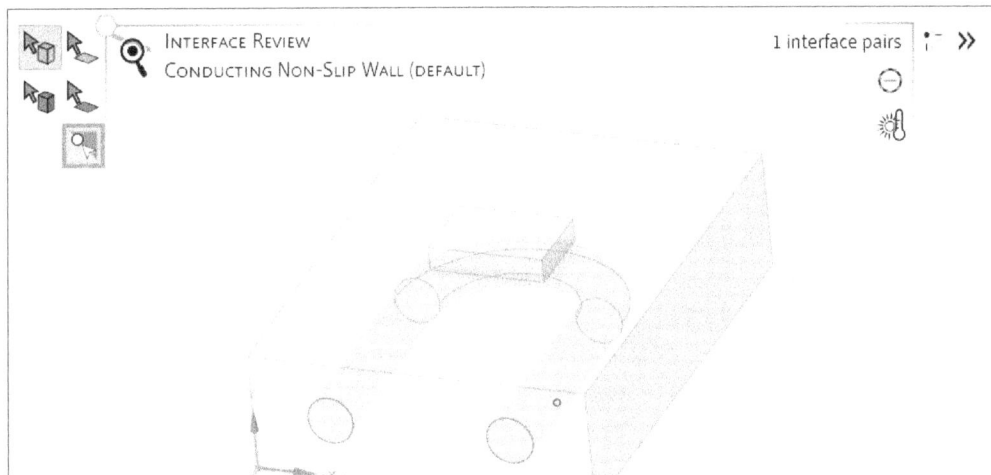

Figure-71. INTERFACE REVIEW HUD toolbar

- Select the **Include Internal Radiation** toggle button from the right **HUD** toolbar to include effect of internal radiation between fluid and solid boundary in the analysis.
- Expand the right **HUD** toolbar to define advanced parameters for interaction. The options will be displayed as shown in Figure-72.

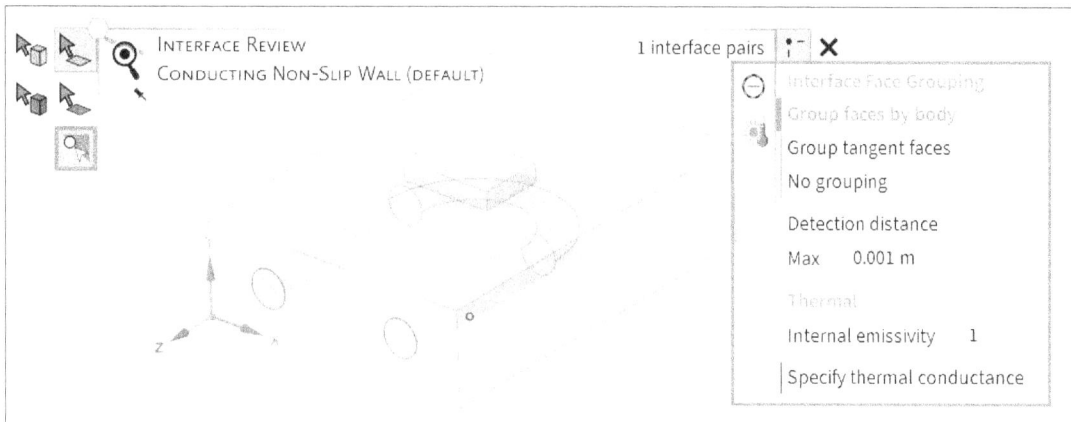

Figure-72. Expanded right HUD toolbar

- Specify desired value in the **Internal emissivity** edit box to define internal radiation rate.
- Select the **Specify thermal conductance** toggle option from the toolbar to define coefficient of heat transfer via conductance from fluid to solid boundary.
- Set the other parameters as discussed earlier and press **ENTER** to apply them. Press **ESC** to exit the tool.

SIMULATION OPTIONS

The options in the **Simulation Options** drop-down are used to define various macro parameters of analysis like whether simulation is static or time dependent, analysis modeling method, and so on; refer to Figure-73. Various options of this drop-down are discussed next.

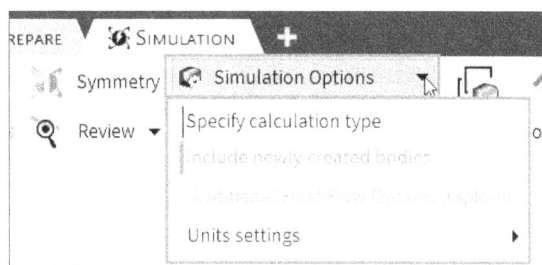

Figure-73. Simulation Options drop-down

Specifying Calculation Type

The **Specify calculation type** tool in **Simulation Options** drop-down is used to define whether the analysis is time dependent or static. After selecting this toggle option from the drop-down, select the **Static or steady-state** option from the drop-down if you want to analyze the model at equilibrium state where effect of time does not affect result of analysis. In most of the situations of structural analysis, static analysis is enough to get insight of failure of structural components.

Select the **Time-dependent** option from the drop-down to perform analysis setup where we are interested in specific duration of time for results. For example, when you start peddling a cycle then you will be exerting more force as compared to when cycle is running at a constant speed on flat road. In such cases, first few seconds of peddling will exert the maximum load on peddle rather than rest of the duration. So in such cases, we perform Time-dependent study to find variables for specific time duration. After selecting the **Time-dependent** option from the drop-down, select the **Specify time-dependent simulation duration** option to define time duration for which analysis will be performed and results will be generated.

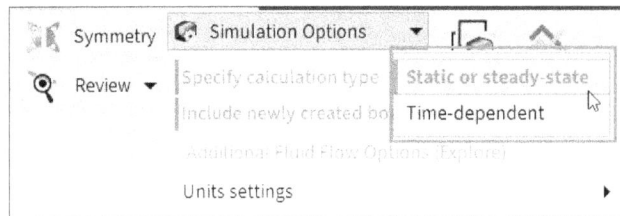

Figure-74. Specify calculation type drop-down

Including Newly created bodies

- Select the **Include newly created bodies** toggle option if you want to include bodies in current analysis as soon as you toggle it to be included in the analysis from the **Model Tree**.

Additional Fluid Flow Options (Explore)

- Select the **Additional Fluid Flow Options (Explore)** toggle option from the **Simulation Options** drop-down to define method for solving fluid dynamics analysis; refer to Figure-75. The **Specify modeling method** option will be displayed in the drop-down below the option. Select the **Laminar** option from **Specify modeling method** drop-down if fluid in your model flows in parallel layers without ripples and disturbances. This kind of flow is generally desired in lubricating mechanical parts. Select the **Turbulent k-omega SST** option from the drop-down to solve free stream flow problems like flow around airfoils, flow separations, and so on. This is the default model used for industrial flow analyses. In this model name, k means turbulent kinetic energy, ω (omega) means kinetic to thermal dissipation rate, and SST means shear stress transport. Since, the model accounts for shear stress variation at different layers of fluid, it is well suited for analyses which have complex boundary layers and you want accurate results. Select the **Smagorinsky (LES)** option from the drop-down if you want to solve a large scale motion model. Here, LES means Large Eddy Simulation. In this model, large energy carrying swirls (eddies) are solved which have greater impact on the flow and then small eddies are used to refine the results. You will learn about these models in later chapters with more detail.

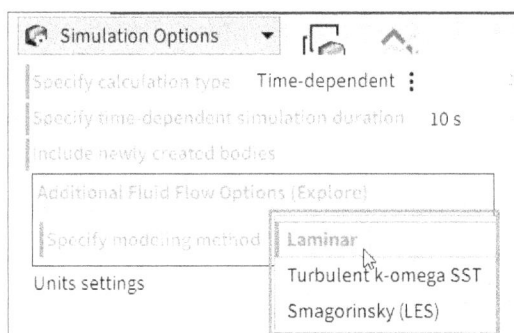
Figure-75. Modeling method drop-down

Defining Additional Fluid Flow Options (Refine)

When you are working in **REFINE** stage then the options for **Additional Fluid Flow** are displayed as shown in Figure-76.

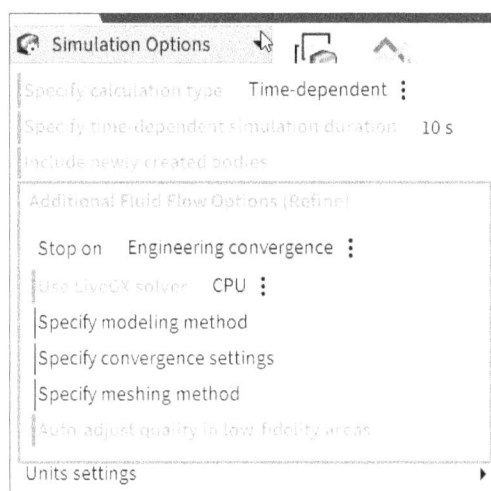
Figure-76. Additional Fluid Flow Options

- Select desired option from the **Stop on** drop-down to define convergence condition at which software will stop calculating and show results. When we perform any FEA then we try to reach the optimal solution by solving iterations of same equations using more and more refined mesh elements and properties. This process can go on indefinitely. We use convergence criteria as stopper for this process and find solution that is accurate upto required level. Select the **Engineering convergence** option from the drop-down if you want to stop calculation and show results when further refinement of mesh or iterations does not significantly affect the result of analysis. Select the **Numerical convergence** option from the drop-down if you want to consider analysis complete when residual of governing equations falls below a certain value and results are consistent on solving further. Select the **Monitored value convergence** option from the drop-down if you want to check stabilization of a physical parameter generated by result like flow rate, pressure, temperature, and so on to mark the completion of analysis.

- Select the **Use LiveGX solver** toggle option from the **Simulation Options** drop-down to advanced solver of Ansys which analyze the fluid flow problems in real time. After selecting toggle option, select the **GPU** or **CPU** option to use respective hardware for solving analysis.

- Select the **Specify modeling method** toggle option from the **Simulation Options** drop-down to define method of modeling CFD analysis. The **Laminar**, **Turbulent k-omega SST,** and **Smagorinsky (LES)** of **Specify modeling method** drop-down have been discussed earlier. Select the **Turbulent k-epsilon standard** option from the drop-down if you want to solve general fully developed turbulent flows. This option is well suited to study flows in ducts, pipes, and large field aerodynamics. Select the **Turbulent k-epsilon realizable** option from the drop-down if you want to solve CFD analysis where flow separation occurs or there is swirling in flow. This option is well suited for studying rotating flows and turbo machines. Select the **Turbulent k-omega standard** option from the drop-down if you want accuracy in solving near wall flow or there is transition in fluid flow. This option is well suited for low Reynolds number flow and transitional flow at boundary layer. Select the **Turbulent Spalart-Allmaras** option from the drop-down to solve external aerodynamics analyses efficiently. You will learn more about these methods later.
- Select the **Specify convergence settings** toggle option from the **Simulation Options** drop-down to define maximum number of iterations and targeted accuracy level for numerical, engineering, and monitored convergences in respective edit boxes; refer to Figure-77.

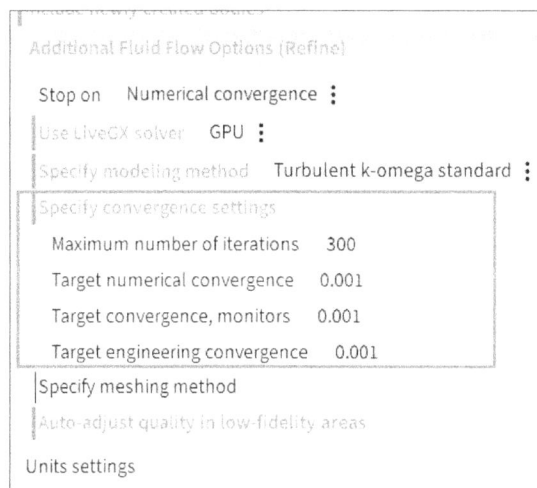

Figure-77. Specify convergence settings options

- Select the **Specify meshing method** toggle option from the drop-down to define whether you want to use Tetrahedra mesh or Polyhedra mesh. By default, Tetrahedral meshing is used.
- Select the **Auto-adjust quality in low-fidelity areas** toggle option from the drop-down if you want the system to automatically adjust mesh size in model where accuracy is low as compared to rest of the model.
- Click on the **Units settings** option to check and modify units for analysis; refer to Figure-78.

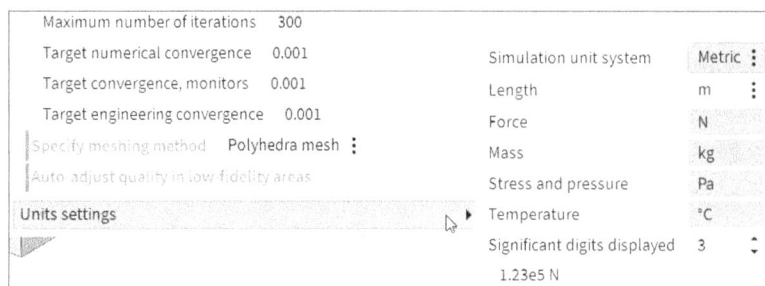

Figure-78. Unit settings cascading menu

NATURAL FREQUENCY SETUP

The **Natural Frequency** toggle option is used to include modal analysis when solving structural analysis to find out natural frequencies and mode shapes of the part. This toggle option is available when you are performing structural analysis. You will work on Modal analysis later.

TOPOLOGY OPTIMIZATION

The **Topology Optimization** toggle option is used to activate topology study for reducing weight, volume, stress, and so on while still fulfilling structural analysis parameters. The procedures to perform the analyses is discussed in subsequent chapters.

SELF-ASSESSMENT

Q1. What is the primary purpose of pre-processing in Ansys Discovery?
A. To generate graphical representations of results
B. To perform calculations based on input parameters
C. To prepare the model for running analysis
D. To generate final reports

Q2. Which of the following is NOT a part of pre-processing?
A. Assigning material properties
B. Applying loads and constraints
C. Creating contour plots
D. Defining meshing parameters

Q3. What is the function of meshing in pre-processing?
A. To refine graphical representations
B. To define material properties
C. To break down the model into smaller elements for analysis
D. To generate simulation reports

Q4. In the processing stage of analysis, which equation is solved in a linear static analysis?
A. $F = m * a$
B. $K * u = F$
C. $P = V * I$
D. $E = mc^2$

Q5. What is the main objective of post-processing in simulation?
A. Preparing the model for analysis
B. Interpreting, analyzing, and visualizing results
C. Defining mesh quality
D. Assigning loads and constraints

Q6. Which of the following is NOT a post-processing activity?
A. Generating contour plots
B. Understanding stress distribution
C. Applying boundary conditions
D. Generating reports

Q7. Which tab contains tools to apply analysis-related parameters in Ansys Discovery?
A. MATERIALS
B. SIMULATION
C. PHYSICS
D. QUICK START

Q8. What does the QuickScope tool help define?
A. Mesh refinement levels
B. Scope of components included in the analysis
C. Load and boundary conditions
D. Post-processing settings

Q9. In the QuickScope tool, what does the "Invert" toggle option do?
A. Inverts the material properties
B. Swaps selected and deselected objects
C. Applies opposite boundary conditions
D. Deletes the selected components

Q10. What is the primary function of the "Materials" drop-down in Ansys Discovery?
A. To assign materials to bodies in the simulation
B. To generate stress-strain curves
C. To define meshing parameters
D. To apply boundary conditions

Q11. What additional material properties are required for fatigue analysis?
A. Density and Poisson's ratio
B. Elastic modulus and yielding strength
C. S-N or E-N curve parameters
D. Thermal conductivity and emissivity

Q12. What is the function of applying supports in Ansys Discovery?
A. To define boundary conditions
B. To refine meshing quality
C. To modify material properties
D. To create vector plots

Q13. Which support type restricts all six degrees of freedom?
A. Sliding Support
B. Hinged Support
C. Fixed Support
D. Ball Support

Q14. Which support allows only rotational motion around the cylindrical face axis?
A. Ball Support
B. Hinged Support
C. Sliding Support
D. Fixed Support

Q15. What is the function of Displaced Support?
A. To allow free translation and rotation in all directions
B. To restrict movement along all axes
C. To manually define movement constraints in specific directions
D. To delete applied boundary conditions

Q16. What is the primary function of the Force tool in simulation?
A. To apply force and torque to selected faces
B. To measure displacement in a model
C. To simulate heat transfer in a material
D. To analyze electrical conductivity

Q17. What option should be selected to apply force uniformly over an entire selected face?
A. Total force
B. Distributed force
C. Remote force
D. Point force

Q18. Which unit is commonly used to specify force per unit area in the simulation?
A. Newton-meter (N-m)
B. Meter per second (m/s)
C. Megapascal (MPa)
D. Kilogram (kg)

Q19. What is the main difference between torque and horsepower in cars?
A. Torque is related to speed, while horsepower is related to acceleration
B. Torque rotates the crankshaft, while horsepower determines work capacity
C. Torque is used only in electric vehicles, while horsepower is for gasoline engines
D. Torque increases fuel efficiency, while horsepower decreases it

Q20. Which HUD toolbar option allows defining X, Y, and Z components of force separately?
A. Use force components
B. Distributed force
C. Remote force
D. Total force

Q21. What does selecting the Use remote point option do when applying force?
A. It applies force from a distant point rather than directly on the face
B. It distributes force evenly over the face
C. It prevents force from affecting the object
D. It allows force to change dynamically during simulation

Q22. What is the SI unit of torque?
A. Newton-second (N-s)
B. Newton-meter (N-m)
C. Joule (J)
D. Pascal (Pa)

Q23. Which type of force is torque most closely associated with?
A. Translational force
B. Rotational force
C. Compressive force
D. Tensile force

Q24. How can the Pressure tool be used to apply force per unit area?
A. By selecting the Total force option
B. By entering a value in the input box of the HUD toolbar
C. By using the Remote point toggle
D. By specifying force components

Q25. When applying displacement, what option should be selected for both linear and rotational movement?
A. Translation
B. Rotation
C. Combined
D. Remote

Q26. What is the function of the Moment tool?
A. To apply loads that bend/twist an object about an axis
B. To measure reaction forces in a structure
C. To analyze thermal expansion of a material
D. To determine impact resistance of a body

Q27. When applying mass load, what does the Use remote point option do?
A. It allows mass to be applied at a specified location without including a physical part
B. It distributes the mass equally over the selected face
C. It applies a force instead of mass
D. It eliminates gravitational effects on the mass

Q28. In which scenario is the Velocity tool commonly used?
A. To simulate static stress in a part
B. To apply kinematic loads for testing collision or object flow
C. To measure heat transfer rate in a body
D. To analyze electrical resistance

Q29. What is the purpose of applying Bearing Loads in simulation?
A. To simulate the effect of rotational friction in a joint
B. To represent radial and axial loads in a bearing housing
C. To determine the torque generated by a shaft
D. To measure the pressure exerted by a fastener

Q30. What does the Bolt Preload tool primarily do?
A. Applies pre-tension load along the axis of bolted faces
B. Measures the expansion of bolts under load
C. Determines the coefficient of friction in bolted connections
D. Simulates the fatigue strength of bolted joints

Q31. What is the function of the Contact tool in simulation?
A. To define how elements of two components interact at common surfaces
B. To measure temperature differences between parts
C. To apply an electrical load to connected components
D. To analyze the sound propagation between materials

Q32. Which contact type ensures that two connected faces act as a single unit under load?
A. Sliding contact
B. Bonded contact
C. Prevented contact
D. Frictional contact

Q33. What does selecting the Prevented contact option do in Ansys?
A. Allows sliding between two faces
B. Keeps two selected faces separated with no stress transfer
C. Bonds two faces together like a weld
D. Simulates frictional interaction between two faces

Q34. What type of joint should be applied to allow only rotational motion between two connected faces?
A. Fixed joint
B. Hinged joint
C. Sliding joint
D. Flexible joint

Q35. What are the two types of CFD analyses that can be performed?
A. Internal Flow Analysis and Boundary Flow Analysis
B. External Flow Analysis and Internal Flow Analysis
C. Aerodynamic Flow Analysis and Hydraulic Flow Analysis
D. Laminar Flow Analysis and Turbulent Flow Analysis

Q36. Where can the tools to define fluid flow analysis type be found in Ansys Discovery?
A. In the PHYSICS panel of the SIMULATION tab
B. In the QUICK START panel of the SIMULATION tab
C. In the MODEL panel of the DESIGN tab
D. In the SETTINGS panel of the RENDER tab

Q37. What is the first step in defining an internal flow analysis in Ansys Discovery?
A. Selecting the fluid material
B. Selecting the inlet face
C. Selecting the outlet face
D. Setting the boundary conditions

Q38. What happens after selecting the inlet and outlet faces in an internal flow analysis?
A. The software generates a fluid volume and applies default boundary conditions
B. The user needs to manually define the fluid domain
C. The user must enter the velocity of the fluid manually
D. The inlet and outlet faces remain unconnected

Q39. In an external flow analysis, what is selected to define the inlet direction?
A. A plane from the model
B. A coordinate axis
C. An arrow from the graphics area
D. A point in the simulation space

Q40. What does selecting the ground plane in an external flow analysis help define?
A. The velocity of the fluid
B. The gravity direction
C. The inlet pressure
D. The turbulent flow

Q41. When do the tools of the Fluid Flow drop-down become active?
A. When structural boundary conditions are applied
B. When the model is closed
C. When the user deletes all boundary conditions from the Structural drop-down
D. When the model is saved in a new file format

Q42. What parameter does the Swirling Flow toggle option allow the user to define?
A. Mass flow rate
B. Rotational speed in RPM
C. Pressure gradient
D. Inlet velocity

Q43. What happens if atmospheric pressure (1 atm) is applied at the outlet of a model?
A. The inlet pressure increases
B. The model simulates an open environment condition
C. The fluid stops flowing
D. The pressure remains constant throughout the flow

Q44. Which tool is used to create an intake or exhaust fan boundary condition?
A. Flow tool
B. Fan tool
C. Wall tool
D. Rotating Fluid Zone tool

Q45. Which value must be specified for a fan boundary condition to define the rate of fluid flow?
A. Velocity
B. Volume flow rate
C. Pressure coefficient
D. Temperature gradient

Q46. What is the function of the Wall tool in CFD analysis?
A. To define different types of fluid boundaries
B. To apply force to a fluid
C. To change the temperature of the fluid
D. To define inlet and outlet points

Q47. What happens when the Stationary option is selected in the Wall tool?
A. The wall allows fluid flow without friction
B. The velocity of the fluid becomes zero
C. The wall moves in a defined direction
D. The temperature of the fluid increases

Q48. What is the purpose of the Convection option in thermal conditions?
A. To prevent heat transfer through the wall
B. To define the external temperature and heat transfer rate
C. To increase the internal temperature
D. To change the emissivity of the wall

Q49. Which parameter is used to define heat transfer occurring via radiation?
A. Heat flux
B. External emissivity
C. Convection coefficient
D. Thermal conductivity

Q50. What does the Porous tool simulate in CFD analysis?
A. A completely closed boundary
B. A rotating pump
C. A filter-like material allowing restricted fluid flow
D. A moving fluid boundary

Q51. What is the function of the Rotating Fluid Zone tool?
A. To simulate a fan boundary condition
B. To create a cylindrical zone where fluid rotates at a defined speed
C. To apply external aerodynamic effects
D. To define porous boundaries

Q52. What is a requirement for a rotating fluid zone in Ansys Discovery?
A. It must be applied to a rectangular object
B. It can share an interface with another rotating fluid zone
C. It must be located on a body with cylindrical or conical outer surfaces
D. It must also be defined as a porous zone

Q53. Which tool is used to assign a specified temperature to a selected face?
A. Convection
B. Temperature
C. Heat
D. Insulated

Q54. What unit is used for the convection coefficient in the Convection tool?
A. W/m.K
B. W/m².K
C. J/K
D. W/K

Q55. Which option should be selected in the Heat tool to define the heat rate per unit area?
A. Total heat
B. Heat per unit area
C. Convection heat
D. Insulated heat

Q56. What does the Insulated tool do?
A. Prevents heat from passing through selected faces
B. Allows heat conduction at a specified rate
C. Increases the heat transfer rate
D. Defines a temperature gradient

Q57. When is the Electromagnetics drop-down active?
A. During the SETUP stage
B. When the REFINE stage is selected
C. When a mesh is generated
D. After running the simulation

Q58. What does the Circuit Port tool represent?
A. A single resistor in a circuit
B. The overall effect of a circuit board in the model
C. A thermal boundary condition
D. The flow of heat through a solid

Q59. What does selecting the Passive port option in the Circuit Port tool do?
A. Creates a power source
B. Defines a receiver that does not generate power
C. Applies a temperature constraint
D. Converts the model into an electrical conductor

Q60. What is another name for the Mode Port tool?
A. Heat port
B. Wave port
C. Thermal port
D. Insulated port

Q61. What is the purpose of the Symmetry tool?
A. To create a mirror copy of a model
B. To define a symmetry condition for simplifying analysis
C. To generate duplicate parts
D. To define a temperature constraint

Q62. What does the Default Solid-Solid Contacts tool do?
A. Applies a symmetry condition
B. Defines heat conduction rates
C. Checks and applies default contact settings between solids
D. Removes contact between solid faces

Q63. Which option in the Default Solid-Solid Contacts tool allows converting highlighted contacts into sliding type?
A. Convert to joint
B. Convert to permanently bonded
C. Convert to sliding
D. Exclude from detection

Q64. What does the Include Internal Radiation toggle button in the Default Fluid-Solid Interface tool do?
A. Ignores radiation effects in fluid-solid interactions
B. Includes the effect of internal radiation between fluid and solid boundaries
C. Removes heat transfer conditions
D. Changes the thermal conductivity of the material

Q65. What is the purpose of the Specify calculation type tool?
A. To define whether the analysis is static or time-dependent
B. To determine material properties
C. To specify mesh refinement levels
D. To apply an insulation condition

Q66. Which option should be selected for modeling fluid flow around airfoils?
A. Laminar
B. Turbulent k-omega SST
C. Smagorinsky (LES)
D. Static

Q67. What does the Monitored value convergence option do in Additional Fluid Flow Options?
A. Stops the analysis based on a selected physical parameter's stabilization
B. Stops when numerical equations fall below a certain residual
C. Stops when further mesh refinement does not significantly affect results
D. Runs the simulation indefinitely

Q68. What hardware options are available for the Use LiveGX solver?
A. RAM or SSD
B. HDD or GPU
C. GPU or CPU
D. CPU or SSD

FOR STUDENT NOTES

Chapter 7

Ansys Discovery-
Structural Analyses

Topics Covered

The major topics covered in this chapter are:

- *Introduction to Structural Analyses Theory*
- *Performing Linear Static Structural Analysis*
- *Generating Results*
- *Linear Static Analysis On Assembly*
- *Non Linear Static Analysis*
- *Theory Behind Non-Linear Analysis*
- *Performing Non-Linear Static Analysis*
- *Load Cases/Parameter Study*
- *Performing Natural Frequency Analysis*
- *Performing Topology Optimization*

INTRODUCTION

In previous chapter, you have learned about various tools used to perform different types of analyses. The tools discussed earlier can be used to perform structural analysis, thermal analysis, fluid dynamics study, and so on. After learning about the basics of these tools, you will now learn to apply these tools for structural analyses in this chapter.

STRUCTURAL ANALYSES

Structural Analysis, as the name suggests, is analysis of structure of a model when different types of loads are applied on it. These loads can be structural like force, torque, pressure, etc., they can be thermal like radiation, convection, flux, etc. or they can be electromagnetic, or combinations of all these types of loads. In all these cases, we are only interested in structural deformations of the model when we are performing structural analysis. We are not concerned about temperature change, voltage change, or other changes. Various types of structural analyses that can be performed in Ansys Discovery are given next.

Static Structural Analysis

Static means stable/fixed. The Static Structural analysis is used to check the effect of fixed value loads on a model at equilibrium state. In simple words, this analysis is used when value of load is not changing with time. This analysis is performed to check whether the object of interest will be able to sustain applied loads or not. We also use this analysis to find out the factor of safety for the design.

Tip : The factor of safety is the dimensionless value which determines the multiple of current loads that can be sustained by design. For a simplified example, assume current force applied on our design is 500 N and design can sustain 1500 N force then factor of safety will be 1500/500 = 3.

There are two types of Static Structural Analyses: Linear Static Structural Analysis and Non-linear Static Structural Analysis. The Linear Static Structural Analysis is used when there is linear relation between stress and strain. For example if you apply 500 N force on a bar then it deflects by 0.05 mm and when you apply 1000 N force on it then it deflects 0.10 mm.

Linear Static Structural Analysis

In case of linear static structural analysis, apart from fulfilling the conditions of static structural analysis, various conditions that are fulfilled by linear static analyses:

• Material behaves elastically which means relation between stress and strain of the material is linear. When expressed in mathematical formula:

$$\sigma = \varepsilon \cdot E$$

Here, σ is stress which is Force per area, ε is strain which is deformation with respect to original size, and E is the Young's Modulus.

- The deformations in structure under loads are below the threshold which can change the structural design of model significantly. In simple words, consider a case of ball pressed by force along Z axis. For the linear static structural analysis to be applied for this case, the ball should not deform so much that it becomes ellipsoid instead of sphere. Because if it becomes ellipsoid then its structure has changed and hence there will be no longer linear relation between stress and strain.
- The deformation in structure is elastic which means after removing load, the structure returns to original form.

So, based on assumptions required for performing linear static structural analysis, it can be concluded that Linear static structural analysis is performed to check early stage design. At this stage, you want to test the general behavior of model under applied loads to decide which type of analysis should be performed next for further study and improve the design. In terms of computing resources, it is one of the simplest and less resource-hungry analysis.

Non-Linear Static Structural Analysis

The Non-linear static structural analysis is used when there is change in geometry of model or there is not a linear relation between stress and strain of the model. Various assumptions of model that require performing a Non-linear static structural analysis are given next.

- Material does not follow Hooke's Law ($\sigma = \varepsilon \cdot E$) and it can deform plastically like aluminum, steel, etc. or hyper-elastically like rubber. This criteria is collectively called Material Non-linearity.
- There is large deformation in the model under load and it has caused permanent change in structure of model. In such cases, system considers new geometry of model for testing after each iteration for applying loads. Since, the geometry is changed, system need to update the load distribution as well by using P-delta factor which counters P-delta effect in the analytical model. P-delta effect is the phenomena in which value of moment generated by same load changes after each small deformation because the anchor point of loading is getting changed.

Tip : In case of analyses, Iteration is repetition of solving mathematical equations of the analysis to reach a more accurate result at each step. In case of non-linear static analysis, total load applied on model is divided into small incremental loading steps which will become equation to net total load at the end of iterations. At each iteration, an incremental load is applied which generates new stiffness matrix due to non-linearity and at the next iteration, next incremental load will be applied to newly formed stiffness matrix; and the process will keep repeating until either desired accuracy in solution is achieved or desired number of steps have been performed.

- There is a connection or contact in the model that changes type when load reaches certain threshold. This can represent breaking of contacts or one part sliding over another, or it may represent flying object striking another object hence developing new type of contacts.
- Load applied on the model changes magnitude and/or direction both internally or externally.

- The load is applied in parts at different time intervals. Note that applying loads in parts does not make the analysis fall in dynamic analysis category. Such type of analyses are still in Static analyses category because effect of time dependent self-acting components like damping and inertia are not considered in these analyses.
- There is a crack formation in the model due to load or there is structural instability like buckling in the model.

Prestressed Static Analysis

The Prestressed static analysis is a type of static analysis where prestress forces have already been applied to the model before applying static load for the analysis. There are many examples of this type of analysis like bolt tightened forcibly, wires holding the bridge, concrete columns, car tires filled with pressured air, and so on.

Tip : You can identify a design analysis problem as prestressed static analysis when the loads applied on the model do not change with time, there is already a fixed load applied on the model when defining design conditions.

Buckling Analysis

Buckling occurs when a structure suddenly deforms under compressive axial loads, converting stored axial energy into bending energy without changing the applied load. This typically happens in slender structures, such as thin-walled components, under small axial loads. The critical load at which buckling begins is crucial because failure occurs due to loss of geometric stability rather than material strength.

In linearized buckling analysis, an eigenvalue problem is solved to predict the critical buckling load and the associated mode shapes (various forms the structure may take when buckling). Designers focus on the lowest mode shape (mode 1) because it corresponds to the lowest critical load. Multiple buckling modes are calculated to identify weak spots in the design, enabling engineers to make modifications to prevent buckling.

In non-linear buckling analysis, more advanced methods are needed to study behavior beyond the critical buckling load.

An everyday example is pressing on an empty soda can: applying force gradually causes the can to suddenly collapse—a visual analogy for buckling. For engineers, buckling often happens at stress levels far below what the material can normally handle, so specialized checks are necessary to prevent this catastrophic failure. Buckling analysis is particularly important in industries like civil engineering (columns, bridges) and mechanical engineering (aircraft, automotive design).

Modal Analysis

Every structure vibrates at specific frequencies called natural or resonant frequencies, each associated with a unique mode shape. When the frequency of an external dynamic load matches a structure's natural frequency, resonance occurs, causing large displacements and stresses. Damping helps to control this response, preventing infinite motion in real systems.

While real structures have infinite natural frequencies, finite element models calculate a finite number based on the degrees of freedom. Typically, only the first few modes are necessary. Natural frequencies are influenced by the structure's geometry, material, and supports. Modal analysis help design systems to avoid resonance and assess responses in dynamic environments. Compressive loads reduce natural frequencies, while tensile loads increase them, as seen with tension in a violin string. To optimize a design for vibration stability, increasing rigidity or reducing weight raises the natural frequency, while increasing weight or reducing rigidity lowers it.

Transient Structural Analysis

The Transient Structural analysis is performed to simulate effect of time varying loads on the model where dynamic forces like damping, impact loads, and so on are also included in the setup. Note that most of the materials provide some amount of damping when load is applied on them. There are various types of transient structural analyses like Modal analysis, Random vibration analysis, explicit dynamics, implicit dynamics, and so on. You will learn about these analyses later.

PERFORMING LINEAR STATIC STRUCTURAL ANALYSIS

Linear static analysis is performed to calculate stresses, displacements, strains, reaction forces, and error estimates under various loading conditions. On applying the loads, the body gets deformed and the loads are transmitted throughout the body. The deformation and other effects of body are studied under this analysis.

Assumptions

Some assumptions are made in this type of analysis, like:

1. The loads applied do not vary with time.
2. All loads are applied slowly and gradually until they reach to the full magnitude and after reaching the full magnitude, the loads remain constant. Thereby, neglecting impact, inertial, and damping forces.
3. The materials applied to the components satisfy the Hooke's law.
4. The change in stiffness due to loading is neglected.
5. Boundary conditions do not vary during the application of loads. Loads must be constant in magnitude, direction, and distribution.

Geometry Assumptions

1. The part model must represent the required CAD geometry.
2. Only the internal fillets in the area of interest will be included in the study.
3. Shells are created when thickness of the part is small in comparison to its width and length.
4. Thickness of the shell is assumed to be constant.
5. If the dimensions of a particular part are not critical and do not affect the analysis results, some approximations can be made in modeling the particular part.
6. Primary members of structure are long and thin like a beam then idealization is required.
7. Local behavior at the joints of beams or other discontinuities are not of primary interest so no special modeling of these area is required.

8. Decorative or external features will be assumed insignificant to the stiffness and the performance of the part, and will be omitted from the model.

Material Assumptions

1. Material remain in the linear regime. It is understood that either stress levels exceeding yield or excessive displacements will constitute a component failure. That is non linear behavior cannot be accepted.
2. Nominal material properties adequately represent the physical system.
3. Material properties are not affected by load rate.
4. Material properties can be assumed isotropic (Orthotropic) and homogeneous.
5. Part is free of voids or surface imperfections that can produce stress risers and skew local results.
6. Actual non linear behavior of the system can be extrapolated from the linear material results.
7. Weld material and the heat affected zone will be assumed to have same material properties as the base material.
8. Temperature variations may have a significant impact on the properties of the materials used. Change in material properties is neglected.

Boundary Conditions Assumptions

1. Choosing proper BC's require experience.
2. Using BC's to represent parts and effects that are not or cannot be modeled leads to the assumption that the effects of these un-modeled entities can truly be simulated or has no effect on the model being analyzed.
3. For a given situation, there would be many ways of applying boundary conditions. But these various alternatives can be wrong if the user does not understand the assumptions they represent.
4. Symmetry/ anti-symmetry/ reflective symmetry/ cyclic symmetry conditions if exists can be used to minimize the model size and complexity.
5. Displacements may be lower than they would be if the boundary conditions being more appropriate. Stress magnitudes may be higher or lower depending on the constraint used.

Fasteners Assumptions

1. Residual stress due to fabrication, pre-loading of bolts, welding and/or other manufacturing, or assembly processes are neglected.
2. Bolt loading is primarily axial in nature.
3. Bolt head or washer surface torque loading is primarily axial in nature.
4. Surface torque loading due to friction will produce only local effects.
5. Bolts, spot welds, welds, rivets, and/or fasteners which connect two components are considered perfect and acts as rigid joint.
6. Stress relaxation of fasteners or other assembly components will not be considered. Load on threaded portion of the part is evenly distributed on engaged threads.
7. Failure of fasteners will not be reflected in the analysis.

General Assumptions

1. If the results in the particular area are of interest, then mesh convergence will be limited to this area.
2. No slippage between interfacing components will be assumed.

3. Any sliding contact interfaces will be assumed frictionless.
4. System damping will be normally small and assumed constant across all frequencies of interest unless otherwise available from published literature or actual tests.
5. Stiffness of bearings in radial or axial directions will be considered infinite.
6. Elements with poor or less than optimal geometry are only allowed in areas that are not of concern and do not affect the overall performance of the model.

When a system under load is analyzed with linear static analysis, the linear finite element equilibrium equations are solved to calculate the displacement components at all nodes. These results are used to calculate the strained components. These strain results along with stress-strain relationship helps to calculate stresses.

Simple Case of Linear Static Analysis

Consider a rectangular bar of size 100mm x 40mm x 500mm which is subjected to a uniformly distributed total load of 980 N on top face and fixed at its both end points; refer to Figure-1. You need to find out the maximum deflection that will occur in the beam due to load. Material used in the bar is AISI 1020 Carbon Steel (Annealed).

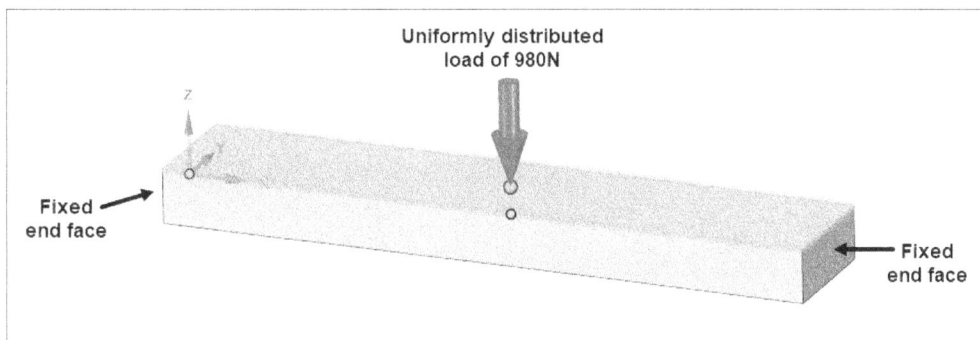

Figure-1. Linear Static Structural Analysis example

Steps involved in performing linear static analysis in Ansys Discovery:

- Creating or importing geometry/model required to perform analysis.
- Applying Material Properties
- Applying Boundary Conditions like constraints and loads
- Defining mesh element size
- Solving the Simulation
- Analyzing the Results

Creating Geometry Model

Generally, we import models in Ansys Discovery from other CAD software because modeling is not the main intention of Ansys Discovery. But in this case, the model is simple so we will create it inside Ansys Discovery software.

- Start Ansys Discovery if not started yet and start a new document. The sketching environment of Ansys Discovery will be active by default with XY plane selected as sketching plane.
- Click on the **Rectangle** tool from the **SKETCH** panel in the **DESIGN** tab of the **Ribbon**. The **RECTANGLE HUD** toolbar will be activated and you will be asked to specify start point of rectangle.

- Select the **Draw from center** toggle option from the right **HUD** toolbar to start creating rectangle with its center at origin and draw the rectangle as shown in Figure-2.

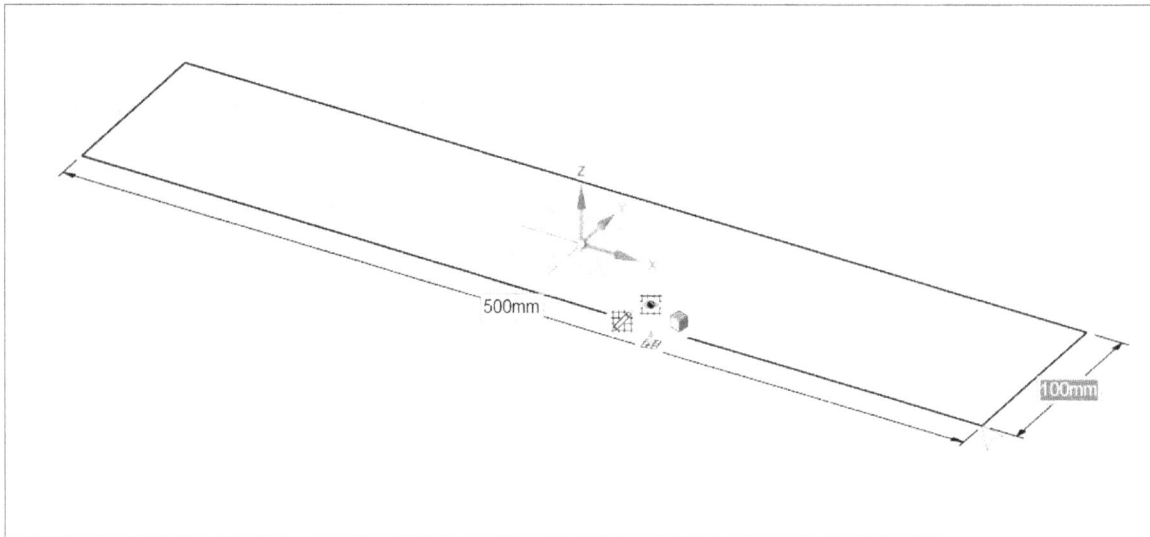

Figure-2. Creating rectangle

- Click on the **Return to 3D** button from the **Sketch Mini-toolbar** to switch to **Pull** tool for generating 3D model. The **Pull** tool will be activated and you will be asked to select the section to be extruded.
- Select the sketch section created earlier from mid, drag it upward, and enter the value as **40** in the input box; refer to Figure-3. Press **ESC** twice to exit the tool.

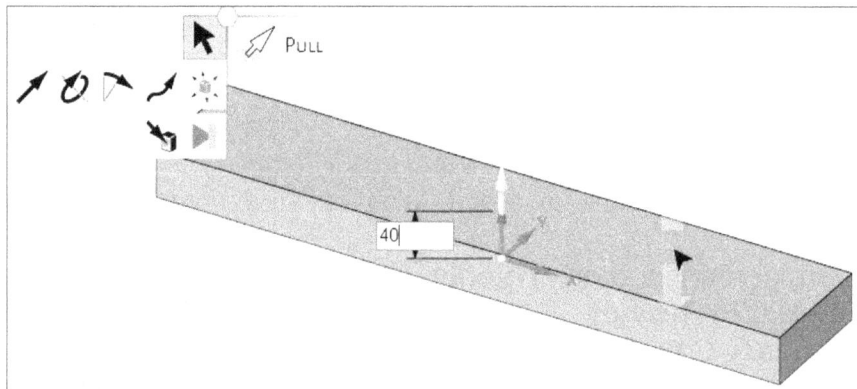

Figure-3. Specifying value for pull

Applying Material Properties

- Select the **EXPLORE** option from the **Stage Selector** at the bottom and select the **SIMULATION** tab in the **Ribbon**. The tools to apply simulation will be displayed.
- Click on the **Materials** tool from the **PHYSICS** panel in the **SIMULATION** tab of the **Ribbon**. The material will be applied to the model and will be listed in the **Physics Tree**.
- Double-click on the currently applied material from the **Physics Tree**. The options to change material will be displayed; refer to Figure-4.

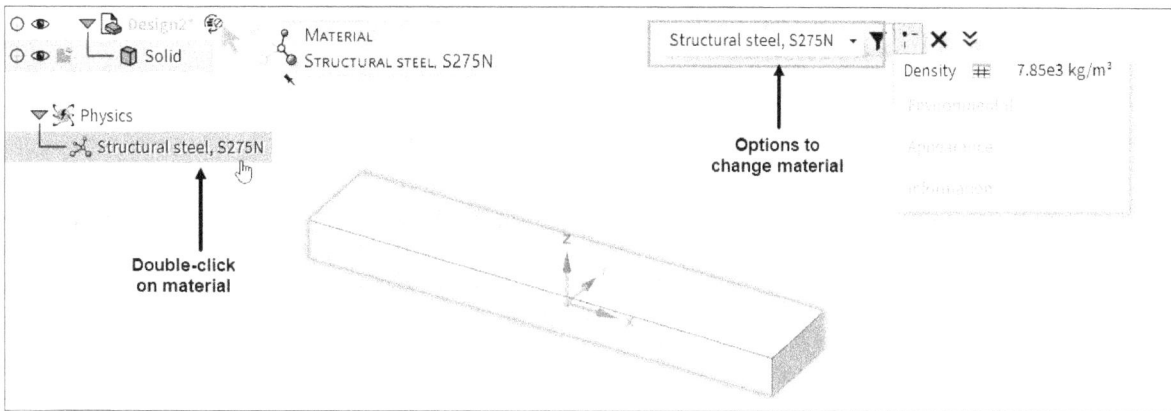

Figure-4. Options to change material

- Select the **Carbon steel, 1020, annealed** option from the **Materials** drop-down in the right **HUD** toolbar. Press **ESC** twice to exit the tool.

Applying Boundary Conditions

- Click on the **Structural** tool from the **PHYSICS** panel in the **SIMULATION** tab of the **Ribbon**. The **STRUCTURAL SUPPORT HUD** toolbar will be displayed.
- Select the **Fixed** option from the **Support** drop-down in the **HUD** toolbar and select two side faces of the model as shown in Figure-5.

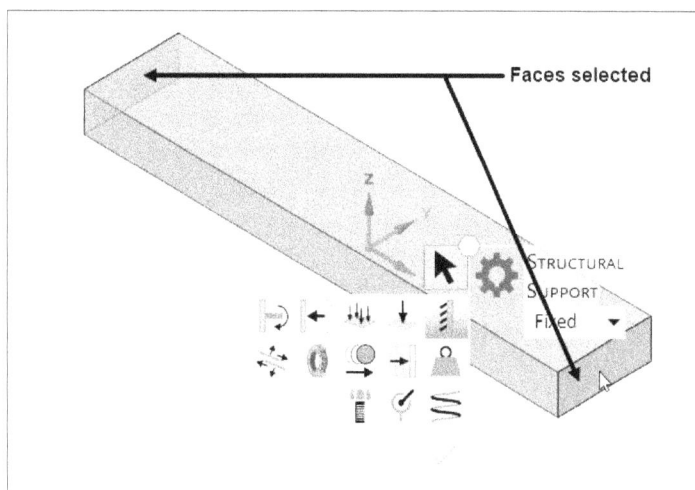

Figure-5. Faces selected for support

- Click on the **OK** button from the **HUD** toolbar to apply constraint.
- Click on the **Force** button from the **HUD** toolbar and apply the force on top face as shown in Figure-6.

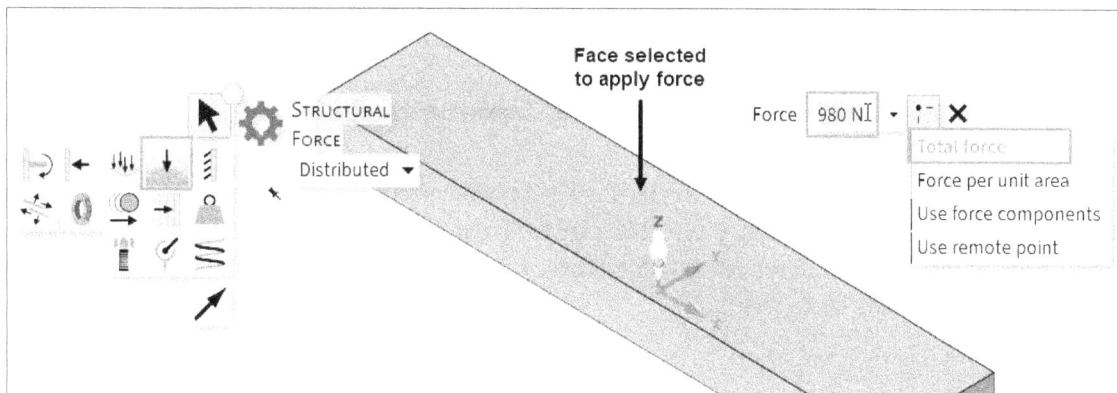

Figure-6. Applying face

- After specifying value in the **Force** edit box, press **ENTER** to apply the force.
- Press **ESC** twice to exit the tool.

Defining Mesh Size

The mesh size is defined by fidelity parameter in Ansys Discovery. You will learn about meshing in details later. The procedure to check and modify size of mesh is given next.

- Select the **Size Preview** toggle button from the **FIDELITY** panel in the **SIMULATION** tab of the **Ribbon** and hover the cursor on the model to check mesh size; refer to Figure-7.

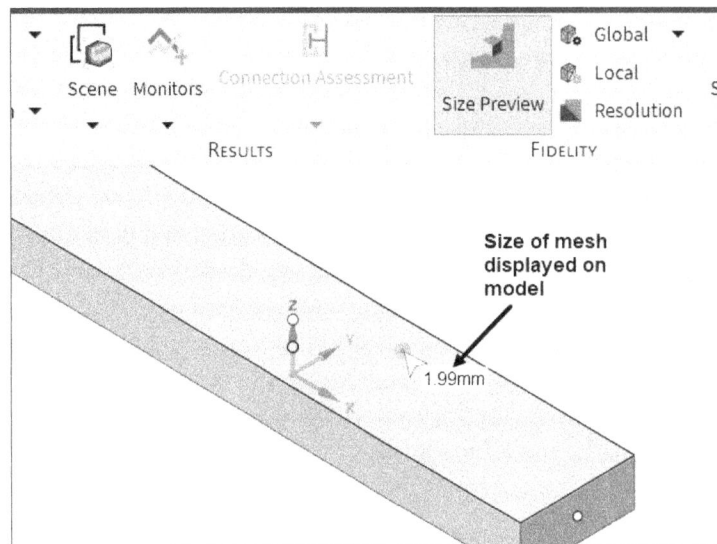

Figure-7. Checking mesh size

- Hover the cursor on **Explore** button in **Stage Selector**. The fidelity (mesh element size) value will be displayed with an interactive slider to change this value; refer to Figure-8.

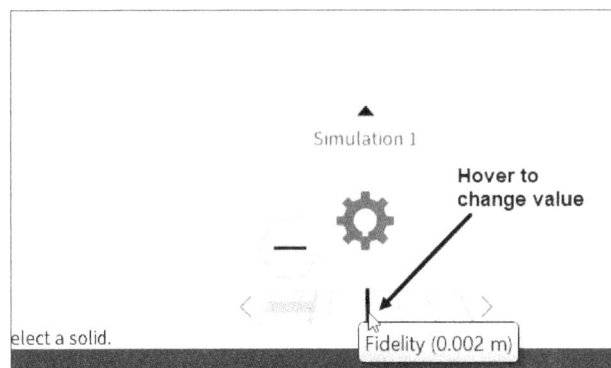

Figure-8. Fidelity value

- Drag the **Fidelity** slider towards right to decrease the mesh element size for increasing solution accuracy at the cost of more computing. Drag the **Fidelity** slider towards left to increase the mesh size for getting faster solution while decreasing solution accuracy.

Solving Simulation and Analyzing Results

- Click on the **Solve** tool from the **Simulation** drop-down in the **STUDY** panel of the **SIMULATION** tab of the **Ribbon**. The process will start and an indication of processing will be displayed with colored halo encapsules around simulation name in **Stage Selector**.

- Once the system completes solving the analysis then displacement result will be displayed in the graphics area. Make sure green colored halo encapsules around the simulation name in **Stage Selector** which marks completion of the analysis; refer to Figure-9.

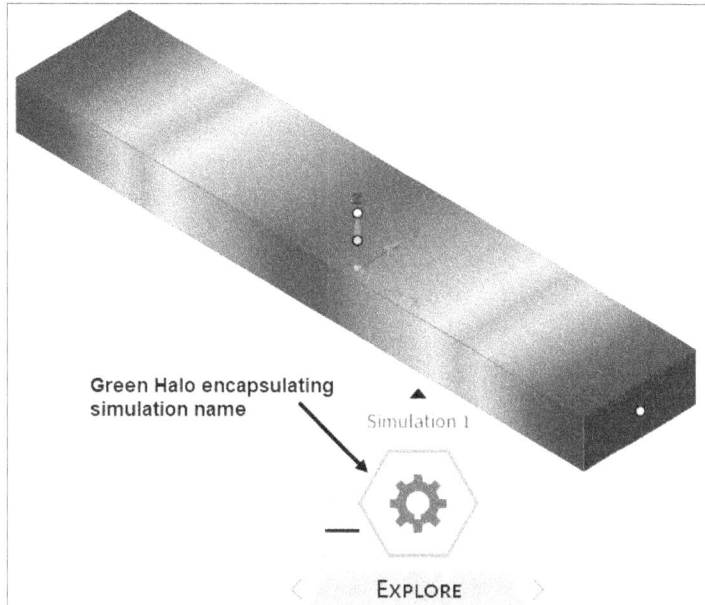

Figure-9. Result generated

- Move the cursor on the model to check value of displacement at current position in the scale of **Legend** area; refer to Figure-10. You can change the unit of result parameter from the **Unit** drop-down at the top right corner of **Legend** area; refer to Figure-11.

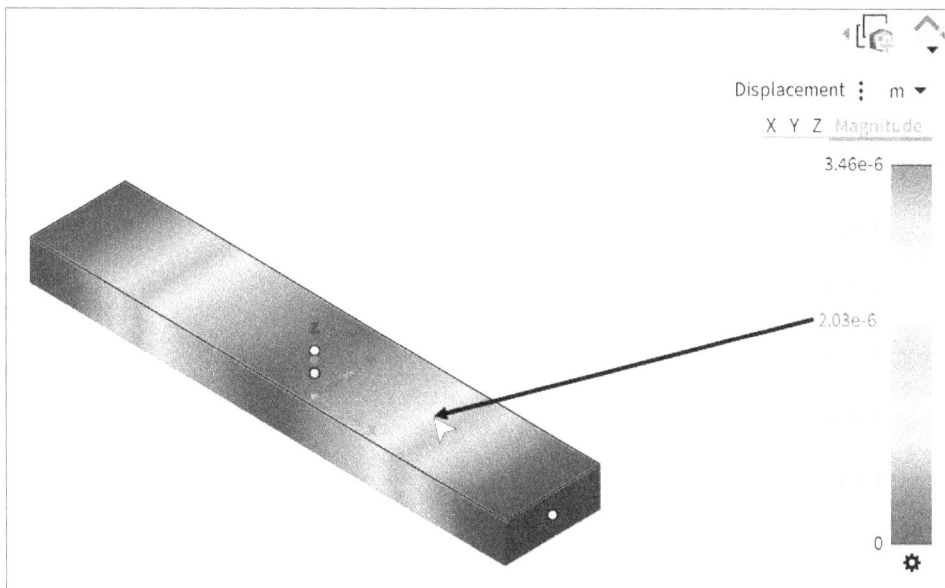

Figure-10. Displacement at current position

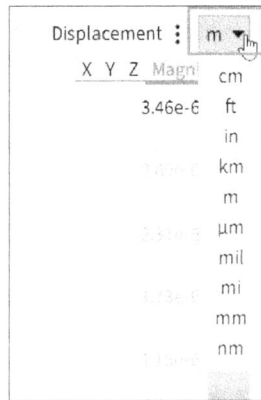

Figure-11. Unit drop-down

You will now learn about rest of the tools in **Ribbon** that were not discussed in previous chapter, because generating analysis results was not explained at that point.

GENERATING RESULTS

After performing analysis, results are graphically displayed in the graphics area of the software. Various sections of user interface of software are used to generate and post-process results; refer to Figure-12. These sections are discussed next.

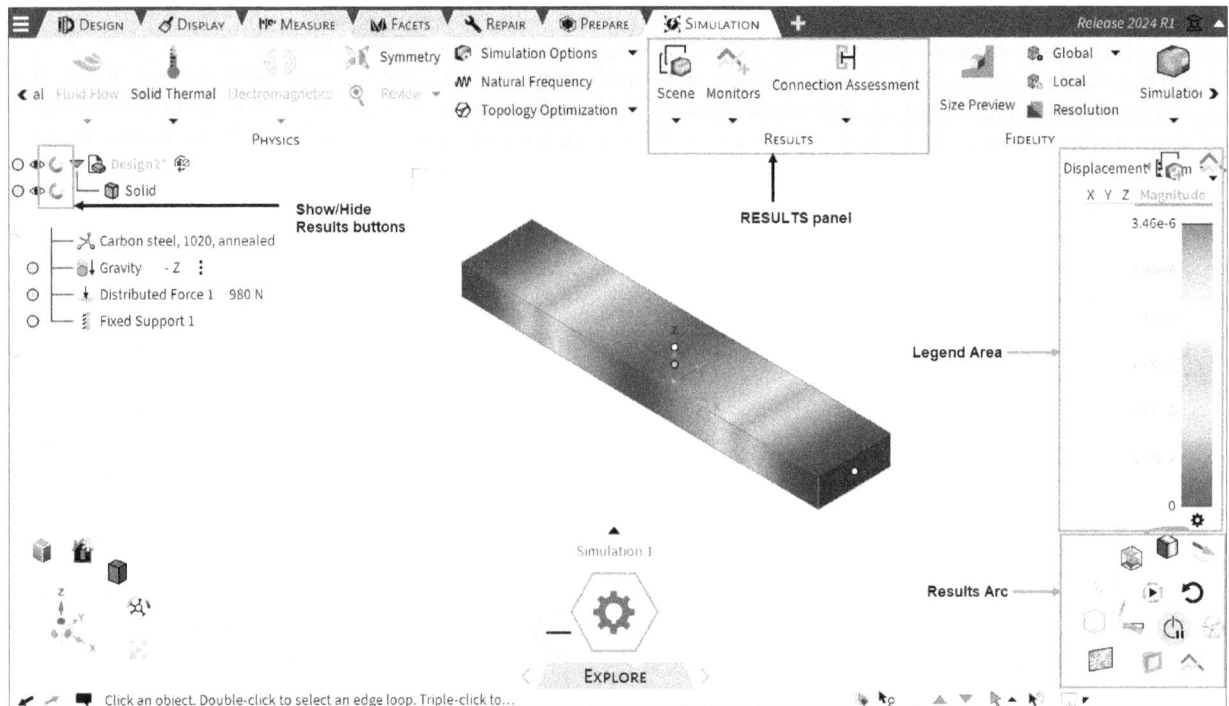

Figure-12. Options for Results

- Click on the Show/Hide results toggle button in the **Design Tree** to switch between showing and hiding results of analysis on model in color gradient in graphics area.

Legend Area

- Select desired option from the **Variables** drop-down in the **Legend** area to define result parameter to be displayed in the graphics area; refer to Figure-13. For example, select the **Stress** option from the drop-down to check stress in the model at different locations.

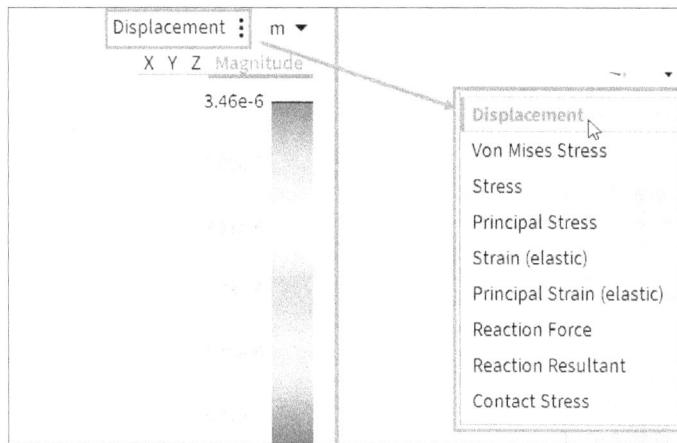

Figure-13. Variables drop-down

- Select desired option from the **Unit** drop-down at the top right corner in the **Legend** area to define the unit to be used displaying analysis result.
- If you want to change maximum and minimum values of the range scale then click on them in the **Legend** area and enter desired values; refer to Figure-14.

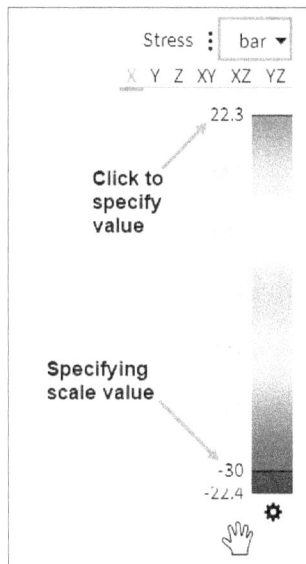

Figure-14. Changing value of scale

- Hover the cursor on **Settings** option from bottom in the **Legend** area to define style and range parameters for the scale of analysis (legend); refer to Figure-15. Select the **Reset user defined ranges** button from the **Settings** flyout to reset values to default in the scale. Select the **Toggle gradient style** toggle button from the **Settings** flyout to display scale in clear value boundaries. Select the **Show min/max locations** toggle option from the flyout to highlight minimum and maximum values of the result variable in the graphics area. Select the **Display out-of-range results as muted colors** toggle option to mute colors which are out of range; refer to Figure-16.

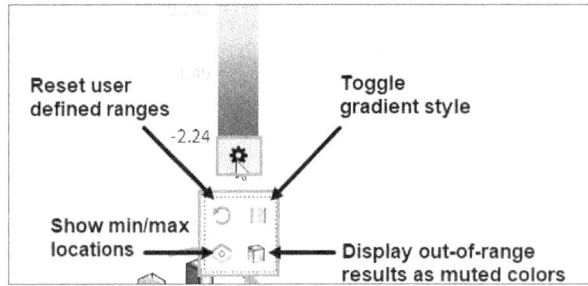

Figure-15. Setting flyout in Legend

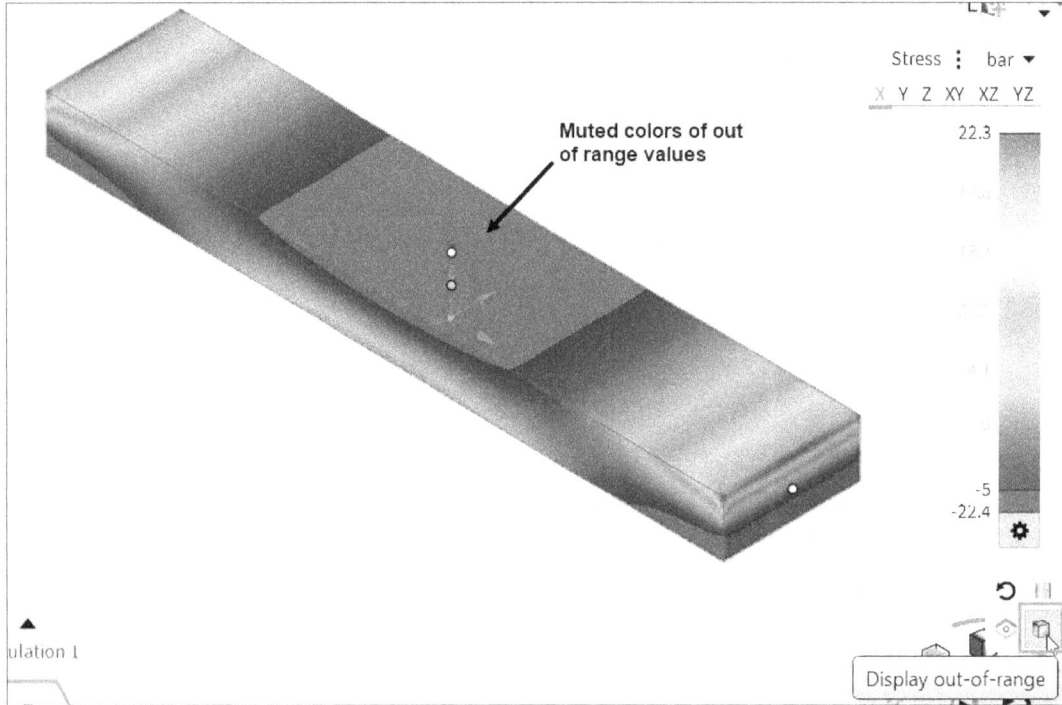

Figure-16. Showing muted colors

Results Arc

The buttons of **Results Arc** are used to set different types of results in graphics area; refer to Figure-17. Functions of various buttons of the arc are discussed next.

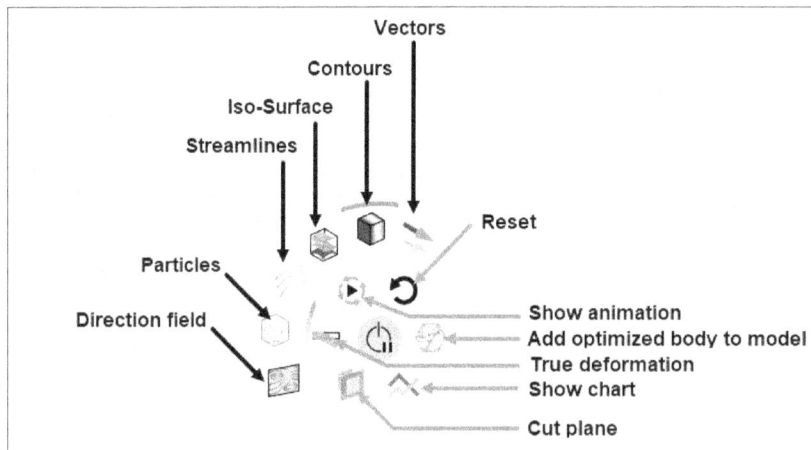

Figure-17. Results Arc

- Click on the center button of **Results Arc** to start solving and pause the analysis.

Vectors

Vectors are arrows colored according to result parameters. These arrows are used to analyze results.

- Hover the cursor on **Vectors** toggle button from the **Results Arc**. The options to define number of arrows and size of arrows will be displayed in a flyout; refer to Figure-18. Use the sliders to increase/decrease the number of arrows and size.

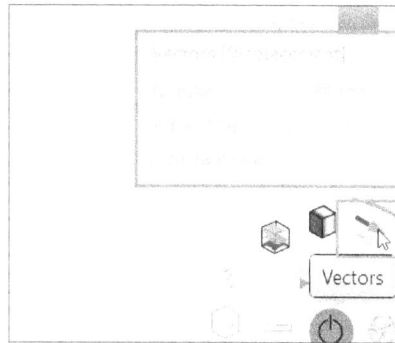

Figure-18. Vectors options

- Select the **Constant size** toggle button from the flyout to keep sizes of all arrows equal.
- After setting desired parameters, click on the **Vectors** toggle button from the **Results Arc**. The results will be displayed in graphics area; refer to Figure-19. Make sure to toggle off all the other result toggle buttons in the **Results Arc**.

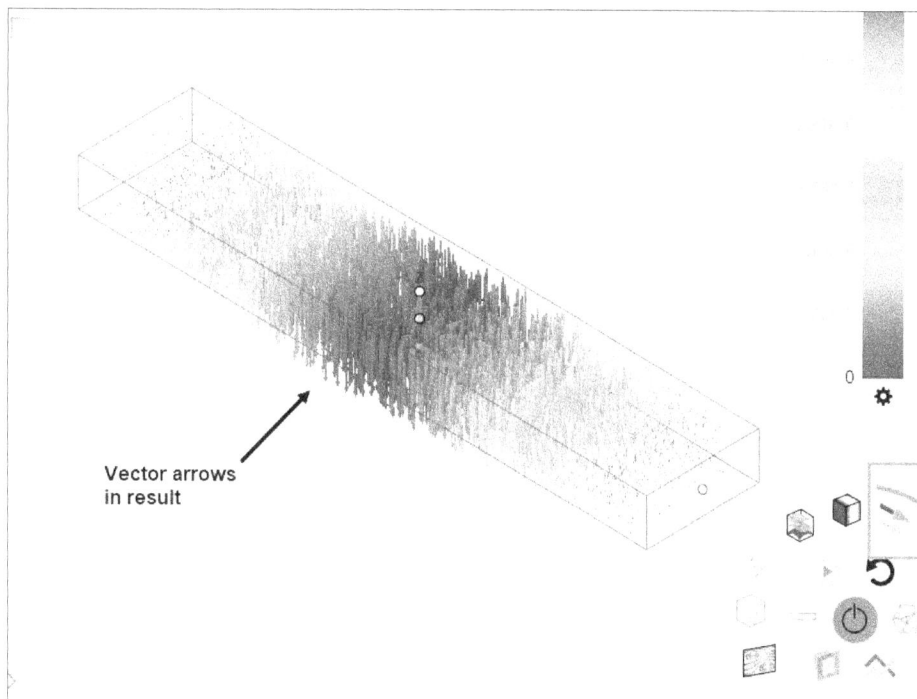

Figure-19. Vector arrows in results

Contours

Contours are used to represent results of analysis by painting faces of model in colors depending on result parameters.

- Hover the cursor on **Contours** toggle button to define parameters for contours result. The **Contours** flyout will be displayed; refer to Figure-20.

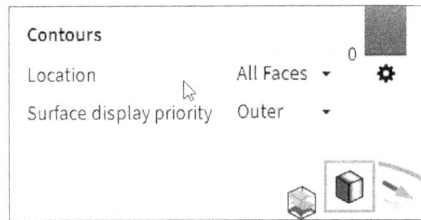
Figure-20. Contours flyout

- Select desired option from the **Location** drop-down in the flyout to define location where contours will be displayed for results. Select the **All Bodies** option from the **Location** drop-down if you want to display results over full body while making solid boundaries transparent. If the **All Faces** option is selected in the **Location** drop-down then you can select desired option from the **Surface display priority** drop-down. Click on the toggle button to apply result. The effects of various options on same result variable are shown in Figure-21.

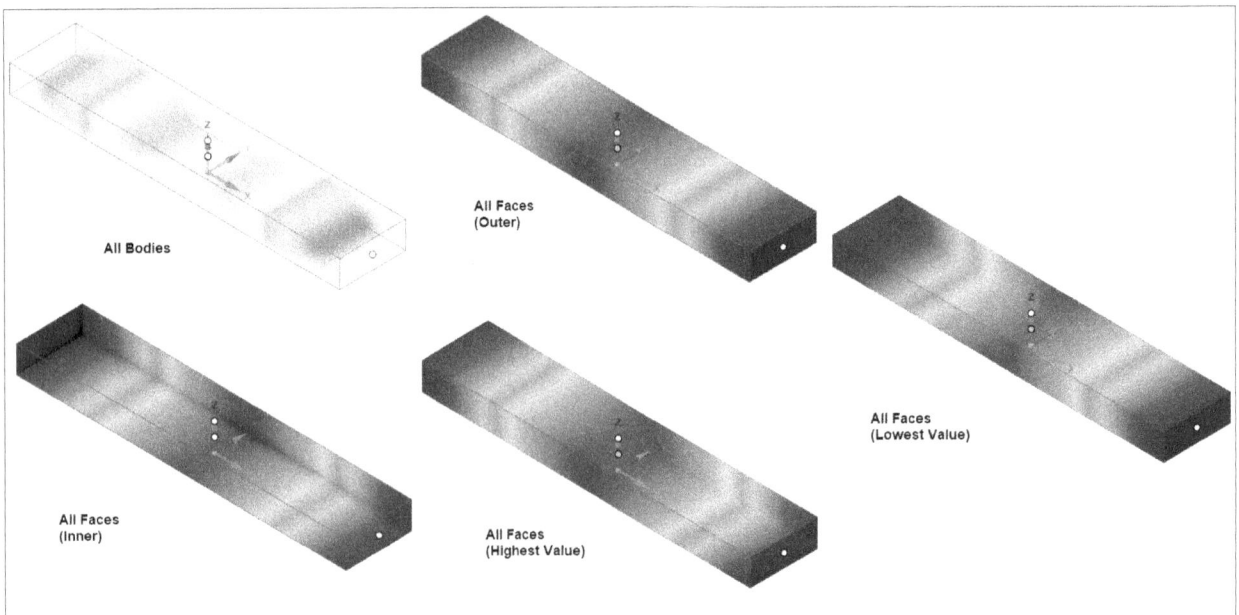
Figure-21. Contours result types

Iso-Surface

The **Iso-Surfaces** are used to represent result parameters as surfaces. Every surface layer depicts same result parameter.

- Hover the cursor on the **Iso-Surface** toggle button from the **Results Arc** and set desired result value to be displayed on the model by an iso surface.

Figure-22. Iso-Surface options

- You can also change the value by using slider displayed on clicking down button. Select the **Iso-surface** toggle button from the **Results Arc** to check the results; refer to Figure-23.

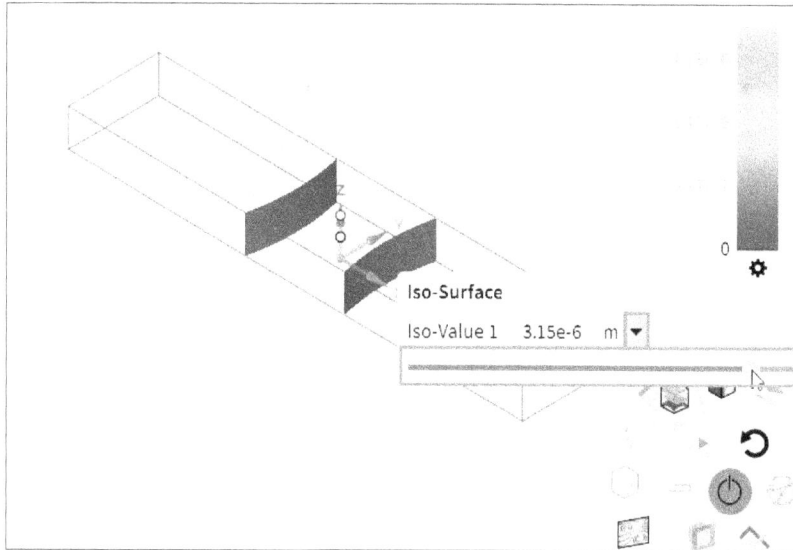

Figure-23. Iso-surface result

Streamlines

Streamlines are used to assess flow of fluids in results after performing CFD (Computational Fluid Dynamics) analysis. You will learn about performing CFD analysis later. The procedure to use streamlines is given next.

- Click on the **Streamlines** toggle button from the **Results Arc**. The result in graphics area will be displayed as shown in Figure-24.

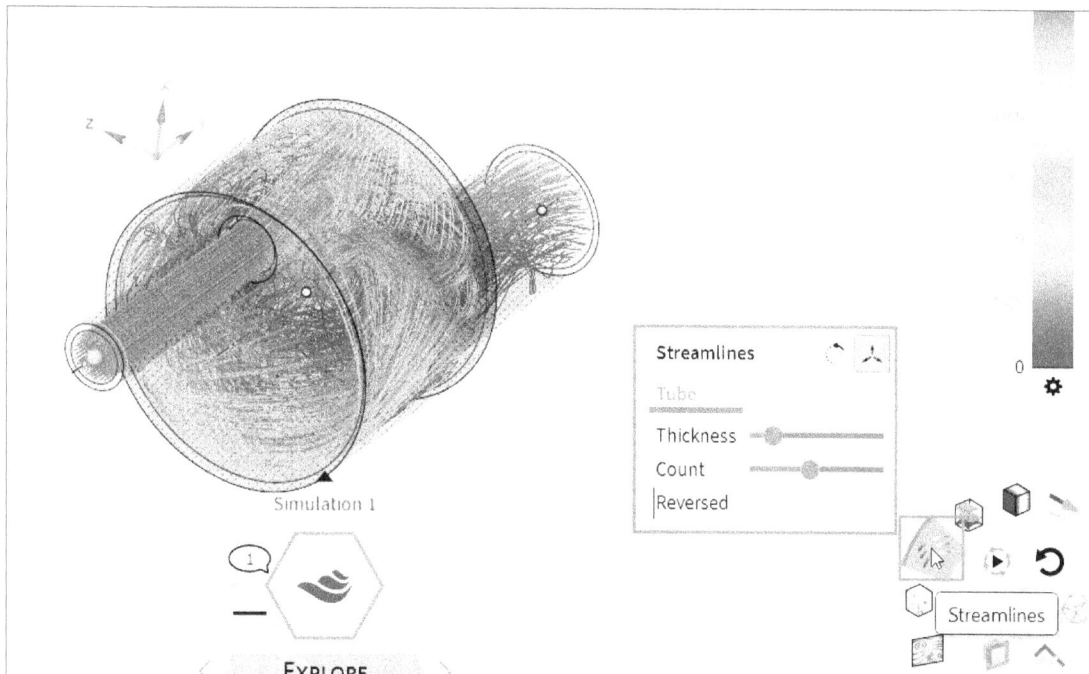

Figure-24. Streamlines in results

- Move the **Thickness** and **Count** sliders to increase/decrease the size and numbers of streamline.
- Select the **Reversed** toggle option to flip the direction of streamlines flow.

Particles

The **Particles** result is used to check flow of fluids after performing CFD analysis. The procedure to use this tool is given next.

- Select the **Particles** toggle button from the **Results Arc**. The results will be displayed in the graphics area; refer to Figure-25.

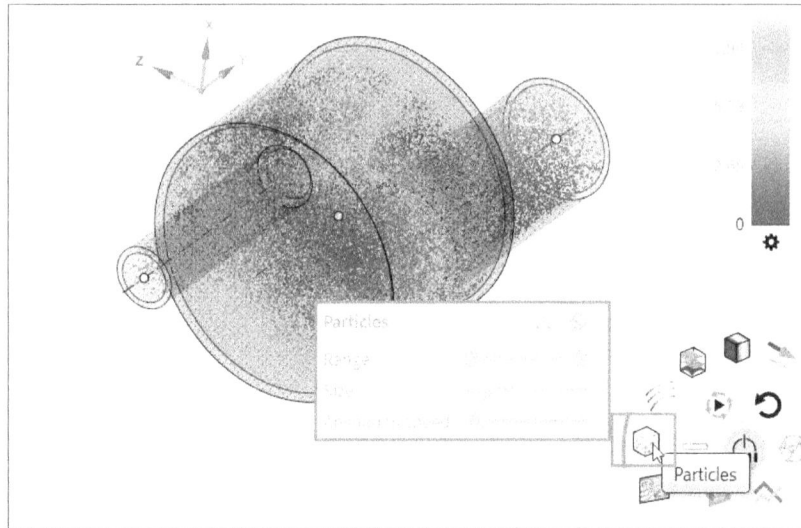

Figure-25. Particles options

- Move the **Range** sliders to define upper and lower limits of parameter for which particles will be displayed in the graphics area.
- Use the **Size** slider to increase/decrease the diameter of particles.
- Use the **Animation speed** slider to define how fast the particles will move from start point to end point in the model.

Direction field

The **Direction field** result is used to check the flow of fluid at center plane of the model. The procedure to use this tool is given next.

- Select the **Direction field** toggle button from the **Results Arc**. The results will be displayed in the graphics area; refer to Figure-26.

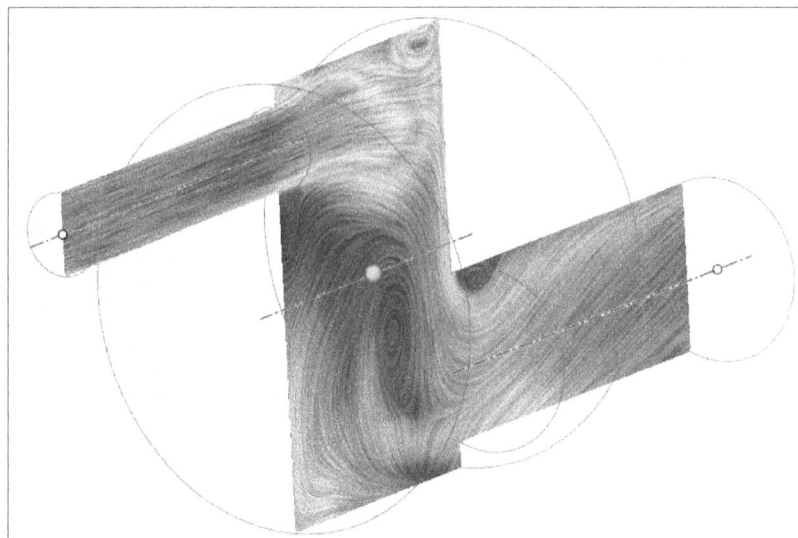

Figure-26. Direction field result

Checking Deformation

The **Deformation** result is used to check the change in geometry in enlarged/ diminished scale depending on displacement value. The procedure to generate the result is given next.

- Select the **Deformation** toggle button from the **Results Arc**. The deformed shape of object will be displayed in analysis result; refer to Figure-27.

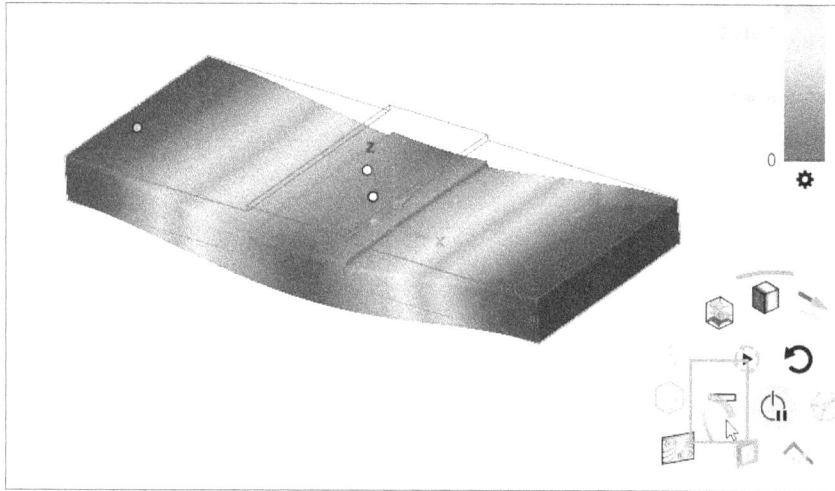

Figure-27. Deformation result

Checking Animation of Result

The **Show Animation** toggle button in **Results Arc** is used to check animation of deformation result in the graphics area.

Cut Plane

The **Cut Plane** toggle button in **Results Arc** is used to check analysis result at center plane of the model; refer to Figure-28.

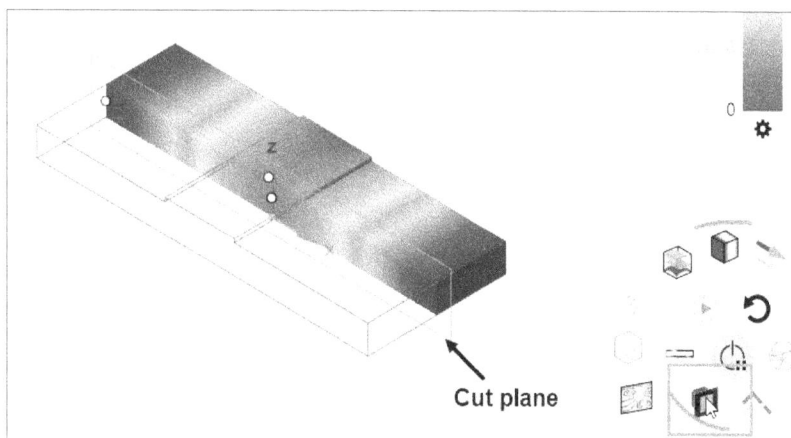

Cut plane

Figure-28. Cut plane results

Showing Charts

Select the **Show chart** toggle button from the **Results Arc** to check various result parameters of the analysis in the form of charts.

Reset

Select the **Reset** button from the **Results Arc** to reset results displayed in graphics area. Note that only visualization of results is reset not the parametric result generated by analysis.

Adding Optimized Body to Model

The **Add optimized body to mold** tool from the **Results Arc** is used to create facet body generated by results of **Topology Optimization** study; refer to Figure-29. Note that this tool will be available in the **Results Arc** only after performing **Topology Optimization** study.

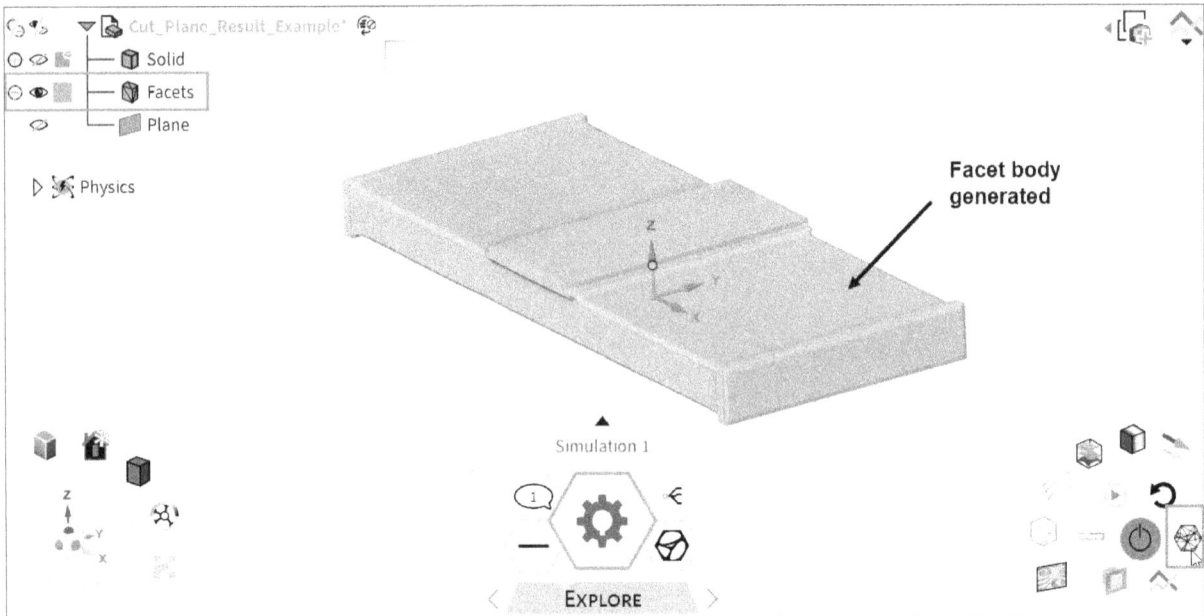

Figure-29. Adding optimized body to mold

LINEAR STATIC ANALYSIS ON ASSEMBLY

As discussed earlier in this chapter, linear static analysis is performed when there is a linear relationship between stress and strain. You previously worked on a simple single-body model for this analysis. Now, you will work with an assembly of multiple components connected by various contacts and connections, and perform linear static analysis.

Problem: Check the factor of safety for the model shown in Figure-30. Also, identify which part fails first under specified load conditions. Consider changing the design of part to achieve same better factor of safety.

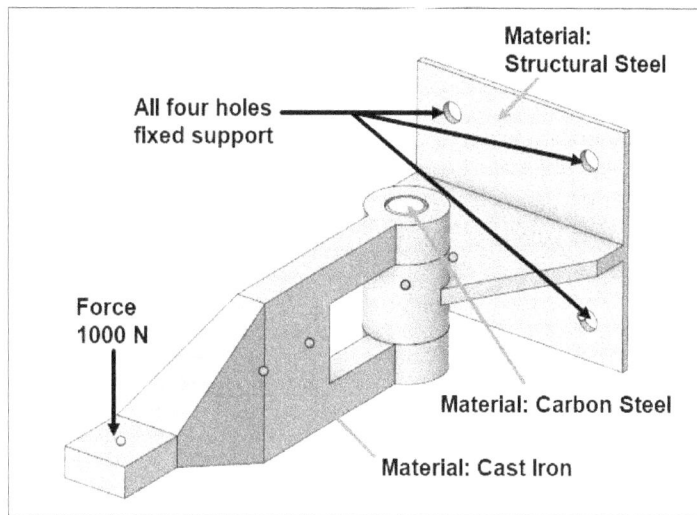

Figure-30. Linear Static Structural Analysis assembly example

Steps involved in performing linear static analysis on assembly in Ansys Discovery:

- Creating or importing geometry/model required to perform analysis.
- Applying Material Properties
- Applying Boundary Conditions like constraints and loads
- Managing Contacts and Connections
- Solving the Simulation
- Analyzing the Results

Importing Assembly in Ansys

- Start Ansys Discovery and start a new document in the application as discussed earlier.
- Click on the **Insert Geometry** tool from the **File Menu**. The **Insert Geometry** dialog box will be displayed and you will be asked to select the model to be used for analysis.
- Select the assembly file provided for this example in resource folder of the book and click on the **Open** button; refer to Figure-31. The model will be displayed in the graphics area.

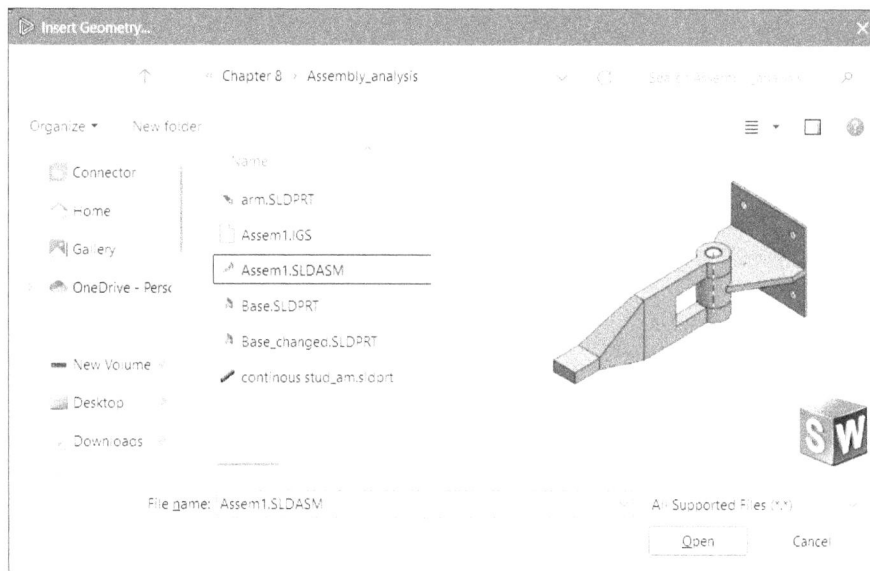

Figure-31. Insert Geometry dialog box

- Select the rotation handle about X axis of model and rotate it by 90 degree so that the face on which we will apply the force later is aligned with Z axis of the analysis environment.

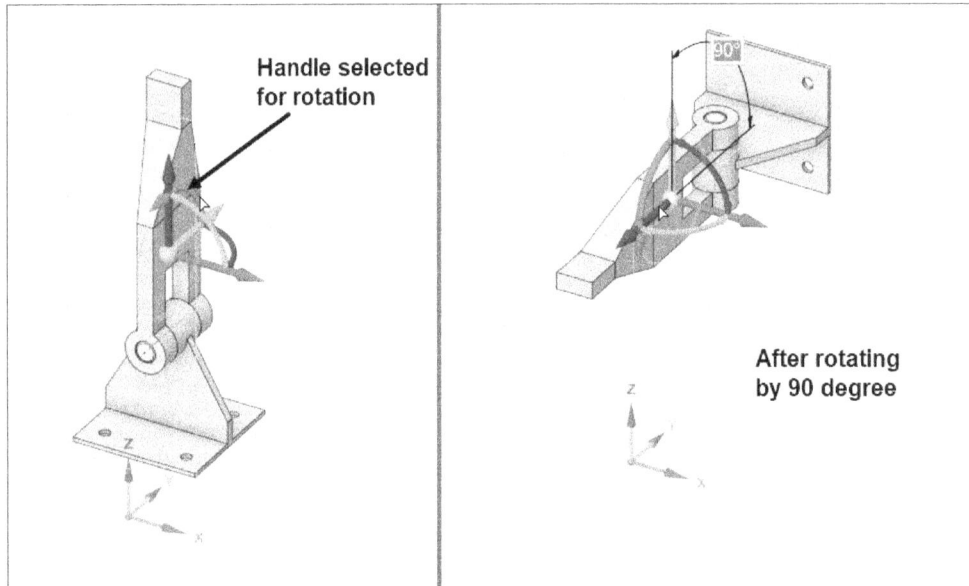

Figure-32. Rotating model for alignment

- After rotating the model, click in empty area and press **ESC** to exit the tool.

Applying Material

- Select the **SIMULATION** tab in the **Ribbon** and switch to **EXPLORE** stage from the bottom in the application area.
- Right-click on the material from the **Physics Tree** and select the **Edit** option from the shortcut menu. The **MATERIAL HUD Toolbar** will be displayed with drop-down to select the material.

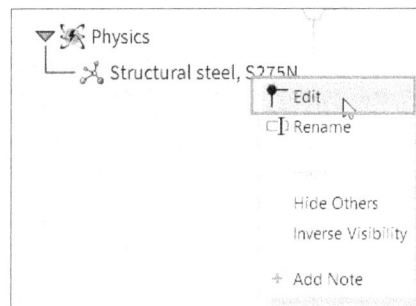

Figure-33. Edit material option

- Select the Base plate from the model and select the **Structural steel, S275N** option from the **Material** drop-down, if not selected; refer to Figure-34.

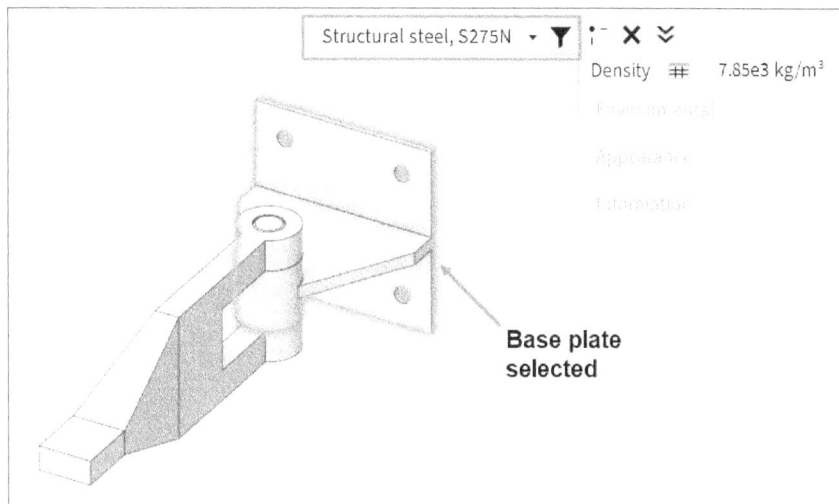

Figure-34. Material applied to Base plate

- Select the Pin from model and apply **Carbon Steel, 1020, annealed** material from the **Material** drop-down; refer to Figure-35.

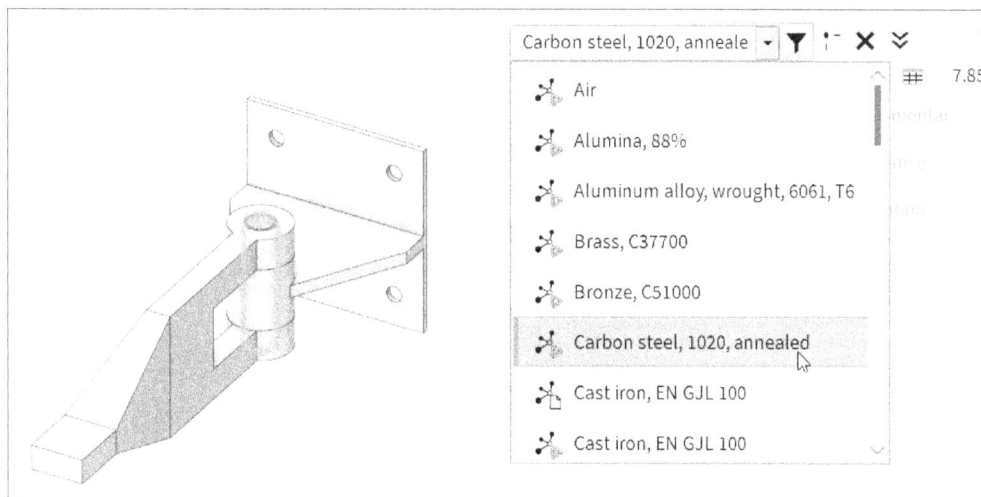

Figure-35. Applying Carbon steel to pin

- Similarly, apply **Cast iron, EN GJL 100** material to Hinge arm in the model.
- After applying the materials, press **ESC** to exit the tool.

Applying Constraints and Loads

- Click on the **Support** tool from the **Structural** drop-down in the **SIMULATION** tab of the **Ribbon**. The **STRUCTURAL SUPPORT HUD** toolbar will be displayed.
- Select the **Fixed** option from the drop-down in toolbar and select inner faces of holes in the base plate while holding the **CTRL** key; refer to Figure-36. Click on the **OK** button from the **HUD** toolbar to apply the constraint.

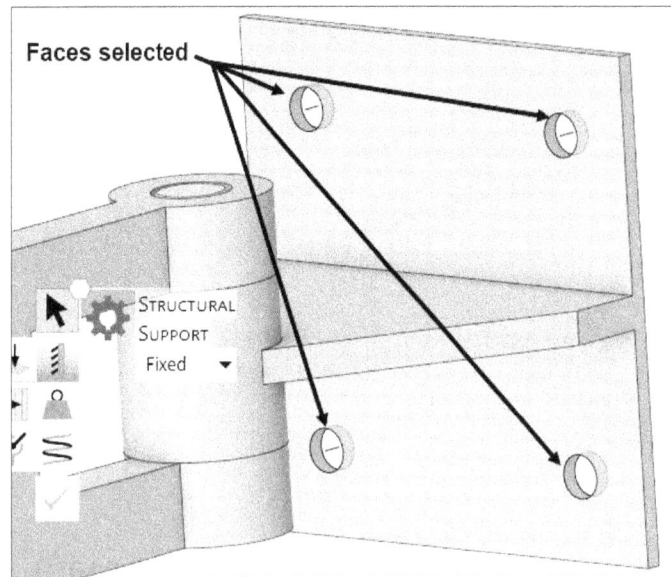
Figure-36. Faces selected for applying support

- Click on the **Force** toggle button from the **HUD** toolbar, select the **Distributed** option from the drop-down in toolbar, and select the face of model as shown in Figure-37. The input box to specify force value will be displayed.

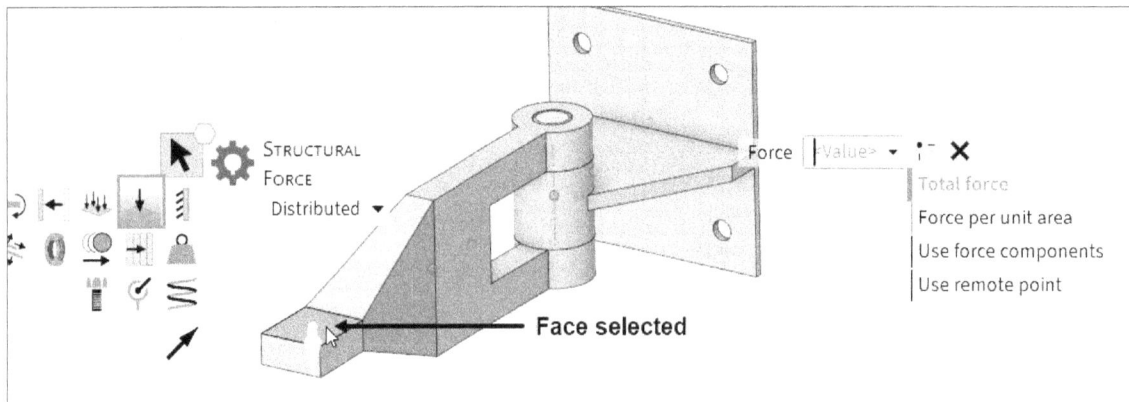
Figure-37. Face selected for force

- Type **1000 N** in the input box and press **ENTER** to apply the force. Press **ESC** twice to exit the tool.

Note: The yellow arrow shows the direction of load application. If it has been upward instead of downward, you would have need to specify -1000 N in the input box to apply a downward load.

Analyzing Contacts and Connections

By default, Bonded contacts are applied to all the boundaries between two components of the assembly. Selection of the contacts are situational and should represent your assumptions and real-world conditions. In this case of model, if you want to check the effect of force at static condition and you are assuming that force will act only along Z axis then default bonded condition is sufficient for our analysis. This is because we are not interested in large deformation or movements of assembly components with respect to each other. Also, applying bonded connection ensures that maximum reaction force is generated in the arm without it being sliding or rotating before maximum loading condition is achieved. Now, consider a different scenario where we need to find out the degree of angle by which arm will rotate under specified loads and we are performing non-linear analysis then it is important to apply hinge relation between pin, base plate, and arm at their connection faces. Also, the flat faces of arm resting on hub feature of base plate should be allowed to slide for representing real world conditions; refer to Figure-38.

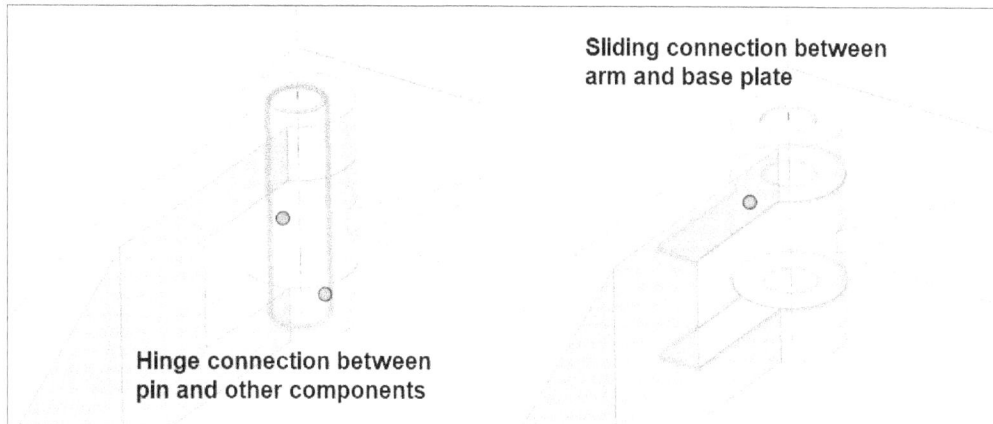

Figure-38. Connections to represent real conditions

Tip: By default, Bonded contacts is applied between all the components. To change contact between any two components, right-click on the **Bonded Contacts (default)** option from the **Physics Tree** and select the **Edit** option. The **CONTACT REVIEW HUD** toolbar will be displayed with all the contacts highlighted in graphics area. Click on the small bubble on contact that you want to change; refer to Figure-39 and then select desired button from right **HUD** toolbar to change the contact.

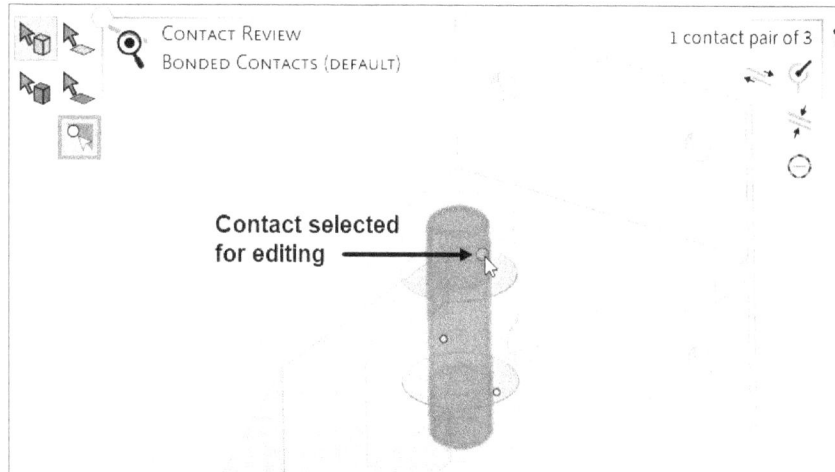

Figure-39. Selecting contact for editing

Solving and Analyzing Results

- Click on the **Solve** button from center of **Results Arc** at the bottom right corner of application window. System will start solving the analysis and once process is complete, the results will be displayed; refer to Figure-40.

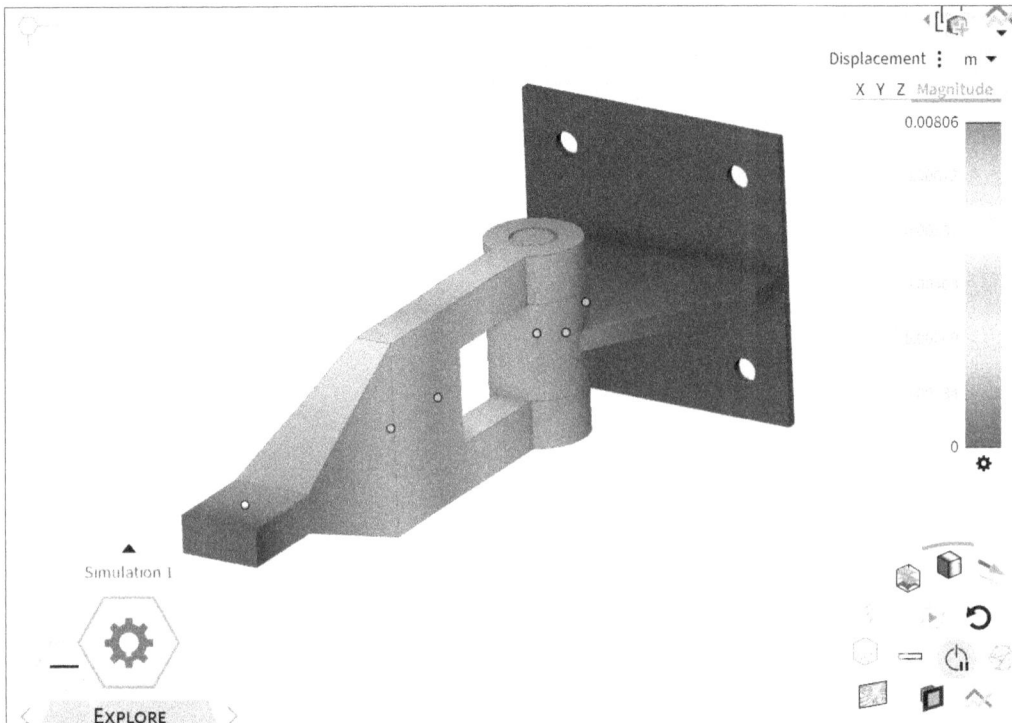

Figure-40. Result of analysis

• Expand the **MONITORS** drop-down and select the **Factor of Safety** option from the drop-down. The **Factor of Safety** (**FoS**) parameter will be displayed in the form of a chart; refer to Figure-41. As you can see from the results, the FoS of model is 0.8 which is not safe. There are various methods to increase Factor of Safety of this model which can include material change, structure change, design change, shifting load to different location by design, and so on. But before we move to these changes, we need to identify the locations in model that are causing failure. To do so, change the result parameter to Von Mises Stress to show check distribution of stress over model; refer to Figure-42.

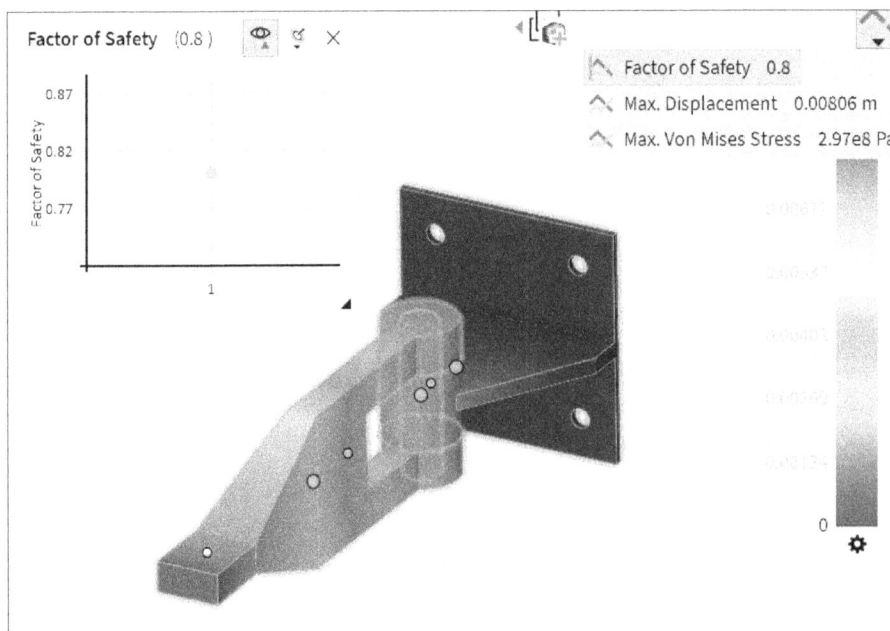

Figure-41. Factor of Safety for model

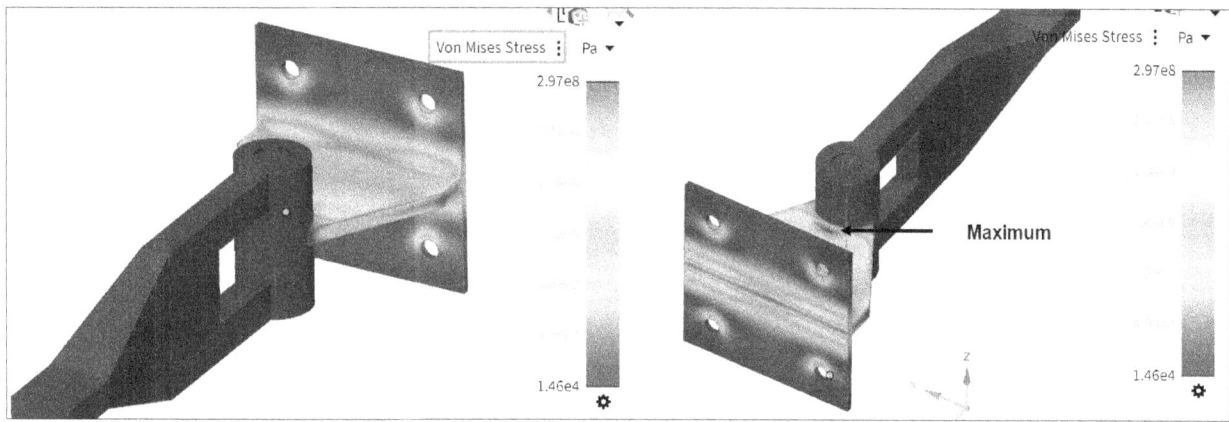

Figure-42. Von mises stress of model

As you can check in the above figure, the maximum stress occurs where cylindrical tube is attached to the thin base plate. You can also confirm this by showing min/ max locations in result as discussed earlier. If you are familiar with engineering methods of strengthening parts using design then you can tell that ribs are the perfect partner here to increase strength with minimum weight addition. The procedure to modify part is given next.

Strengthening Part using Design

- Click on the **Pause** button displayed instead of **Solve** button in the **Results Arc** to stop solving the analysis dynamically. We are doing this to solving taking processing of computer while we are still editing the model. Once the model is changed as desired then we can start solving the analysis again.
- Expand the **Model Tree** and right-click on **Base** component from the tree. The shortcut menu will be displayed as shown in Figure-43.

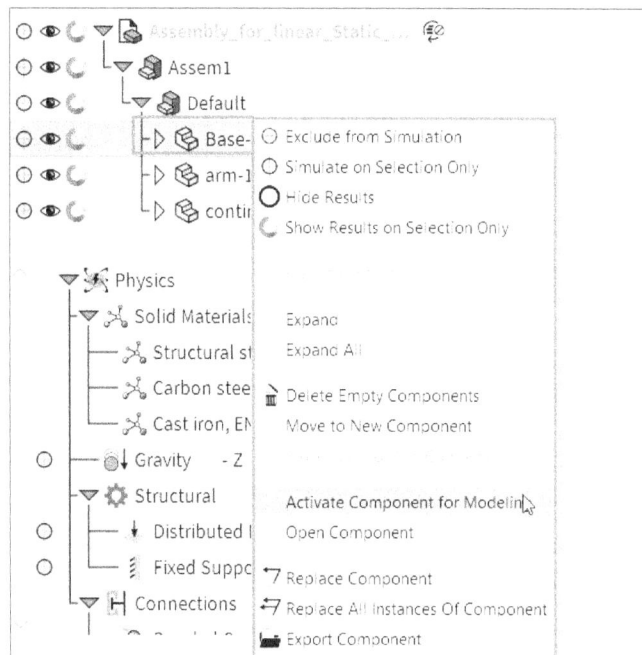

Figure-43. Right-click shortcut menu

- Select the **Activate Component for Modeling** option from the shortcut menu. Now, all the modeling changes will be applied to Base component.
- Toggle off the **Contour** result from the **Results Arc** at bottom right corner of application window to view the model clearly.

- Click on the **PLANE** tool from the **CREATE** panel in the **DESIGN** tab of the **Ribbon** to start creating a vertical plane at center of the model. We will be using this plane to create sketch for rib feature. After activating this tool, select the side face of model as shown in Figure-44 to define placement location for new plane and then press **ESC** to exit the tool.

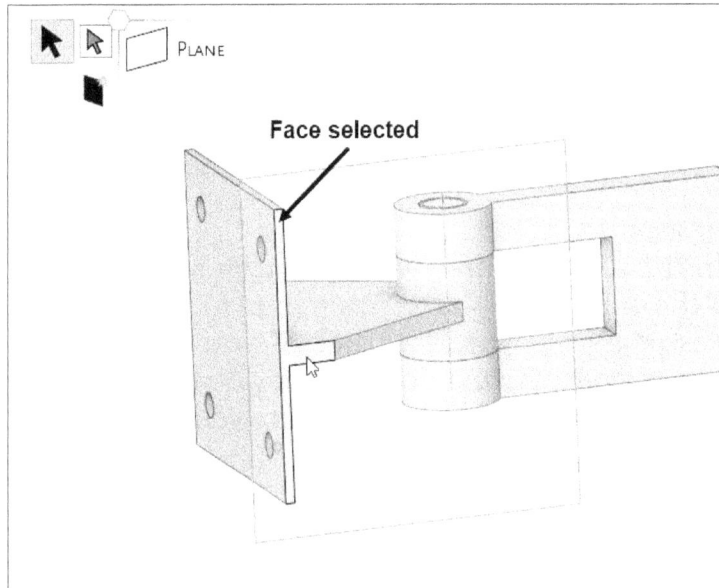

Figure-44. Selecting face for plane

- Click on the **Measure** tool from the **INSPECT** panel in the **MEASURE** tab of the **Ribbon** and measure total width of base plate so that we can find out distance by which new plane should move to be at center of the base; refer to Figure-45. In this case, the total length of edge is **150 mm** so we will move the newly created plane by **75 mm** to put it at center. Press **ESC** to exit the tool.

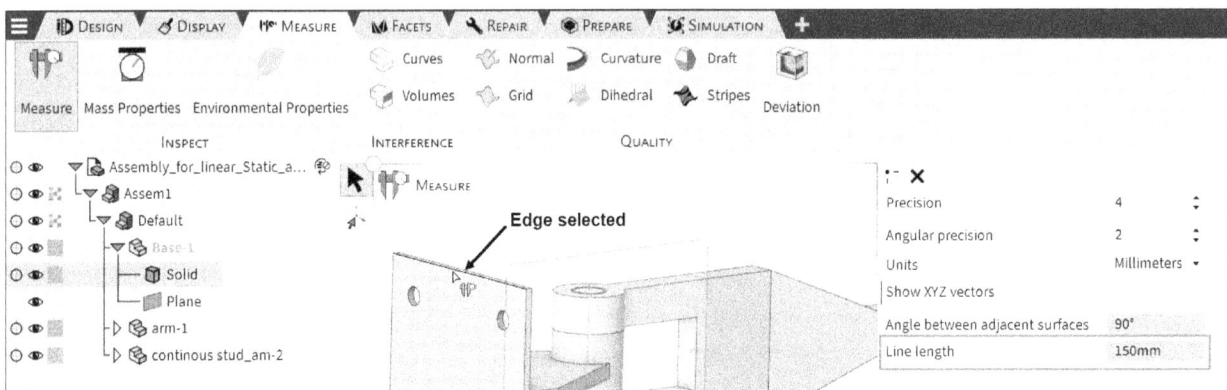

Figure-45. Measuring length of edge

- Click on the **Move** tool from the **EDIT** panel in the **DESIGN** tab of the **Ribbon**. The **MOVE HUD** toolbar will be displayed and you will be asked to select object to be moved.
- Select the plane earlier created and move it by **75 mm** using the blue handle; refer to Figure-46.

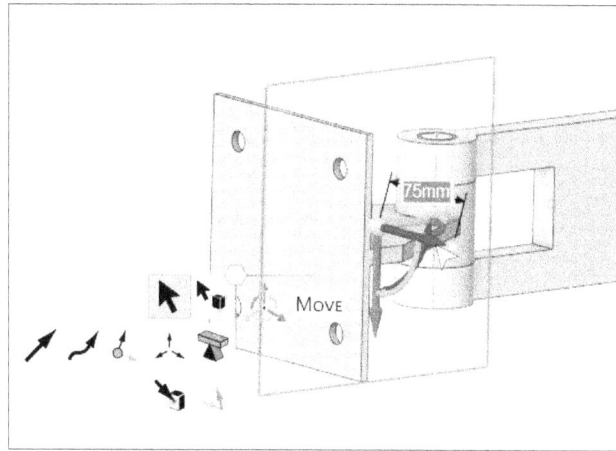

Figure-46. Moving plane to center

- After setting parameters, click in empty space. The plane will move to specified distance. Press **ESC** to exit the tool.
- Select the plane and click on the **Sketch** tool from the **MODE** panel in the **DESIGN** tab of the **Ribbon**. The model will be displayed in sketching environment; refer to Figure-47.

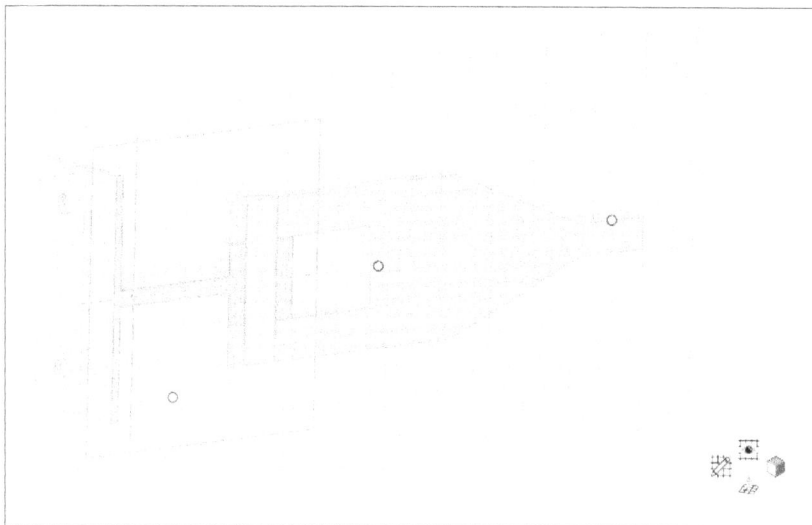

Figure-47. Sketching environment

- Press **v** key from keyboard to make the sketching plane parallel to screen.
- Click on the **Line** tool from the **SKETCH** panel in the **DESIGN** tab of the **Ribbon** and create the closed loops in sketch as shown in Figure-48.

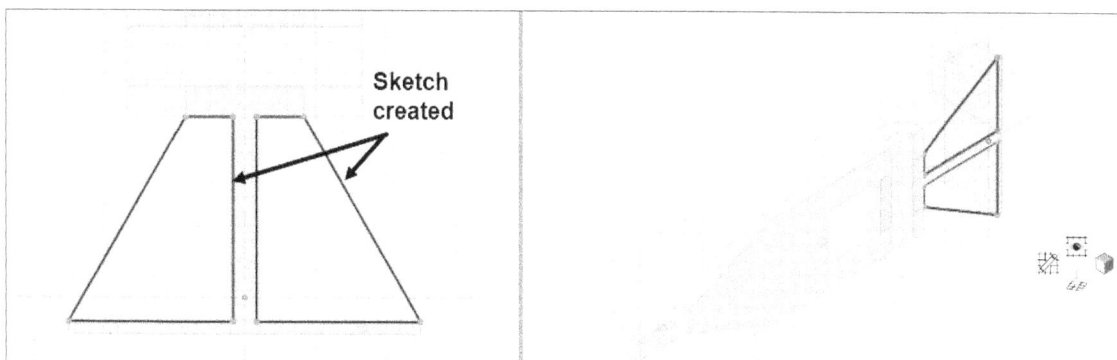

Figure-48. Sketch created

- Click on the **Return to 3D** tool from the **HUD** toolbar. The **PULL HUD** toolbar will be displayed.
- Select the upper sketch section and select the **Pull both sides** toggle button from the **HUD** toolbar to create extrude feature on both sides; refer to Figure-49.

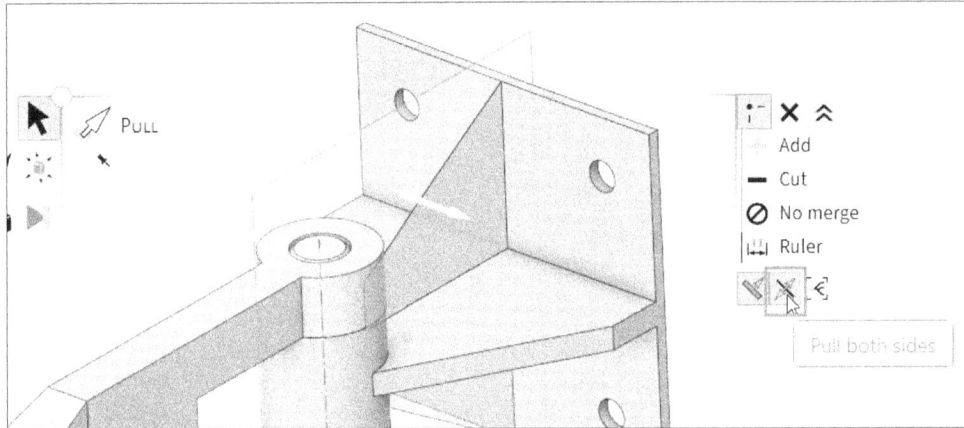

Figure-49. Pull both sides toggle button

- Drag the arrow and enter the thickness as **10** in the dynamic input box. The upper rib feature will be created.
- Similarly, create rib from the bottom section of the sketch. Press **ESC** to exit the tool after creating rib. The model should display as shown in Figure-50.

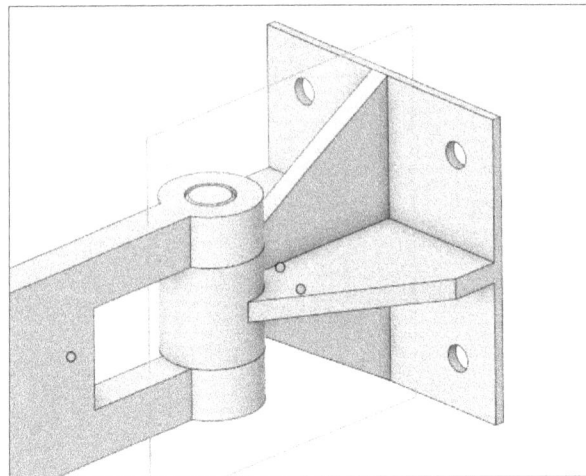

Figure-50. Model after creating rib

- Now, click on the **Solve** tool from the **Results Arc** and check the factor of safety in results; refer to Figure-51. As you can see in results, previous factor of safety was **0.8** for the model but after adding ribs, the factor of safety is **1.51** which is almost doubled.

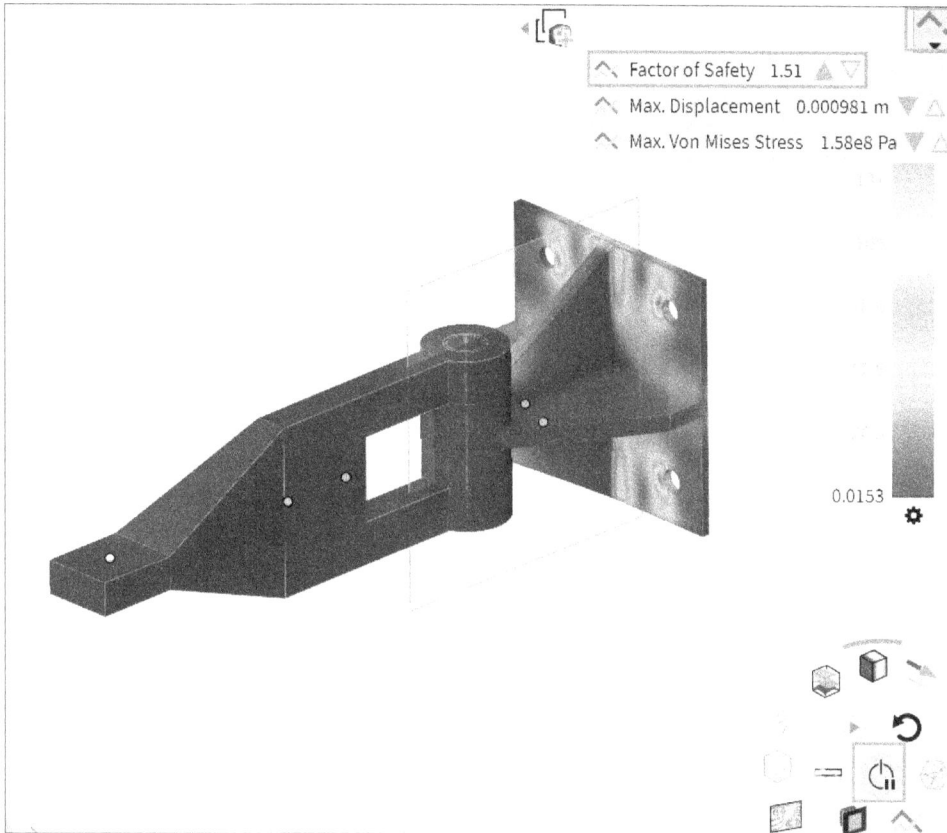

Figure-51. Factor of safety for modified model

- Now, rotate the model and you will find that highest stress is at the holes used for fastening base to wall which is to be expected. You can further increase the strength of model by increasing thickness of base plate by **10 mm** from the side which will be attached to wall so that thickness of holes will increase. This will further raise the factor of safety; refer to Figure-52. You can further find the areas which are of highest stress and increase their strength.

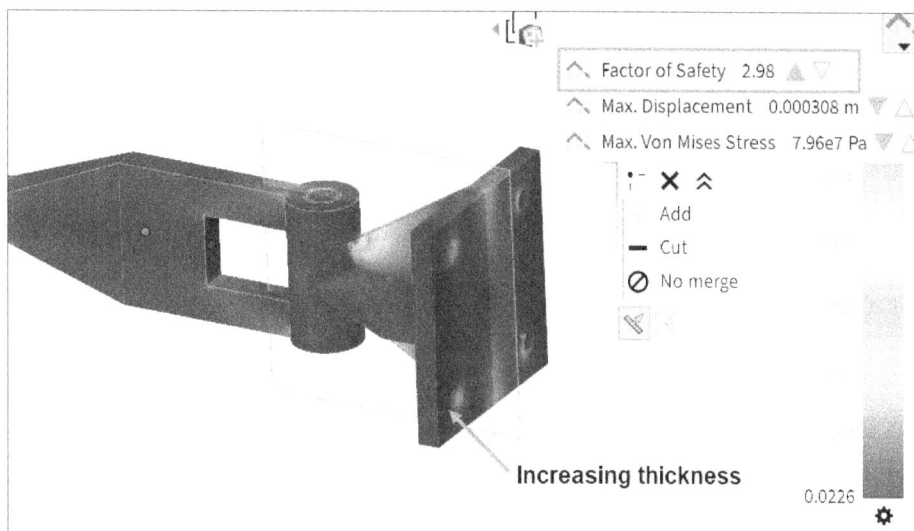

Figure-52. Factor of safety after increasing thickness

NON LINEAR STATIC ANALYSIS

All real structures behave nonlinearly in one way or another at some level of loading. In some cases, linear analysis may be adequate. In many other cases, the linear solution can produce erroneous results because the assumptions upon which it is based are violated. Nonlinearity can be caused by the material behavior, large displacements, and contact conditions.

You can use a nonlinear study to solve a linear problem. The results can be slightly different due to different assumption.

In nonlinear finite element analysis, a major source of nonlinearity is due to the effect of large displacements on the overall geometric configuration of structures. Structures undergoing large displacements can have significant changes in their geometry due to load-induced deformations which can cause the structure to respond nonlinearly in a stiffening and/or a softening manner.

For example, cable-like structures generally display a stiffening behavior on increasing the applied loads while arches may first experience softening followed by stiffening, a behavior widely-known as the snap-through buckling.

Another important source of nonlinearity stems from the nonlinear relationship between the stress and strain which has been recognized in several structural behaviors. Several factors can cause the material behavior to be nonlinear. The dependency of the material stress-strain relation on the load history (as in plasticity problems), load duration (as in creep analysis), and temperature (as in thermoplasticity) are some of these factors.

This class of nonlinearity, known as material nonlinearity, can be idealized to simulate such effects which are pertinent to different applications through the use of constitutive relations.

A special class of nonlinear problems is concerned with the changing nature of the boundary conditions of the structures involved in the analysis during motion. This situation is encountered in the analysis of contact problems.

Pounding of structures, gear-tooth contacts, fitting problems, threaded connections, and impact bodies are several examples requiring the evaluation of the contact boundaries. The evaluation of contact boundaries (nodes, lines, or surfaces) can be achieved by using gap (contact) elements between nodes on the adjacent boundaries.

Yielding of beam-column connections during earthquakes is one of the applications in which material nonlinearity are plausible.

THEORY BEHIND NON-LINEAR ANALYSIS

As you have learnt from introduction, there are three factors which cause non-linearity in the body which are elastic behavior of material, large displacement in body, and variation in contact. In terms of equation, this can be expressed as

$$[K(D)]\{D\} = \{F\}$$

Here, K(D) is stiffness matrix which is function of displacement.

     D is the displacement matrix
and   F is the Force applied

So, from the above equation you can easily understand that the stiffness is changing according to the displacement. Note that displacement is vector and it also has a direction. If we compare the linear static analysis and non-linear static analysis curves then we can find out the different effects of load in both conditions; refer to Figure-53.

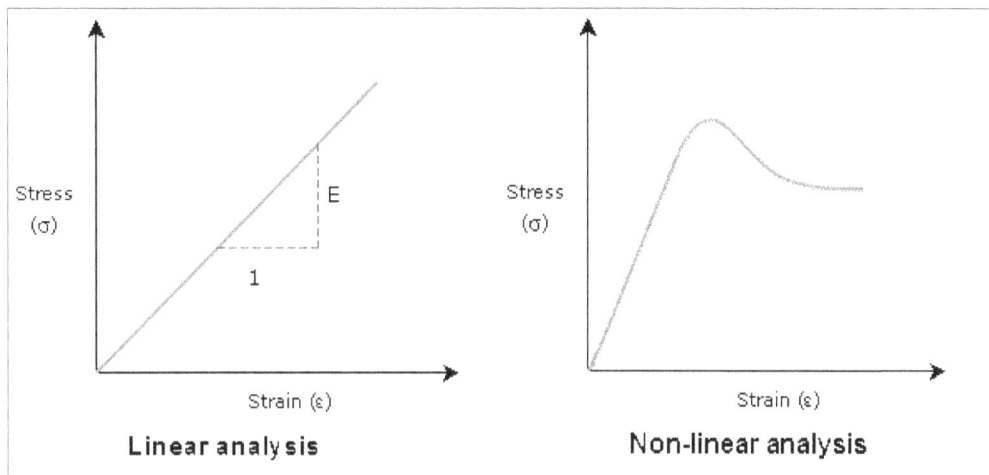

Figure-53. Stress Strain diagrams

In terms of load and displacement, the curve for both analysis can be given by Figure-54.

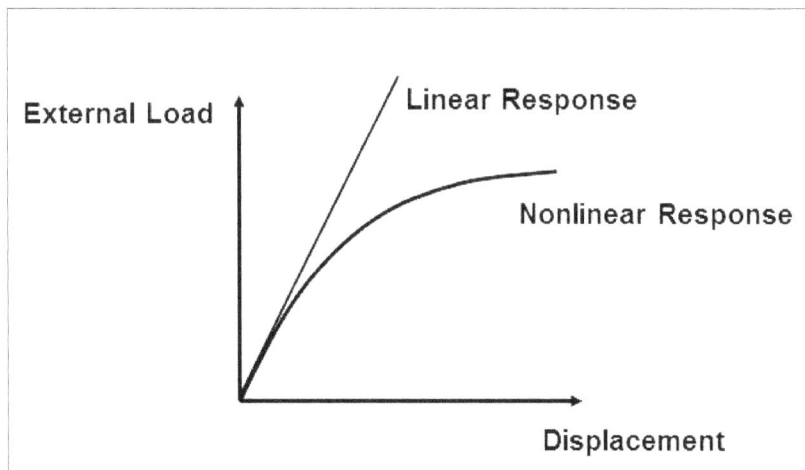

Figure-54. Load displacement curve

The normal incremental iteration does not work good for the non-linear analysis and generate errors as shown in Figure-55. So, Newton-Raphson algorithm is used to solve the non-linear equation.

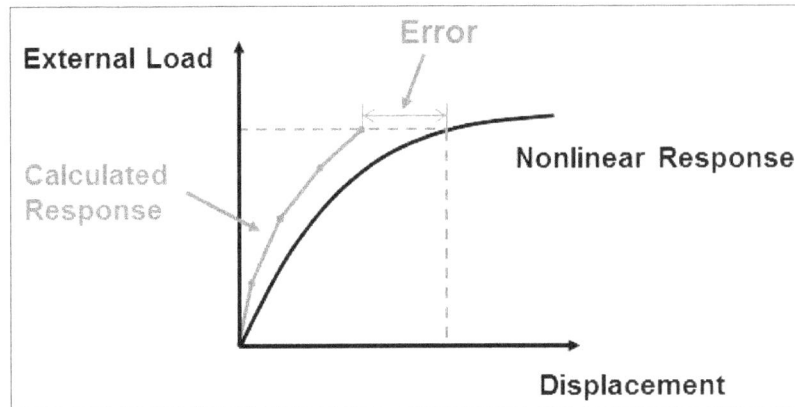

Figure-55. Error in incremental interative method

The equation for Newton-Raphson method is given as:

$$[K_T]\{\Delta u\} = \{F\} - \{F^{nr}\}$$

$[K_T]$ = tangent stiffness matrix
$\{\Delta u\}$ = Displacement increment
$\{F\}$ = external load vector
$\{F^{nr}\}$ = internal force vector

The iteration continues till $\{F\}$ - $\{F^{nr}\}$ (difference between external and internal loads) is within a tolerance; refer to Figure-56.

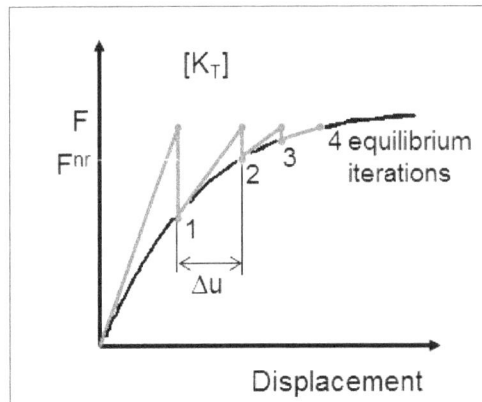

Figure-56. Newton-Raphson method

Thus, a nonlinear solution typically involves the following:

*   One or more load steps to apply the external loads and boundary conditions.(This is true for linear analyses too.)
*   Multiple sub-steps to apply the load gradually. Each sub-step represents one load increment.(A linear analysis needs just one sub-step per load step.)
*   Equilibrium iterations to obtain equilibrium (or convergence) at each sub-step. (Does not apply to linear analyses.)

Role of Time in Non-Linear Analysis

Each load step and sub-step is associated with a value of time. Time in most nonlinear static analyses is simply used as a counter and does not mean actual, chronological time. Figure-57 shows a load-time curve.

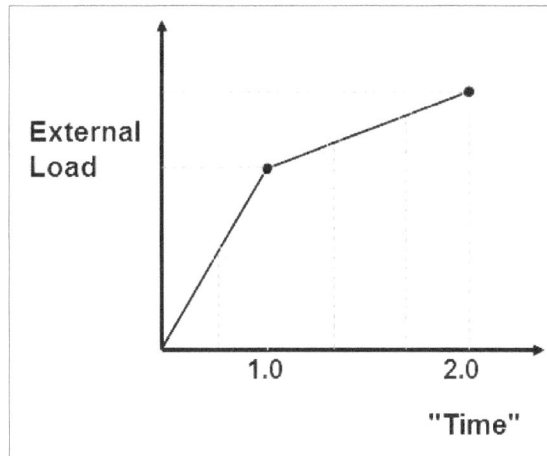

Figure-57. Load time curve

By default, time = 1.0 at the end of load step 1, 2.0 at the end of load step 2, and so on. For rate-independent analyses, you can set it to any desired value for convenience. For example, by setting time equal to the load magnitude, you can easily plot the load-deflection curve.

The "time increment" between each sub-step is the time step Δt. Time step Δt determines the load increment ΔF over a sub-step. The higher the value of Δt, the larger the ΔF, so Δt has a direct effect on the accuracy of the solution. SolidWorks Simulation gives the flexibility to use automatic time stepping algorithm and manual setting of time steps.

Now, we are going to start with Non-linear Static Analysis which means there is no damping or resistance to the force. Note that the last example in previous chapter, an assembly of wall bracket, was a problem of Nonlinear static analysis because there was large displacement in the body. Now, we will learn the procedures of Ansys Discovery to perform Nonlinear static analysis.

PERFORMING NON-LINEAR STATIC ANALYSIS

Problem: Find out the area of model which will be under highest stress due to applied loads and conditions in the model; refer to Figure-58.

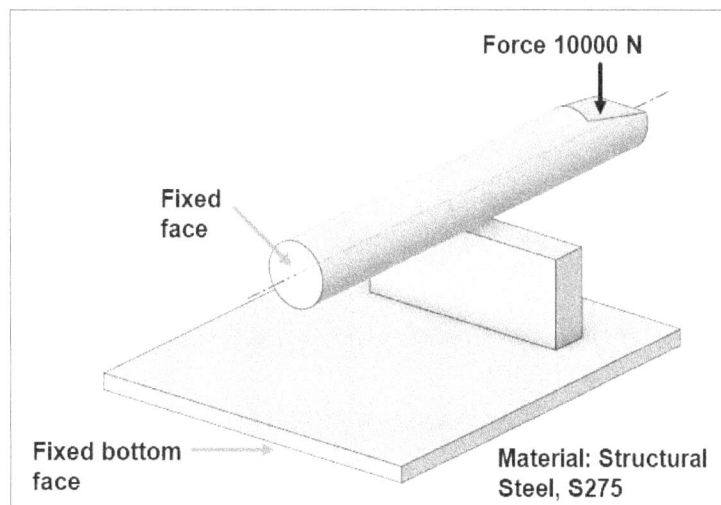

Figure-58. Model for non-linear static analysis

Steps involved in performing non-linear static analysis on assembly in Ansys Discovery:

- Creating or importing geometry/model required to perform analysis.
- Applying Material Properties
- Applying Boundary Conditions like constraints and loads
- Managing Contacts and Connections
- Setting Non-linear static analysis parameters
- Solving the Simulation
- Analyzing the Results

Creating/Importing Model

- Start Ansys Discovery if not started yet.
- Click on **Insert Geometry** tool from the **File** menu. The **Insert Geometry** dialog box will be displayed.
- Select the model for this example from the resource kit; refer to Figure-59 and click on the **Open** button. The model will be displayed in graphics area (you may need to zoom-out using scroll button in mouse).

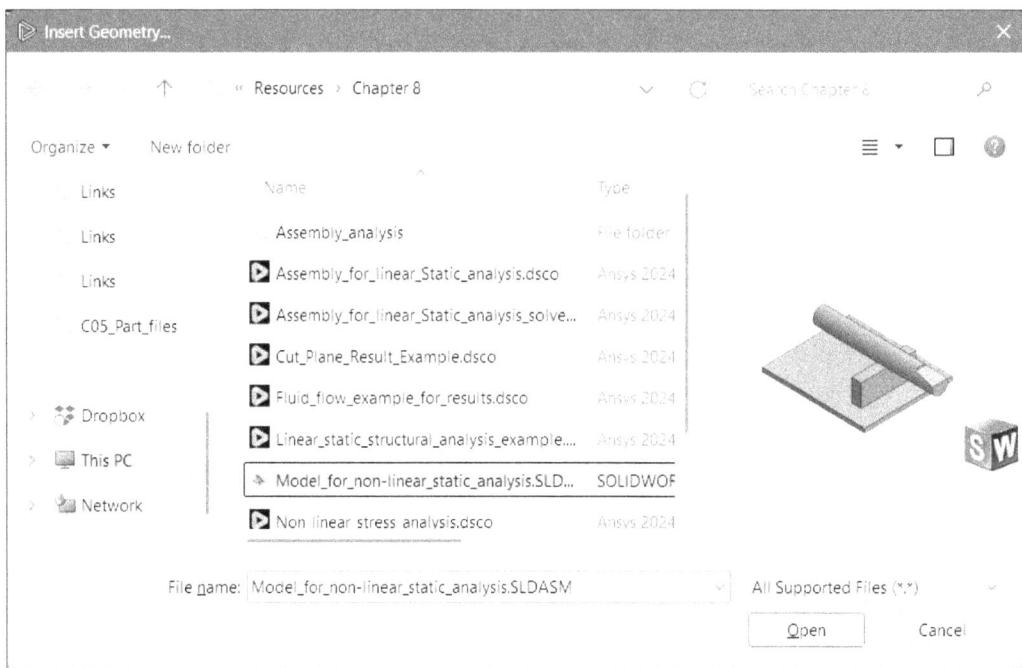

Figure-59. Opening model for non-linear static analysis

- Rotate the model using red rotation handle by 90 degree to align model with default coordinate system; refer to Figure-60. Click in the empty area and press **ESC** to exit the tool.

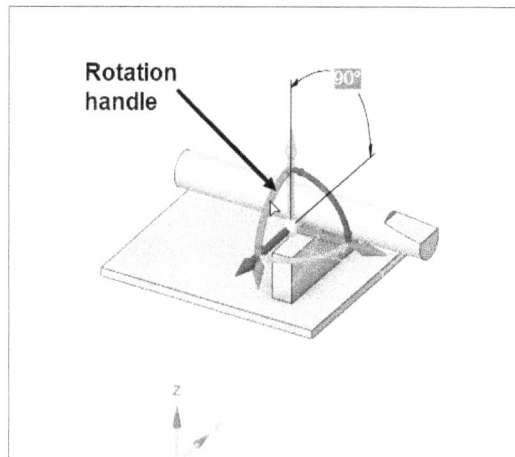

Figure-60. Rotating model

Applying Material

- Switch to **REFINE** stage from the **Stage Selector** at the bottom in the application window because options to define setup for nonlinear analysis are available in the **REFINE** stage of Ansys Discovery.
- Expand the **Physics Tree** to check properties of analysis. You will find that Structural steel, S275N is already applied to the model by default. If this material is not applied by default then apply it using the procedure discussed earlier.

Applying Constraints and Loads

- Click on the **Support** tool from the **Structural** drop-down in the **PHYSICS** panel of the **SIMULATION** tab in the **Ribbon**. The **STRUCTURAL SUPPORT HUD** toolbar will be displayed.
- Select the bottom face of base plate and back face of cylinder while holding the **CTRL** key to apply fixed support and press **ENTER**. The fixed support constraint will be applied.
- Click on the **Force** toggle button from the **HUD** toolbar and select the flat face of the cylinder as shown in Figure-61.

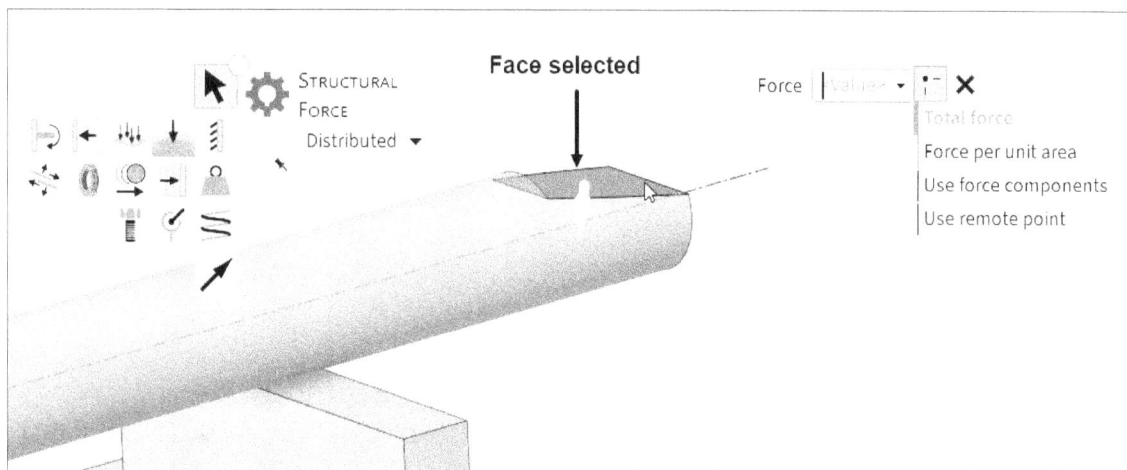

Figure-61. Selecting Face for force

- Type the value **10000 N** in the input box and press **ENTER**. The force will be applied. Press **ESC** to exit the tool.

Modifying Contacts

- In this model, there are three components: Base plate, a rectangular block, and a cylindrical shaft. The shaft will be bending under specified force and it will hit the rectangular block. By default, bonded contact is applied to all the components in the assembly when performing analysis in Ansys Discovery. But, if we keep default contact between rectangular block and cylindrical shaft then there will be no impact load which should happen in real world. To represent real loading conditions, we need to change the contact to sliding type.

- Double-click on **Bonded Contacts (default)** option from the **Physics Tree**. The **CONTACT REVIEW HUD** toolbar will be displayed with contacts highlighted in graphics area; refer to Figure-62.

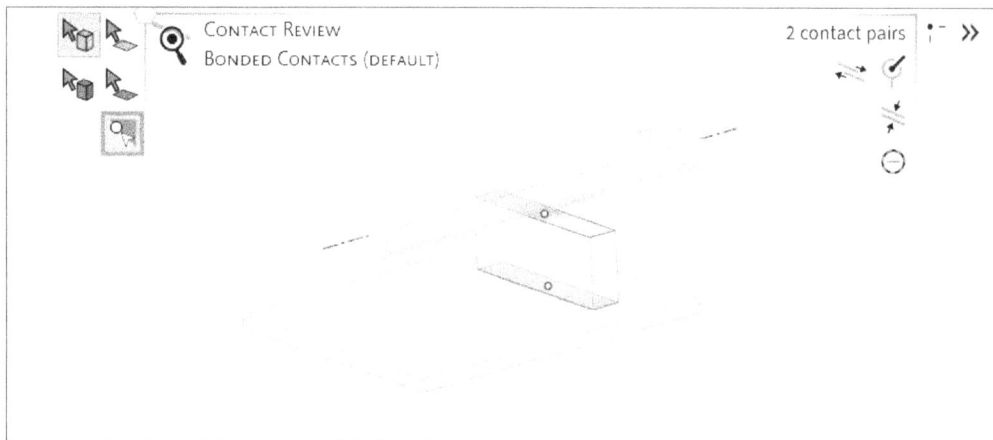

Figure-62. Contacts displayed

- Select the cylindrical shaft from the model to modify contacts for it and select **Convert to sliding** button from the right **HUD** toolbar; refer to Figure-63. The contact for cylindrical shaft will change to sliding.

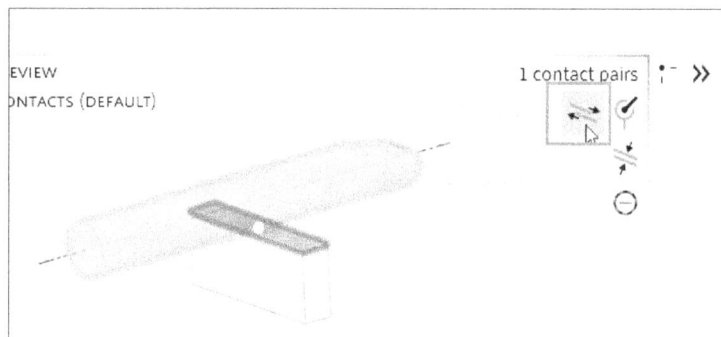

Figure-63. Converting contact to sliding

- Press **ESC** to exit the tool.

Defining Non-Linear Analysis Parameters

- Select the **Specify modeling method** toggle button from the **Simulation Options** drop-down in the **PHYSICS** panel of **SIMULATION** tab in the **Ribbon** and select the **Non-linear** option from the drop-down next to it; refer to Figure-64. This will allow large deformation in the cylindrical shaft which is important aspect of our analysis.

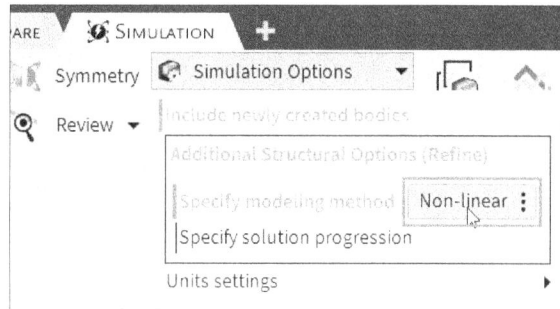

Figure-64. Non-linear option

Performing Analysis and Analyzing Results

- Click on the **Solve** button from the **Results Arc** or **Simulation** drop-down in **STUDY** panel of **SIMULATION** tab in the **Ribbon**. Once the analysis is complete, the results will be displayed in graphics area.
- Select the **Von Mises Stress** option from the **Variables** drop-down in the **Legends** area. The results will be displayed as shown in Figure-65.

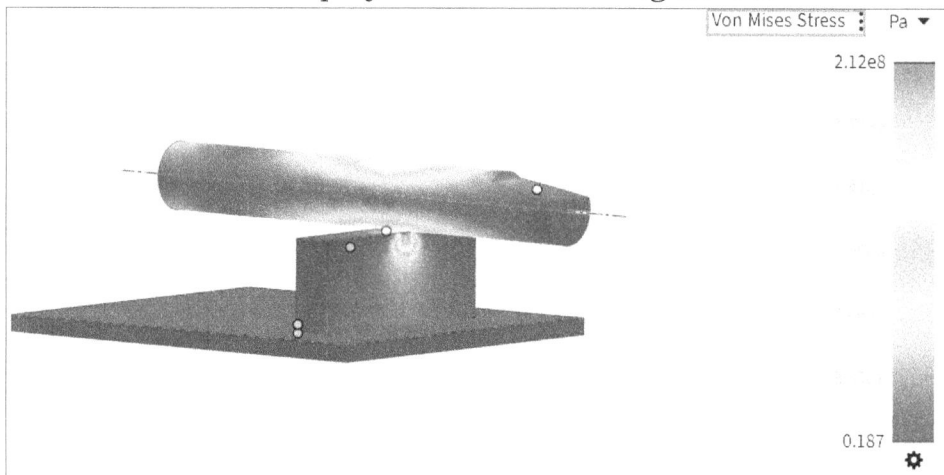

Figure-65. Stress results

- From the results, you can find that maximum stress occurs at the point where cylindrical shaft touches the rectangular block after bending. Check the factor of safety as discussed earlier and improve the design.

LOAD CASES/PARAMETER STUDY

The Load cases are used to check variations of boundary condition parameters for same analysis setup. For example, you have an analysis setup as shown in Figure-66 where you want to check effect 500 N force, 800 N force, and 1000 N force at same location in the same setup. In such cases, you can use load cases to analyze various results. In Ansys Discovery, you can use **Parameters** tool to perform load case/ parameter study. The procedure to perform parameter study is given next.

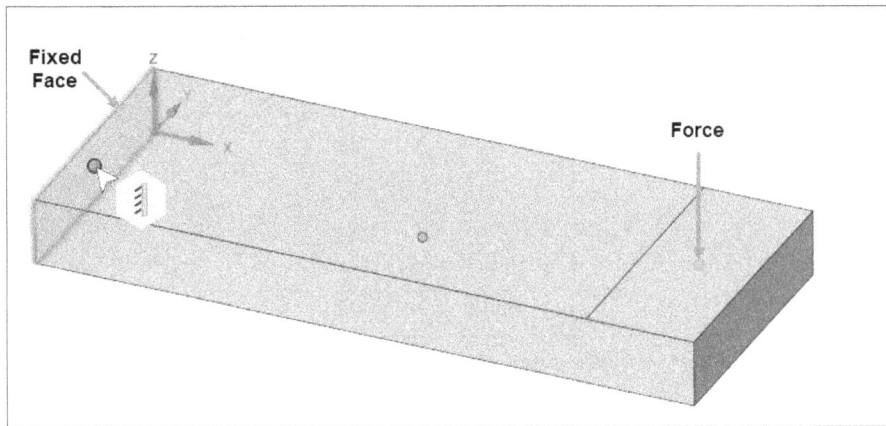

Figure-66. Setup for load case

- Create the support, force, and material setup for the model as shown in Figure-67.

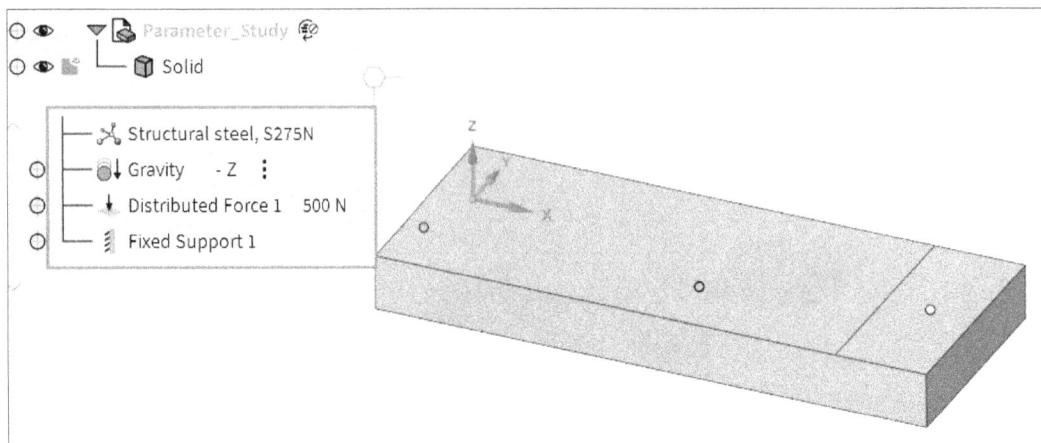

Figure-67. Setup parameters for load case

- After creating setup, click on the **Add as a parameter** button for **Distributed Force 1** from the **Physics Tree**; refer to Figure-68. The force value will be added as analysis parameter.

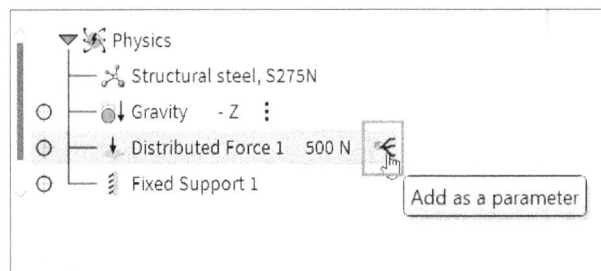

Figure-68. Adding force as parameter

- Click on the **Variations** tool from the **STUDY** panel in the **SIMULATION** tab of **Ribbon**. The **Parameter Study** window will be displayed.
- Click on the **Add a variation** button from the **Parameter Study** window. A new variation will be added; refer to Figure-69.

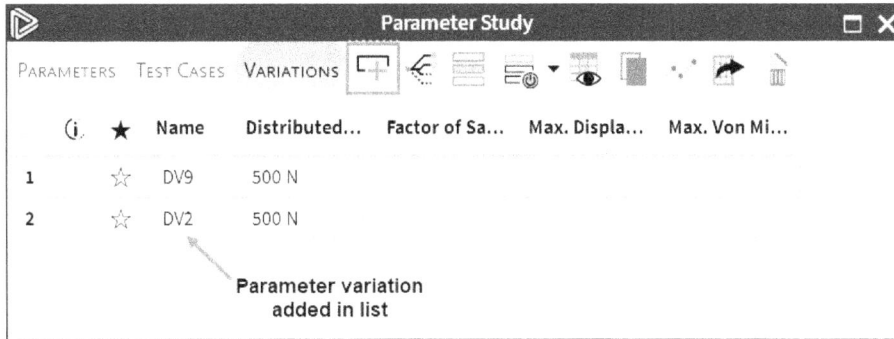

Figure-69. Parameter variation added

- Click in the field of **Distributed Force 1** column for newly added variation and enter the force value as **800 N** in the field.
- Similarly, add the force variation of **1000 N** in the table; refer to Figure-70.

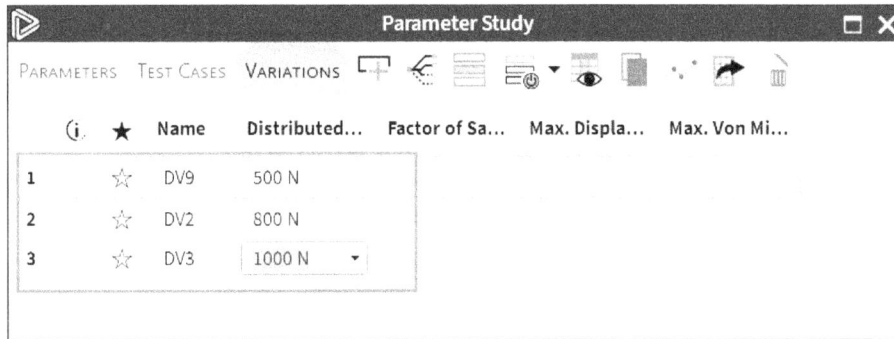

Figure-70. Variations added in window

- After setting desired parameters, click on the **Update All** tool from the **Update** drop-down in the toolbar; refer to Figure-71. The analysis will be solved for all the variations and results will be displayed in the window; refer to Figure-72.

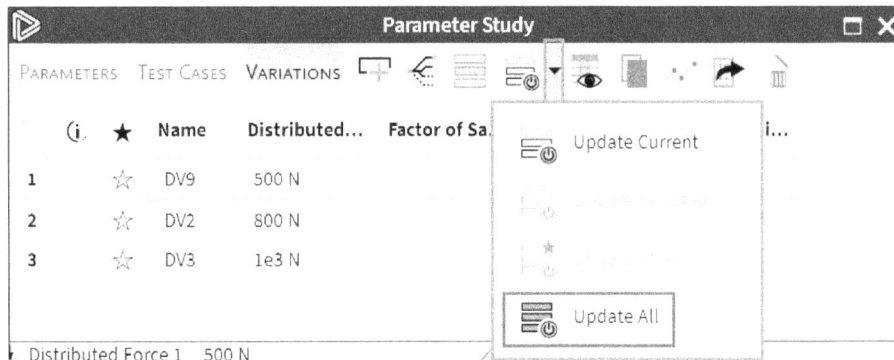

Figure-71. Update All tool

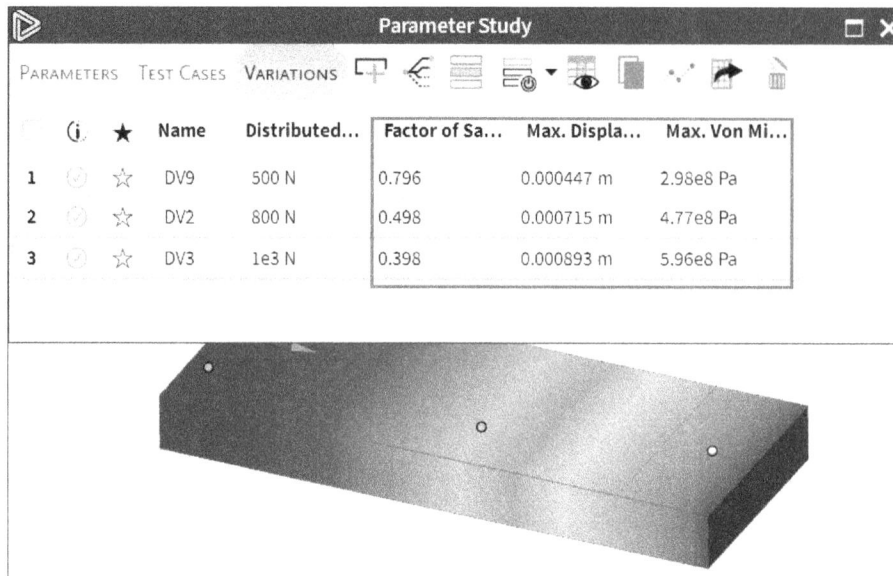

Figure-72. Results in variations

- Click on the **Show/hide variations chart** tool from the **Parameter Study** window to check results on a chart. Close the **Parameter Study** window and chart window by using **Close** button at the top right corner.

PERFORMING NATURAL FREQUENCY ANALYSIS

Natural Frequency analysis also known as Modal analysis is performed to find out resonance frequencies of the model at which current body shows maximum deformation. When creating mechanical parts/assemblies, depending on the application of design part/assembly, it may be mandatory to perform this analysis like in case of a bridge being used by general public or in case of a motor frame of ropeway. In such cases, it is important to make sure that natural frequencies of structures are outside the working frequencies of the assembly. If the assembly frame with motor is vibrating due to motor by a frequency of 100 Hz and natural frequency of frame is 100 Hz then it can cause catastrophic damage under small load. Note that force or other loads are not required to perform natural frequency analysis. The procedure to perform natural frequency analysis is given next.

- Import or create the model to be tested for natural frequencies.
- Switch to the **Explore** stage using the **Stage Selector**.
- By default, the **Structural steel, S275N** material will be applied to the part.
- Click on the **Support** tool from the **Structural** drop-down in the **PHYSICS** panel of **SIMULATION** tab in the **Ribbon**. The **STRUCTURAL SUPPORT HUD** toolbar will be displayed.
- Select the **Fixed** option from the drop-down and apply the constraint as shown in Figure-73. Note that applying constraint to body is important for getting accurate results from the analysis.

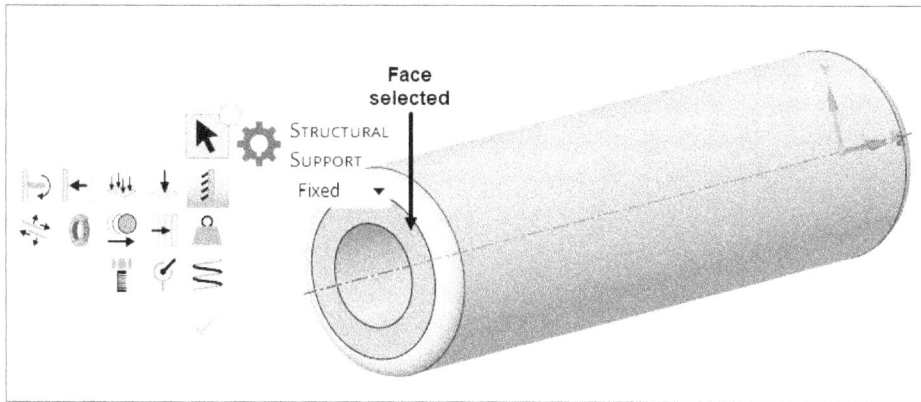

Figure-73. Face selected for fixed support

- Click on the **Appy** button from the **HUD** toolbar or press **ENTER** to apply constraint. Press **ESC** to exit the tool.
- Click on the **Natural Frequency** toggle button from the **PHYSICS** panel in the **Ribbon**. The options in **Physics Tree** will be displayed as shown in Figure-74.

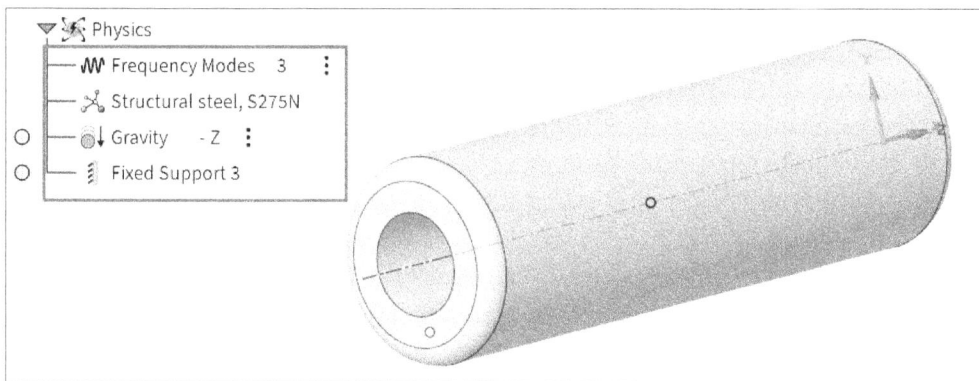

Figure-74. Frequency options for Natural Frequency analysis

- Click in the **Frequency Modes** drop-down of **Physics Tree** and set desired number of natural frequencies to be found by analysis. Generally, we are concerned about first 5 natural frequencies so select the **5** option from the **Frequency Modes** drop-down.
- After setting desired parameters, click on the **Solve** button from the center of **Results Arc** at bottom right corner of application window. After the analysis solving is complete, the result will be displayed in graphics area.
- Select desired mode frequency from the **Mode** drop-down in the **Legends** area; refer to Figure-75. The mode frequencies in results are 1160, 1160, 4460, 5570, and 5580. Note that 1160 Hz is repeated in results because it might be generating two different mode shapes. Mode shapes are the shapes of model that it tends to attain under the influence of respective frequency.

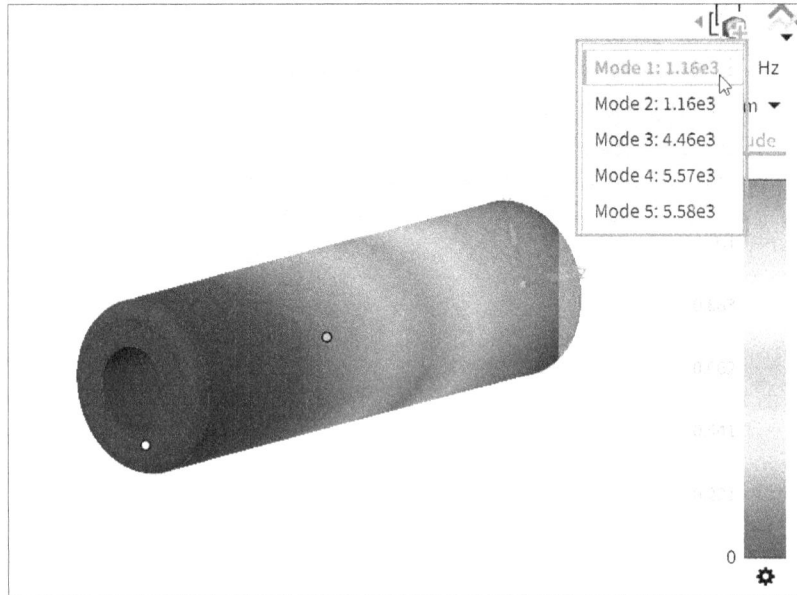

Figure-75. Mode frequencies

PERFORMING TOPOLOGY OPTIMIZATION

The **Topology Optimization** is performed to reduce extra material and enhance the performance of part under specified load conditions. In **Topology Optimization** study, system solves the analysis with sections removed from marked area of part. The procedure to perform topology optimization is given next.

- Create/Open the part on which you want to perform topology study.
- Create the linear static study setup for the model to define the conditions for which topology study will be performed; refer to Figure-76.

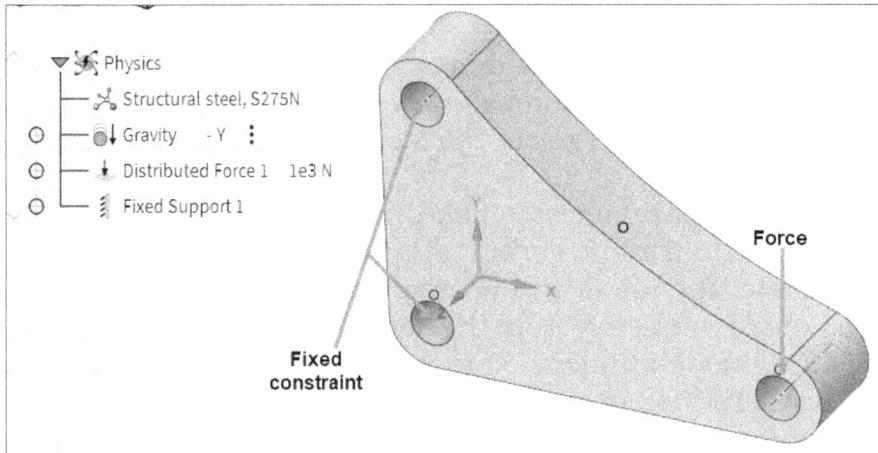

Figure-76. Setup for topology study

- Click on the **Topology Optimization** tool from the **PHYSICS** panel in the **SIMULATION** tab of the **Ribbon**. The options to setup topology optimization study are displayed; refer to Figure-77.

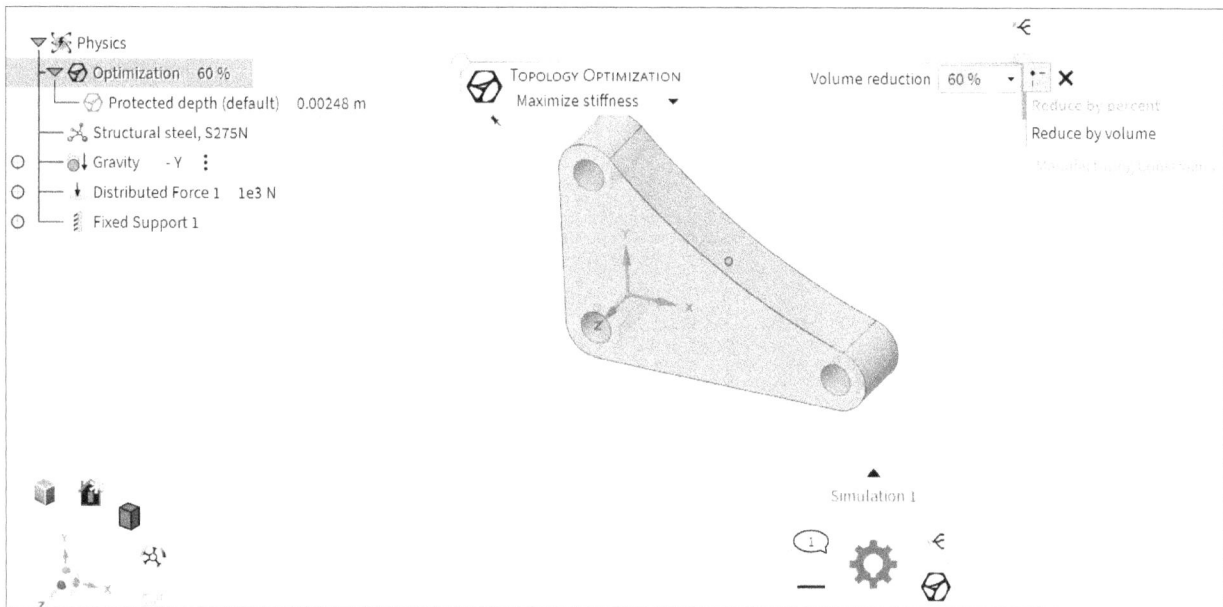

Figure-77. Topology optimization study parameters

- Select desired option from drop-down in the left **HUD** toolbar to define goal of topology optimization. Select the **Maximize stiffness** option from the drop-down to keep stiffness value maximum while reducing volume/mass of the model. Select the **Maximize Natural Frequency** option from the reduce effect of natural frequency on model while raising the value of lowest natural frequency. Select the **Balance Stiffness and Frequency** option from the drop-down to use maximize stiffness of model while keeping natural frequency response of model to minimum. Select the **Target Natural Frequency** option from the drop-down to define the target frequency to be reached by mass/volume increase/reduction. Select the **Minimize volume** option from the drop-down to minimize the volume/mass of part while still maintaining specified stress/factor of safety/stiffness. Select the **Minimize stress** option from the drop-down to keep the stress in model to minimum.

- Set desired value in the **Volume reduction** edit box to define the maximum volume reduction in the part that can be achieved by optimization. Select the **Reduce by percent** toggle button to specify target value in percentage. Select the **Reduce by volume** toggle button to define target volume reduction value.

- If you have selected **Target natural frequency** option in **Goal** drop-down then the **Target frequency** edit box will be displayed at the right in the **HUD** toolbar. Specify desired value of target frequency in the edit box.

- If you have selected the **Minimize volume** option from the **Goal** drop-down then the options in right **HUD** toolbar will be displayed as shown in Figure-78. Select the **Von Mises stress** option from the right **HUD** toolbar to specify maximum Von Mises stress that can occur in the body. Similarly, you can set Principal stress, Factor of safety, or Stiffness reduction as termination condition for optimization study by using respective toggle option from the **HUD** toolbar.

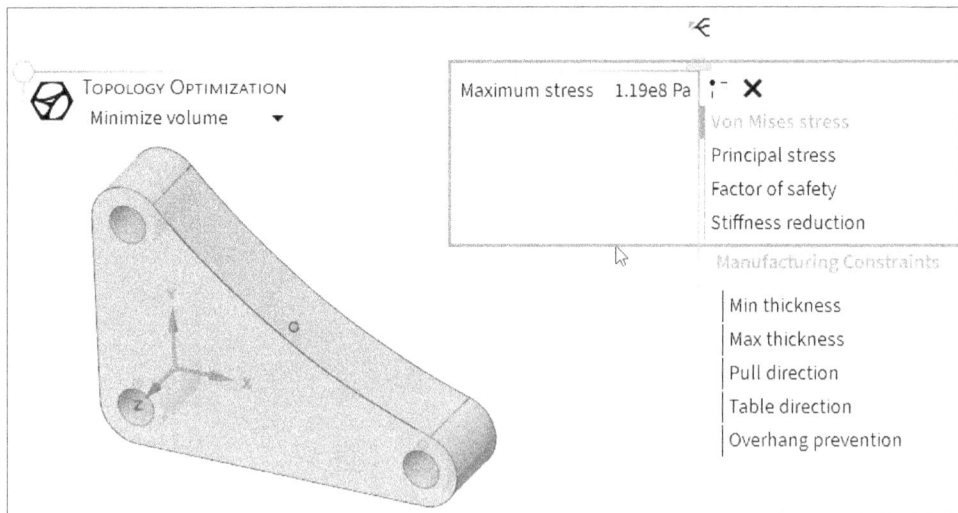

Figure-78. Minimize volume options

- Select desired toggle buttons from the **Manufacturing Constraints** section of the right **HUD** toolbar to define various manufacturing limits of your machine setup.
- Similarly, specify the thickness of projected region in the **Protected depth (default)** edit box in the **Physics Tree**; refer to Figure-79. Note that protected regions are automatically defined around load and constraint faces.

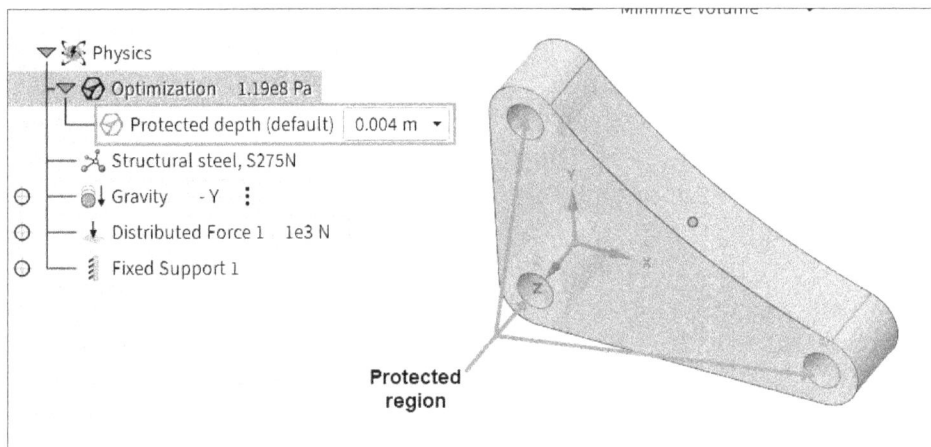

Figure-79. Protected depth edit box

- After setting desired parameters, click on the **Solve** button from center of **Results Arc**. Once the solving of analysis is complete, the result will be displayed in graphics area; refer to Figure-80.

Figure-80. Optimized body

- Click on the **Pause** button from the **Results Arc** to stop solving the analysis further after you have achieved a stable result.
- Click on the **Add optimized body to model** tool from the **Results Arc** to add optimized body as facets body. The Facets body node will be added in the **Model Tree**; refer to Figure-81.

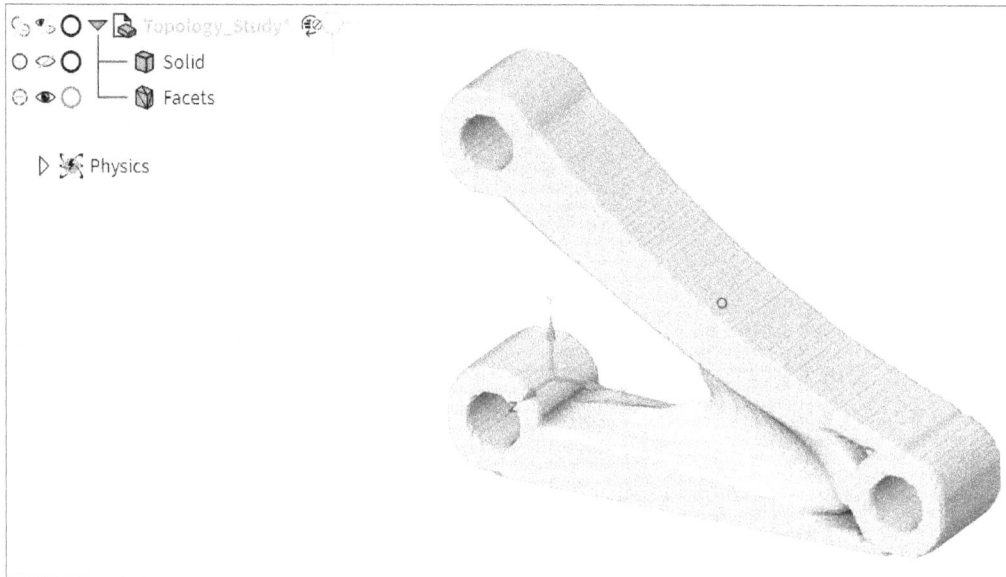

Figure-81. Optimized facets body generated

- Right-click on the **Facets** feature from the **Model Tree** and select the **Using Surfaces** option from the **Convert to Solid** cascading menu of the shortcut menu; refer to Figure-82. The surface model will be generated; refer to Figure-83.

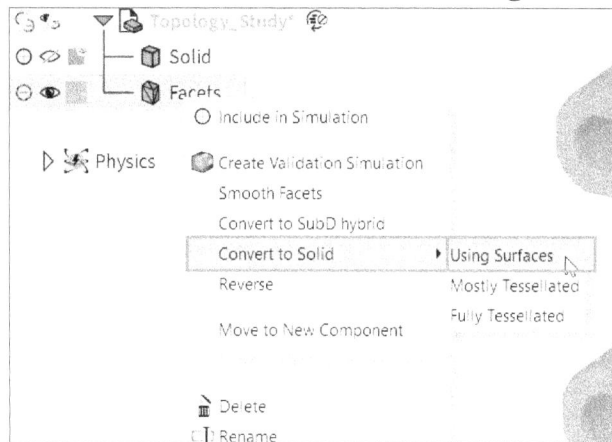

Figure-82. Using Surfaces option

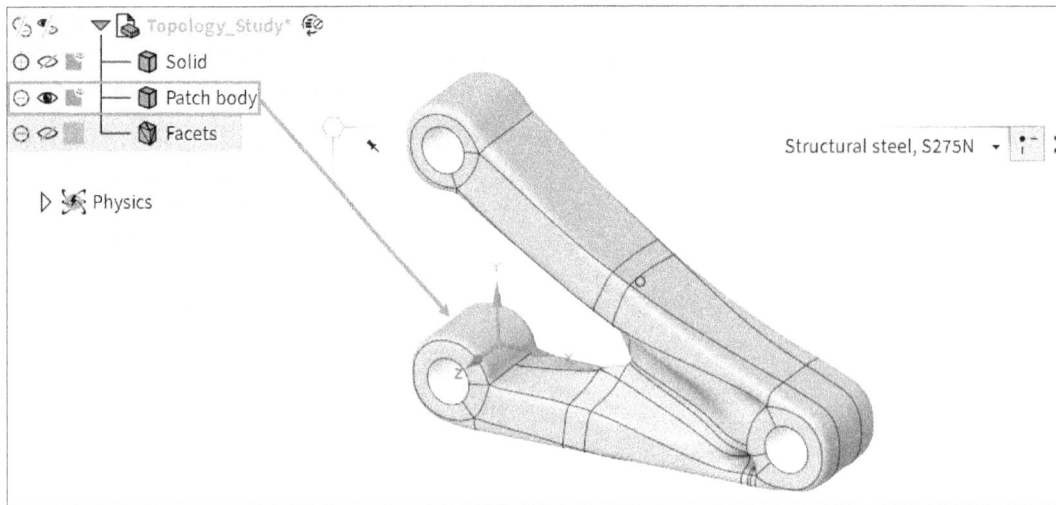

Figure-83. Patch surface body created

- Right-click on the **Patch body** from the **Model Tree** and select the **Move to New Component** option from the shortcut menu; refer to Figure-84. A new component will be created which can be saved as separate body for exporting; refer to Figure-85.

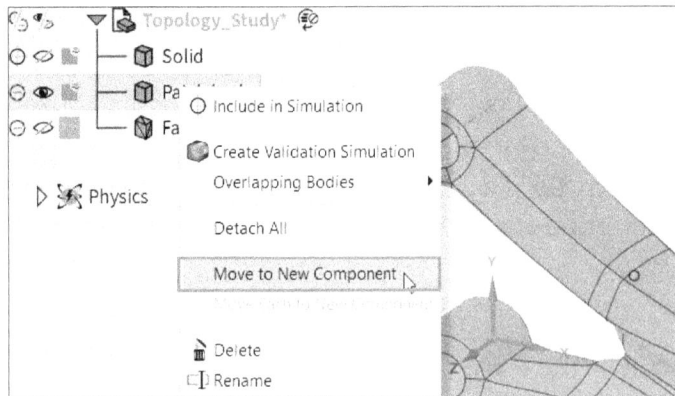

Figure-84. Move to New Component option

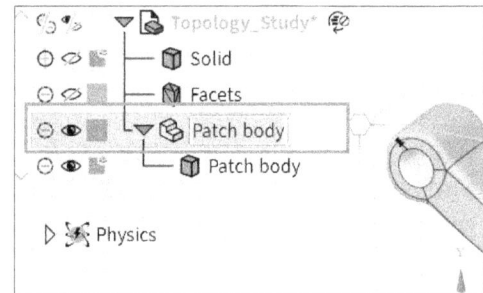

Figure-85. Patch body component

- Right-click on the newly added component and select the **Export Component** option from the shortcut menu; refer to Figure-86. The **Save Discovery Document** dialog box will be displayed. Set desired name of file and click on the **Save** button.

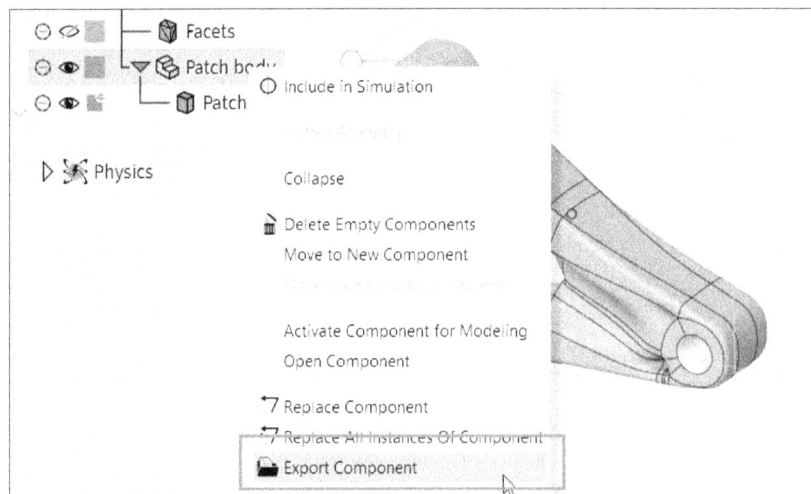

Figure-86. Export Component tool

Note that in Student version of the software, you do not get option to export this model as a CAD geometry in formats like IGES, STP, etc. but if you are using commercial version then you can export this model and refine it in CAD design software.

PRACTICE 1

Consider a rectangular plate with cutout. The dimensions and the boundary conditions of the plate are shown in Figure-87. It is fixed on one end and loaded on the other end. Under the given loading and constraints, plot the deformed shape. Also, determine the principal stresses and the von Mises stresses in the bracket. Thickness of the plate is 0.125 inch and material is **AISI 1020**.

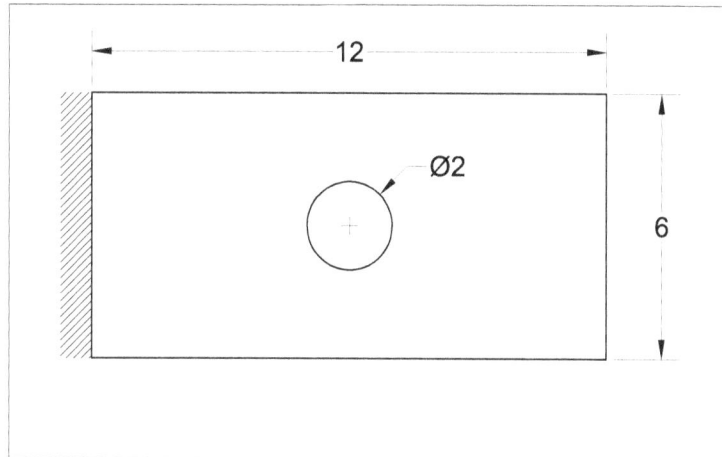

Figure-87. Drawing for Practice 1

PRACTICE 2

Download the model for Practice 2 of this chapter from the resource kit and perform the static analysis using the conditions given in Figure-88. Find out the **Factor of Safety** for the model.

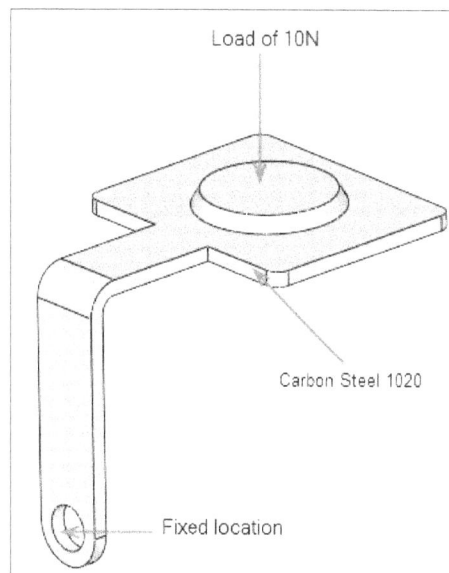

Figure-88. Practice 2

PRACTICE 3

Download the model of chair from the resource kit and perform the static analysis using the conditions given in Figure-89. The material used for manufacturing chair is PET (Polyethylene terephthalate). Check whether it is safe. Also, check whether Rikishi (wrestler) can sit on this chair without any consequences. His weight is 193 Kg.

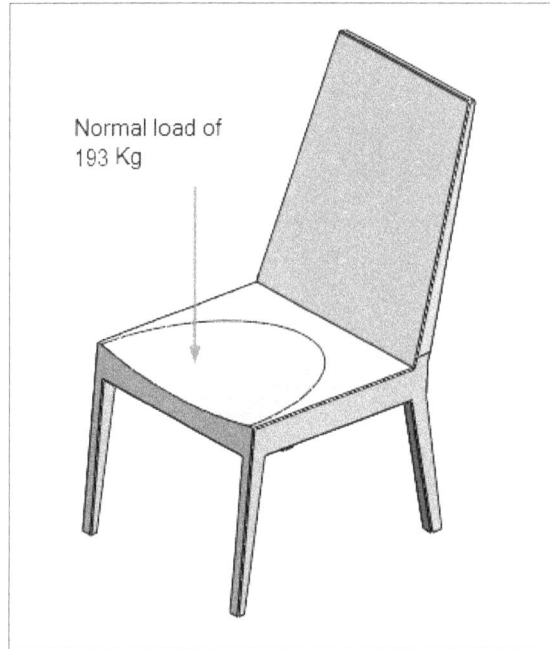

Normal load of 193 Kg

Figure-89. Practice 3

SELF-ASSESSMENT

Q1. What type of analysis is primarily concerned with structural deformations under applied loads?
A. Thermal Analysis
B. Structural Analysis
C. Fluid Dynamics Study
D. Electromagnetic Analysis

Q2. Which of the following is NOT a type of load considered in Structural Analysis?
A. Force
B. Torque
C. Voltage
D. Pressure

Q3. What is the primary characteristic of Static Structural Analysis?
A. Load varies with time
B. Load remains constant over time
C. Only thermal loads are considered
D. Only fluid loads are considered

Q4. The Factor of Safety is defined as:
A. Ratio of applied load to ultimate load
B. Ratio of design load to applied load
C. Ratio of ultimate load to applied load
D. Ratio of stress to strain

Q5. Which condition must be met for Linear Static Structural Analysis?
A. The stress-strain relationship must be non-linear
B. The material must deform plastically
C. The material must behave elastically
D. The structure must undergo large permanent deformations

Q6. What happens to the structure in a Linear Static Structural Analysis when the load is removed?
A. It remains deformed permanently
B. It returns to its original shape
C. It undergoes further deformation
D. It collapses

Q7. Which equation represents the linear relationship between stress and strain in Linear Static Structural Analysis?
A. $\sigma = E/\varepsilon$
B. $\sigma = \varepsilon/E$
C. $\sigma = \varepsilon \cdot E$
D. $\sigma = 1/\varepsilon$

Q8. Why is Non-Linear Static Structural Analysis required?
A. When there is a linear relationship between stress and strain
B. When the material follows Hooke's Law
C. When the structure undergoes large deformations
D. When static loads are constant over time

Q9. What is the P-delta effect?
A. A phenomenon where the load remains constant regardless of deformation
B. A change in applied moment due to small deformations
C. The effect of damping in structural systems
D. The impact of thermal expansion on stress

Q10. What happens in Non-Linear Static Structural Analysis when geometry changes due to deformation?
A. Load distribution remains unchanged
B. The system updates the geometry and load distribution in each iteration
C. The material properties change permanently
D. The model is no longer valid for analysis

Q11. What distinguishes Prestressed Static Analysis from other static analyses?
A. It is performed before loads are applied
B. It includes damping and inertia forces
C. It considers a fixed pre-applied stress before applying further loads
D. It is a type of fluid analysis

Q12. In Buckling Analysis, what is the critical buckling load?
A. The maximum load a structure can support before material failure
B. The load at which the structure suddenly deforms without changing applied load
C. The ultimate tensile load a structure can sustain
D. The force required to initiate yielding

Q13. What is the purpose of Modal Analysis?
A. To determine static load resistance
B. To analyze temperature distribution
C. To find natural frequencies and vibration modes of a structure
D. To determine fluid flow characteristics

Q14. Which factor influences the natural frequency of a structure?
A. Material properties
B. Boundary conditions
C. Load magnitude
D. All of the above

Q15. What distinguishes Transient Structural Analysis from Static Structural Analysis?
A. It ignores damping and impact loads
B. It considers time-varying loads and dynamic effects
C. It only applies to metallic structures
D. It assumes material deformation is always elastic

Q16. Which of the following is NOT an assumption of Linear Static Structural Analysis?
A. Loads do not vary with time
B. Material follows Hooke's Law
C. Stiffness changes due to loading are considered
D. Boundary conditions remain constant

Q17. Which assumption is made about fasteners in Linear Static Structural Analysis?
A. They are considered to be perfect rigid joints
B. Their stress relaxation is always included
C. They always fail under high loads
D. They are never considered in the analysis

Q18. What is assumed about material behavior in Linear Static Structural Analysis?
A. Material remains in the elastic region
B. Plastic deformation is included
C. Creep effects are considered
D. Non-homogeneous material properties are included

Q19. In which situation would a Non-Linear Buckling Analysis be necessary?
A. When analyzing the first mode shape
B. When considering behavior beyond the critical buckling load
C. When all loads are static and time-independent
D. When performing a thermal analysis

Q20. What is the primary goal of Linear Static Structural Analysis?
A. To determine stresses, displacements, and reaction forces
B. To analyze crack formation in structures
C. To evaluate dynamic impact loads
D. To study fluid-structure interaction

Q21. What is the function of the Show/Hide results toggle button in the Design Tree?
A. To delete the analysis results
B. To toggle between showing and hiding analysis results
C. To change the color gradient of the model
D. To reset the analysis results

Q22. Where can you select the desired result parameter to be displayed in the graphics area?
A. Results Arc
B. Design Tree
C. Legend Area
D. Cut Plane

Q23. How can you change the unit used for displaying analysis results?
A. By modifying the Design Tree
B. By selecting the desired unit from the Unit drop-down in the Legend Area
C. By clicking on the Show Animation toggle button
D. By adjusting the Streamlines settings

Q24. What does the Toggle gradient style button do?
A. Displays scale in clear value boundaries
B. Changes the color of analysis results
C. Resets all applied analysis results
D. Removes the graphics display

Q25. What is the function of the center button in the Results Arc?
A. To open the Legend Area
B. To solve and pause the analysis
C. To display deformation results
D. To reset the user-defined ranges

Q26. What are Vectors used for in analysis results?
A. To modify the material properties
B. To highlight the deformation scale
C. To analyze results using colored arrows
D. To delete specific result parameters

Q27. Which option allows keeping all vector arrows equal in size?
A. Constant size toggle button
B. Cut Plane toggle button
C. Show/Hide results toggle button
D. Reset results button

Q28. What is the function of Contours in analysis results?
A. To show analysis results using colored arrows
B. To represent analysis results by painting model faces in different colors
C. To highlight only stress results
D. To animate the analysis results

Q29. What does selecting "All Bodies" from the Location drop-down do in the Contours settings?
A. Displays results over the entire body while making solid boundaries transparent
B. Hides the model from the graphics area
C. Shows only the edges of the model
D. Resets the analysis results

Q30. What do Iso-Surfaces represent in analysis results?
A. The flow of fluid around the model
B. The displacement of objects in different layers
C. The result parameters as surfaces where each layer depicts the same value
D. The speed of animation

Q31. How can you modify the value of the Iso-Surface result parameter?
A. By using the slider in the Iso-Surface settings
B. By clicking on the Design Tree
C. By modifying the Direction Field settings
D. By selecting the Streamlines toggle button

Q32. What does the Streamlines feature represent?
A. The flow of electrical currents in solid models
B. The flow of fluids in CFD analysis results
C. The deformation of a model due to stress
D. The movement of solid particles inside the model

Q33. How can you reverse the direction of fluid flow in Streamlines?
A. By selecting the Reversed toggle option
B. By changing the Color Gradient settings
C. By modifying the Unit drop-down in the Legend Area
D. By adjusting the Contours settings

Q34. What is the function of the Particles result?
A. To check the solid body deformation
B. To check the flow of fluids in CFD analysis
C. To adjust the cut plane of the model
D. To change the legend display settings

Q35. What does the Animation speed slider control in the Particles result?
A. The speed at which particles move in the model
B. The resolution of the analysis results
C. The color of the particle display
D. The cut section of the model

Q36. What does the Direction Field result display?
A. The fluid flow at the center plane of the model
B. The vibration frequency of a model
C. The stress distribution in a mechanical part
D. The electrical resistance of a circuit

Q37. What does the Deformation result show?
A. The force applied to the model
B. The change in geometry due to displacement
C. The fluid flow path around the model
D. The material properties of the model

Q38. What is the purpose of the Show Animation toggle button in the Results Arc?
A. To display the animation of the deformation result
B. To add fluid flow results to the model
C. To reset the parametric analysis results
D. To modify the stress contours

Q39. What does the Cut Plane toggle button do?
A. Displays the model's analysis result at the center plane
B. Adds an additional boundary to the analysis
C. Hides all surfaces except the solid boundaries
D. Creates a new model from the existing analysis

Q40. What happens when you select the Reset button in the Results Arc?
A. The visualization of results is reset, but parametric results remain unchanged
B. The analysis is completely deleted
C. The model is restored to its default shape
D. The animation of the analysis result is restarted

Q41. What does the Add Optimized Body to Model tool do?
A. Creates a facet body generated by Topology Optimization study results
B. Deletes all previous analysis results
C. Increases the resolution of analysis results
D. Reduces the computation time of simulations

Q42. What is the primary condition for performing linear static analysis?
A. Nonlinear relationship between stress and strain
B. Large displacements in the structure
C. Linear relationship between stress and strain
D. Contact conditions varying over time

Q43. In linear static analysis of an assembly, what should be considered apart from a single-body model?
A. Material nonlinearity
B. Various contacts and connections
C. Large deformations
D. Time-dependent loading

Q44. What is the first step in performing linear static analysis on an assembly in Ansys Discovery?
A. Applying boundary conditions
B. Applying material properties
C. Creating or importing geometry/model
D. Solving the simulation

Q45. What material is applied to the Base plate in the given example?
A. Structural steel, S275N
B. Carbon Steel, 1020, annealed
C. Cast iron, EN GJL 100
D. Stainless steel

Q46. Which constraint is applied to the base plate holes in the example?
A. Hinged
B. Fixed
C. Frictionless
D. Roller

Q47. How is force applied in Ansys Discovery according to the example?
A. As a single concentrated load
B. As a distributed force on a selected face
C. As a torque on the base plate
D. As a uniform temperature load

Q48. What is the default contact type between components in Ansys Discovery?
A. Frictional
B. No separation
C. Bonded
D. Free

Q49. Why is a bonded contact suitable for the static analysis in the example?
A. It allows large deformations
B. It prevents relative movement between components
C. It ensures nonlinearity in material behavior
D. It helps determine the angle of rotation of the arm

Q50. What is the initial factor of safety (FoS) obtained in the example?
A. 0.5
B. 0.8
C. 1.0
D. 1.5

Q51. What modification is suggested to improve the factor of safety?
A. Increasing load on the base plate
B. Applying a nonlinear material model
C. Adding ribs to strengthen weak areas
D. Reducing the thickness of the base plate

Q52. After adding ribs, what is the new factor of safety obtained?
A. 0.8
B. 1.2
C. 1.51
D. 2.0

Q53. Which part of the model experiences the highest stress after adding ribs?
A. Base plate edges
B. Holes used for fastening the base to the wall
C. Hinge arm
D. The pin

Q54. What additional modification can further improve the model's strength?
A. Decreasing the thickness of the base plate
B. Using a lower-grade material
C. Increasing the thickness of the base plate at the attachment points
D. Removing bonded contacts

Q55. What can cause nonlinearity in structural analysis?
A. Large displacements
B. Material behavior
C. Contact conditions
D. All of the above

Q56. What is the main difference between linear and nonlinear static analysis?
A. Linear analysis assumes constant stiffness, while nonlinear analysis accounts for changing stiffness
B. Linear analysis accounts for large deformations
C. Nonlinear analysis ignores material properties
D. Linear analysis cannot be performed in Ansys

Q57. Which of the following is an example of material nonlinearity?
A. Beam-column yielding during earthquakes
B. Elastic deformation of a steel beam
C. A rigid body rotating in space
D. Linear stress-strain relationship

Q58. What type of nonlinearity is caused by large displacements?
A. Material nonlinearity
B. Contact nonlinearity
C. Geometric nonlinearity
D. Elastic linearity

Q59. What algorithm is used to solve nonlinear equations in Ansys Discovery?
A. Newton-Raphson method
B. Finite Difference method
C. Gauss-Seidel method
D. Euler's method

Q60. What condition must be met for the Newton-Raphson method to stop iterating?
A. The difference between external and internal loads is within a tolerance
B. The displacement reaches zero
C. The material becomes perfectly elastic
D. The force remains constant

Q61. In nonlinear static analysis, what does time represent?
A. Real-world time
B. A counter for load steps and sub-steps
C. The exact duration of loading
D. The material relaxation time

Q62. What is the purpose of Load Cases in analysis?
A. To change the material properties of a model
B. To check variations of boundary condition parameters for the same analysis setup
C. To modify the geometry of a model
D. To increase the processing speed of the analysis

Q63. In Ansys Discovery, which tool is used to perform Load Case/Parameter Study?
A. Variations
B. Parameters
C. Constraints
D. Physics

Q64. What is the first step in performing a parameter study?
A. Applying force variations
B. Creating support, force, and material setup
C. Running the simulation
D. Exporting the results

Q65. Where can you find the 'Add as a parameter' button in Ansys Discovery?
A. In the Physics Tree
B. In the Ribbon toolbar
C. In the Results Arc
D. In the Model Tree

Q66. What happens when you click on the 'Update All' tool after setting parameters?
A. The software applies material properties
B. The analysis is solved for all variations and results are displayed
C. The model is exported
D. The forces are removed from the setup

Q67. What does the Show/Hide variations chart tool do?
A. Displays the parameter study results on a chart
B. Hides the geometry of the model
C. Removes unnecessary boundary conditions
D. Modifies the applied forces

Q68. What is another name for Natural Frequency Analysis?
A. Stress Analysis
B. Modal Analysis
C. Heat Transfer Analysis
D. Load Case Analysis

Q69. Why is Natural Frequency Analysis important in mechanical designs?
A. To check material properties
B. To ensure natural frequencies do not match working frequencies
C. To improve software performance
D. To minimize the load applied

Q70. What happens if the natural frequency of a frame matches the vibration frequency of a motor?
A. The system will become more efficient
B. The system may suffer catastrophic damage
C. The frame will become more stable
D. The motor will stop working

Q71. What is NOT required to perform Natural Frequency Analysis?
A. Force or other loads
B. Material properties
C. Constraints
D. A solid model

Q72. How many natural frequencies are generally considered in an analysis?
A. 3
B. 5
C. 10
D. 20

Q73. What does a repeated natural frequency in results indicate?
A. The software has an error
B. The model is unstable
C. Different mode shapes may exist at that frequency
D. The frequency should be ignored

Q74. What is the purpose of Topology Optimization?
A. To modify the color of the model
B. To reduce extra material and enhance performance under load conditions
C. To change the material properties of a part
D. To apply forces in different directions

Q75. Which of the following is NOT an option in the Topology Optimization goal drop-down?
A. Maximize stiffness
B. Minimize volume
C. Increase temperature
D. Minimize stress

Q76. If you select the 'Minimize Volume' option in Topology Optimization, which parameter can you define as a termination condition?
A. Von Mises stress
B. Elastic modulus
C. Hardness value
D. Density

Q77. What does the 'Add optimized body to model' tool do?
A. Deletes the original model
B. Adds the optimized body as a facets body
C. Increases the volume of the part
D. Reduces the number of mesh elements

Q78. In Ansys Discovery, what is the final step to export an optimized component as a separate body?
A. Select 'Convert to Solid'
B. Click on 'Move to New Component'
C. Click on 'Export Component'
D. Save the study file

Q79. What limitation exists in the Student version of Ansys Discovery?
A. Cannot apply forces
B. Cannot perform Natural Frequency Analysis
C. Cannot export CAD geometry in formats like IGES or STP
D. Cannot create support constraints

Chapter 8

Ansys Discovery-Thermal, Electromagnetic, and Fluid Flow Analyses

Topics Covered

The major topics covered in this chapter are:

- *Introduction to Thermal Analysis*
- *Performing Thermal Analysis*
- *Introduction To Fluid Flow Analysis*
- *Performing Internal Fluid Flow Analysis*
- *Performing External Fluid Flow Analysis*
- *Performing Electromagnetics Analysis*

INTRODUCTION

Thermal analysis is a method to check the distribution of heat over a body due to applied thermal loads. Note that thermal energy is dynamic in nature and is always flowing through various mediums. There are three mechanisms by which the thermal energy flows:

- Conduction
- Convection
- Radiation

In all three mechanisms, heat energy flows from the medium with higher temperature to the medium with lower temperature. Heat transfer by conduction and convection requires the presence of an intervening medium while heat transfer by radiation does not.

The output from a thermal analysis can be given by:

1. Temperature distribution.
2. Amount of heat loss or gain.
3. Thermal gradients.
4. Thermal fluxes.

This analysis is used in many engineering industries such as automobile, piping, electronic, power generation, and so on.

Important terms related to Thermal Analysis

Before conducting thermal analysis, you should be familiar with the basic concepts and terminologies of thermal analysis. Following are some of the important terms used in thermal analysis:

Heat Transfer Modes

Whenever there is a difference in temperature between two bodies, the heat is transferred from one body to another. Basically, heat is transferred in three ways: Conduction, Convection, and Radiation.

Conduction

In conduction, the heat is transferred by interactions of atoms or molecules of the material. For example, if you heat up a metal rod at one end, the heat will be transferred to the other end by the atoms or molecules of the metal rod.

Convection

In convection, the heat is transferred by the flowing fluid. The fluid can be gas or liquid. Heating up water using an electric water heater is a good example of heat convection. In this case, water takes heat from the heater.

Radiation

In radiation, the heat is transferred in space without any matter. Radiation is the only heat transfer method that takes place in space. Heat coming from the Sun is a good example of radiation. The heat from the Sun is transferred to the earth through radiation.

Thermal Gradient

The thermal gradient is the rate of increase in temperature per unit depth in a material.

Thermal Flux

The Thermal flux is defined as the rate of heat transfer per unit cross-sectional area. It is denoted by q.

Bulk Temperature

It is the temperature of a fluid flowing outside the material. It is denoted by Tb. The Bulk temperature is used in convective heat transfer.

Film Coefficient

It is a measure of the heat transfer through an air film.

Emissivity

The emissivity of a material is the ratio of energy radiated by the material to the energy radiated by a black body at the same temperature. Emissivity is the measure of a material's ability to absorb and radiate heat. It is denoted by e. Emissivity is a numerical value without any unit. For a perfect black body, e = 1. For any other material, e < 1.

Stefan–Boltzmann Constant

The energy radiated by a black body per unit area per unit time divided by the fourth power of the body's temperature is known as the Stefan-Boltzmann constant. It is denoted by σ.

Thermal Conductivity

The thermal conductivity is the property of a material that indicates its ability to conduct heat. It is denoted by K.

Specific Heat

The specific heat is the amount of heat required per unit mass to raise the temperature of the body by one degree Celsius. It is denoted by °C.

PERFORMING THERMAL ANALYSIS

Thermal analysis is mainly performed to get temperature distribution in model under specified load conditions. Using this analysis, you can check whether heat dissipation system designed by you is capable of dissipating the heat from source. For example, take a case of an IC which generates 30W of heat during operation and you want to keep the temperature at maximum 70 degree Celsius. One way to reduce the temperature of IC is by increasing the surface area from where heat will be dissipated by convection. To do this, generally a heat sink is added and if that is not enough then force air circulation is also added. Using this example as setup, the procedure to perform a thermal analysis is given next.

For the given example scenario, we will first try to find out if there is a need to use heat sink for maintaining temperature below 70 degree Celsius in our setup. This can be checked by performing thermal analysis without heat sink. Later, we will include the heat sink if needed and rerun the analysis.

- Create/import the model to be tested for thermal analysis in the application; refer to Figure-1. (Model is available in resources of this chapter.) Reorient the model using drag handles if needed.

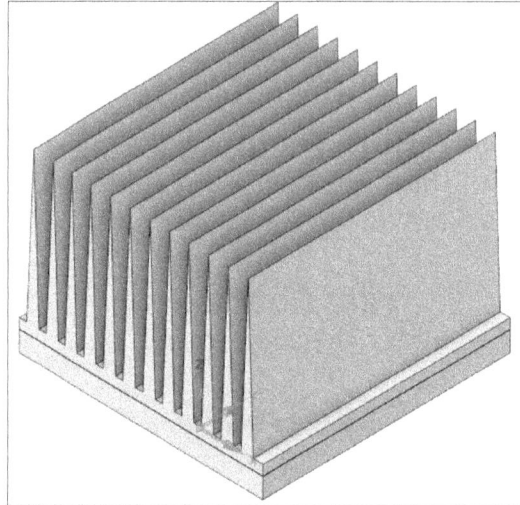

Figure-1. Model for thermal analysis

- Switch to **EXPLORE** stage using the **Stage Selector**.
- Expand the nodes in **Design Tree** and exclude the cooling pad component from the analysis. Also, hide the component using **Show/Hide** button; refer to Figure-2.

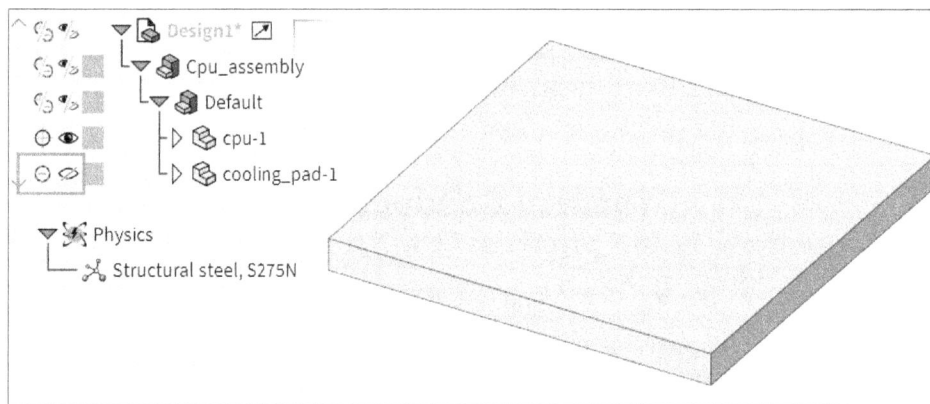

Figure-2. Excluding and hiding cooling pad component

- Right-click on **Structural steel** option from the **Physics Tree** and select the **Edit** option from the shortcut menu. The **HUD** toolbar will be displayed.
- Select the **Aluminum Alloy, wrought, 6061** material from the drop-down to change the material and click in the empty area of graphics window. Press **ESC** to exit the tool.
- Click on the **Heat** tool from the **Solid Thermal** drop-down in the **SIMULATION** tab of the **Ribbon**. You will be asked to select the face to apply boundary condition.
- Select the top face of component marked as CPU/IC (Heat source); refer to Figure-3. The options to define heat value will be displayed in the **HUD** toolbar.
- Type the total heat value as **30 W** in the **Total heat** input box and press **ENTER**. The thermal condition will be applied. Press **ESC** twice to exit the tool.

Note that default value of convection rate will automatically applied to the setup. Generally, 10 W/m².°C is assumed as open air convection rate between solid and air. If you are using forced air circulation then this value can go upto 250 W/m².°C.

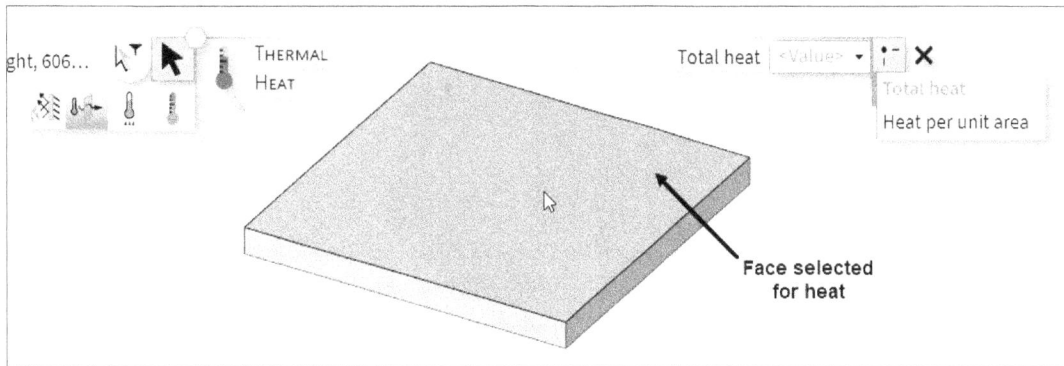

Figure-3. Face selected for applying heat

- Click on the **Solve** button from center of **Results Arc**. Once the solution is achieved, the temperature distribution of body will be displayed in graphics area; refer to Figure-4. As it can be seen from the result below, temperature of chip can go as high as 940 degree Celsius if left without heatsink.

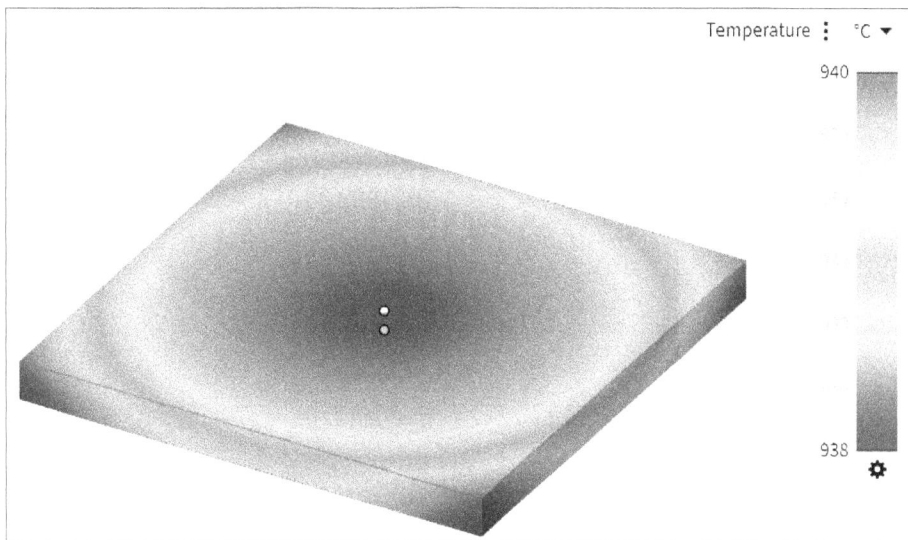

Figure-4. Result without heat sink

- Click on the **Pause** button from center of **Results Arc**. Click on the **Include in simulation** and **Show** buttons for **Cooling pad** component in the **Model Tree**. The heat sink component will be included in the analysis.
- Click on the **Solve** button from the center of **Results Arc**. Once the solution is achieved, the results will be displayed in the graphics area; refer to Figure-5. As you can see from the results, the maximum temperature now reaches to 113 degree Celsius only. Note that this temperature is still higher than 70 degree Celsius allowed. So, forced airflow is needed in the setup. You can represent forced flow by specifying convection value as **20W/m².°C** in the **Physics Tree**.
- After setting the value, the analysis result will be displayed within safe range; refer to Figure-6.

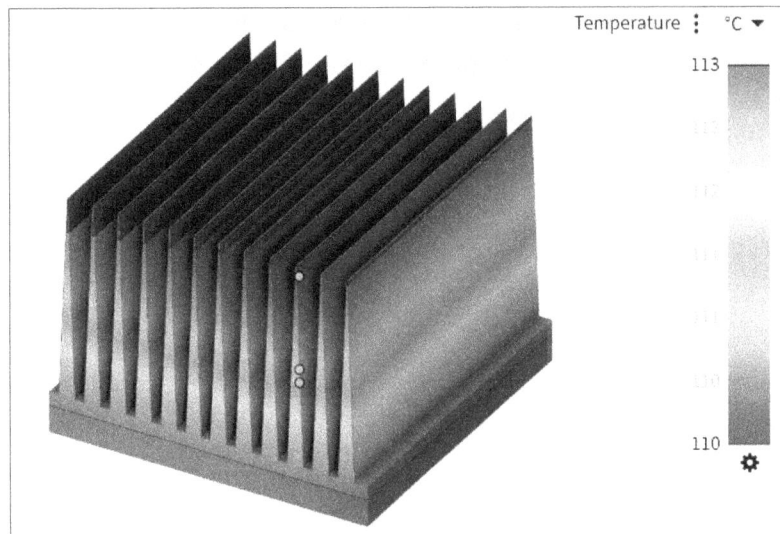

Figure-5. Result after adding heat sink

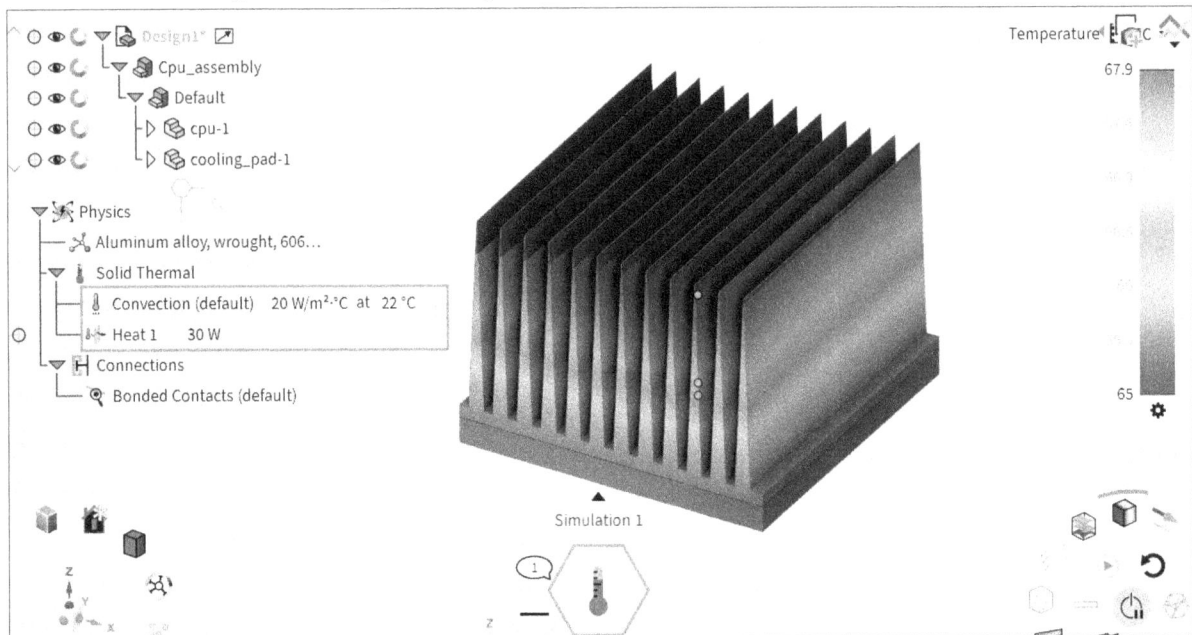

Figure-6. Result after changing convection value

INTRODUCTION TO FLUID FLOW ANALYSIS

The fluid flow analysis is performed to analyze how fluid flows through the specified domain. This analysis is also performed to check how physical properties of fluid and solid in contact with fluid change like temperature, pressure, inertia, and so on. The tools to apply fluid flow boundary conditions are available in **Fluid Flow** drop-down of **SIMULATION** tab in **Ribbon**. In Ansys Discovery, you can perform two types of fluid flow analyses: Internal Flow Analysis and External Flow Analysis. If you want to check the flow of a fluid inside closed region like tube, pipe, container, etc. then you need to perform Internal Flow Analysis and if you want to check the flow of fluid around a closed body like fins of aeroplane, blades of turbine, etc. then you need to perform External Flow Analysis.

Performing Internal Fluid Flow Analysis

You need a closed region in the model with faces to define inlet and outlet of fluid to perform internal fluid flow analysis; refer to Figure-7. The procedure to perform internal fluid flow analysis is given next.

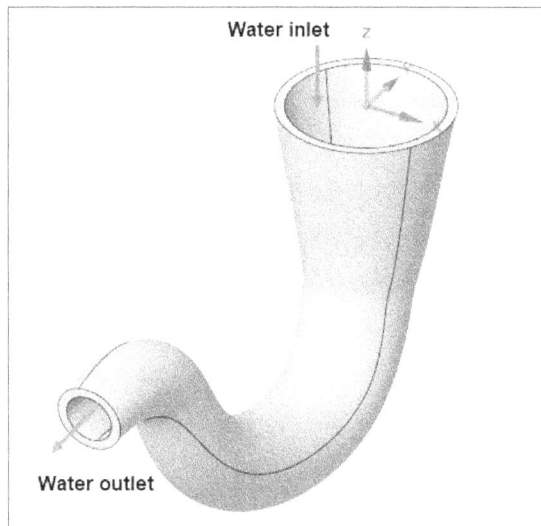

Figure-7. Model for internal fluid flow analysis

- Create/import the model on which you want to perform the analysis; refer to Figure-8. (Model for this problem is available in resources of the book).

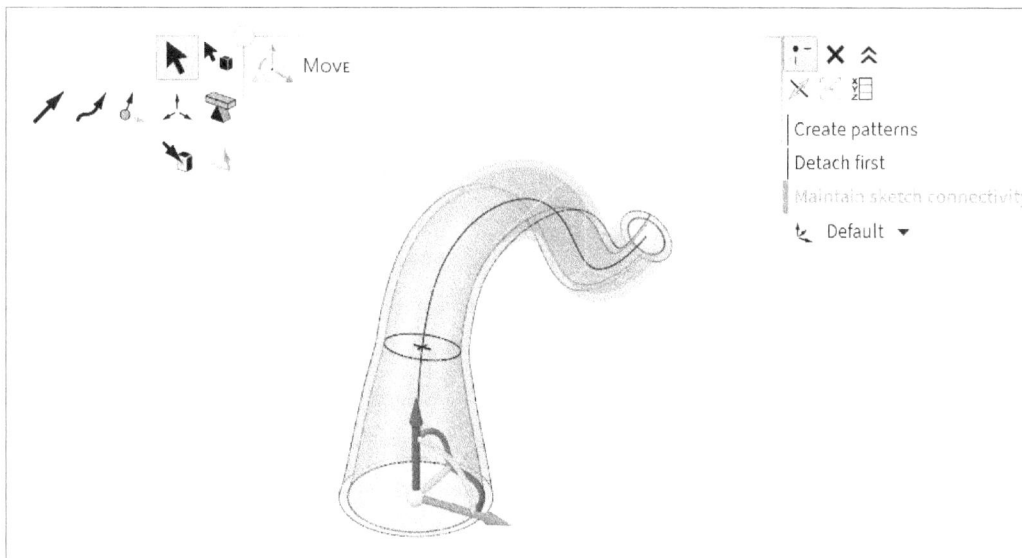

Figure-8. Model imported

- Use the rotation handle and orient the model as shown in Figure-9. Press **ESC** twice to exit reorientation mode.

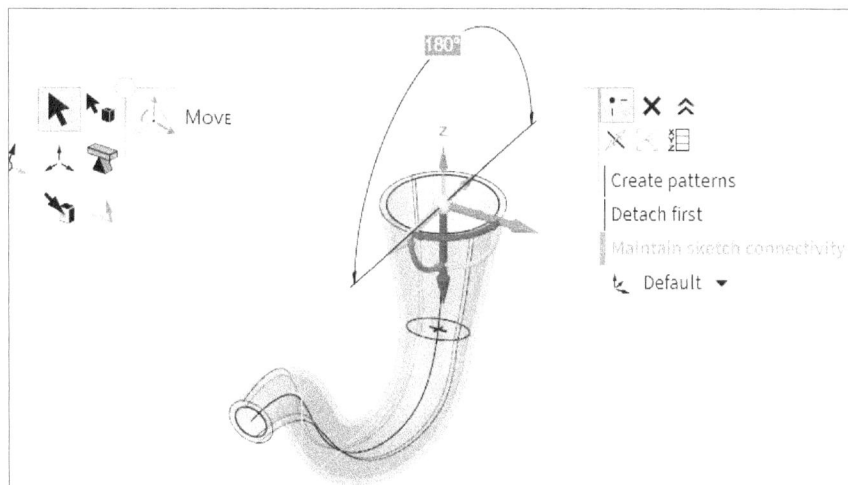

Figure-9. Reorienting the model

- After placing the model, you will notice that walls of the model are transparent. This is because the surfaces in model are not forming a solid body. Click on the **Stitch** tool from the **SOLIDIFY** panel of **REPAIR** tab in the **Ribbon**. The stitchable edge will be highlighted in the graphics area with **HUD** toolbar; refer to Figure-10.

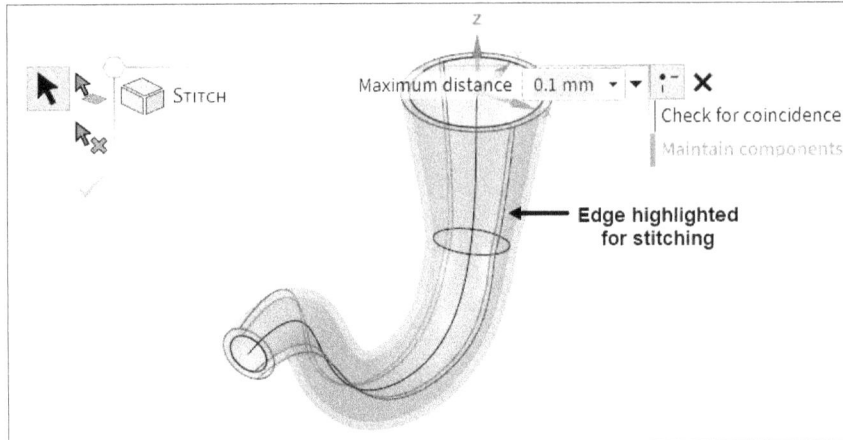

Figure-10. Edge highlighted for stitching

- Select the highlighted edge. The model will be converted to a solid body; refer to Figure-11. Press **ESC** to exit the tool.

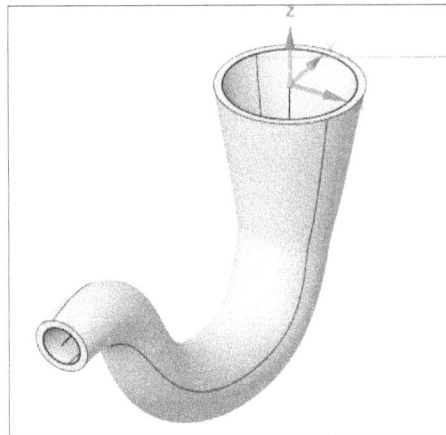

Figure-11. Solidified model

- Switch to **EXPLORE** stage from the **Stage Selector** and click on the **INTERNAL FLOW** tool from the **QUICK START** panel in the **SIMULATION** tab of the **Ribbon**. You will be asked to select face for inlet.
- Select the top face of model as shown in Figure-12 to define inlet face. You will be asked to select face for outlet of fluid.

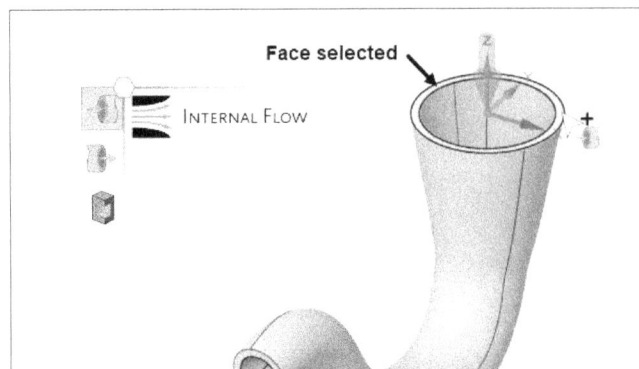

Figure-12. Face selected for inlet

- Select the face of model as shown in Figure-13 to define the outlet. Default boundary conditions will be applied automatically to the model; refer to Figure-14. Press **ESC** to exit the tool.

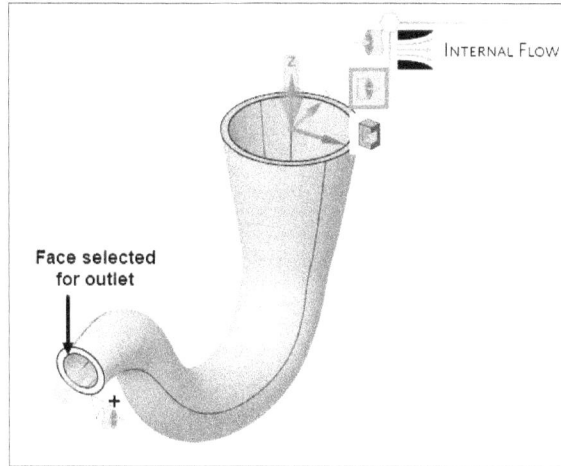

Figure-13. Face selected for outlet

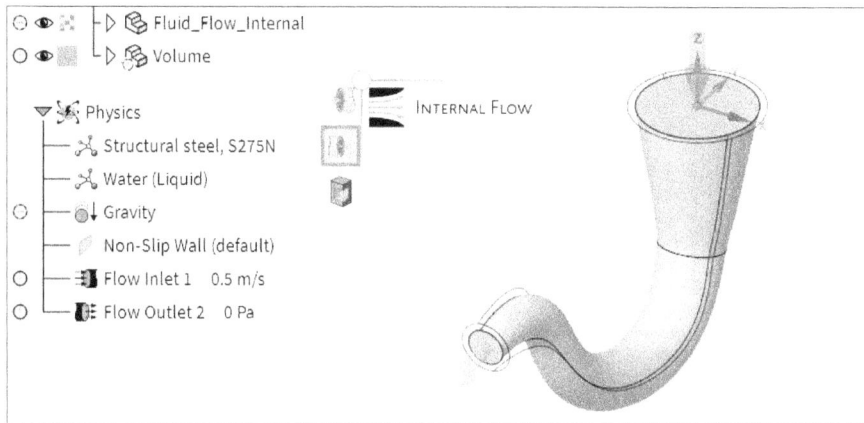

Figure-14. Default boundary conditions

- Click in the **Flow Inlet 1** edit box from the **Physics Tree** and specify the value as **0.2 m/s** to define velocity of fluid at inlet. If you know the mass flow rate instead of velocity then right-click on **Flow Inlet 1** node from the **Physics Tree** and select the **Edit** option from shortcut menu. The **FLUID FLOW HUD** toolbar will be displayed; refer to Figure-15. Select the **Mass flow rate** option from the right **HUD** toolbar and specify desired value in the edit box.

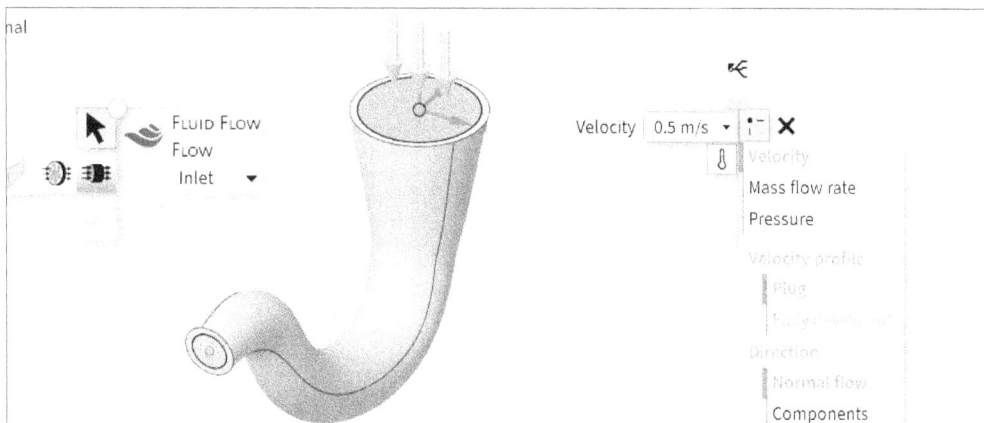

Figure-15. FLUID FLOW HUD toolbar

- Similarly, specify the pressure value as **1 atm** in **Flow Outlet 2** edit box in the **Physics Tree**.
- After setting the parameter, click on the **Solve** button from the center of **Results Arc**. The result of analysis will be displayed as shown in Figure-16.

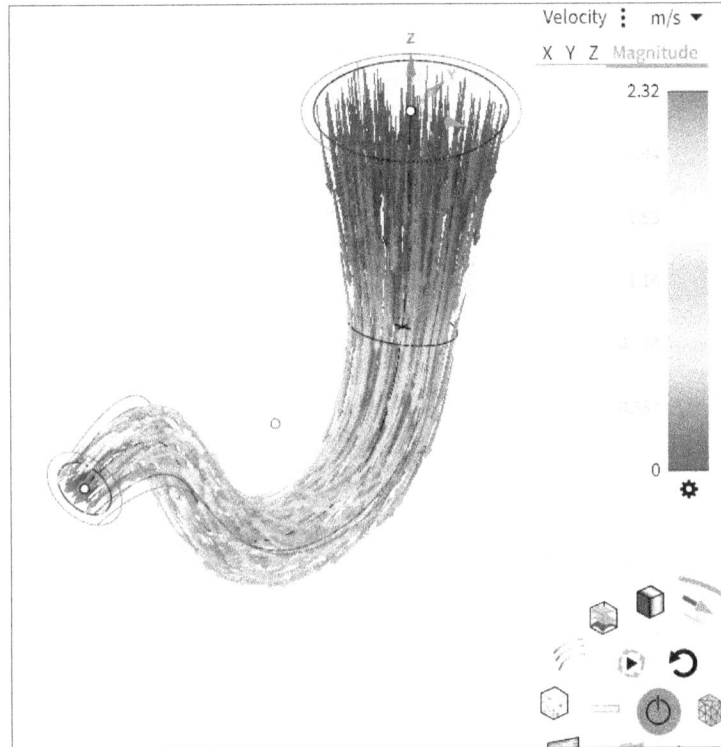

Figure-16. Result of fluid flow analysis

- You can check the pressure drop and maximum velocity from the **MONITORS** drop-down in the results section.

Performing External Fluid Flow Analysis

External fluid flow analysis can be used to check aerodynamics of the model surface. It is important to define the domain within which external flow analysis will be performed otherwise it may take forever to solve the analysis. The procedure to perform the analysis is given next.

- Create/import the model to be used for performing external fluid flow analysis; refer to Figure-17.

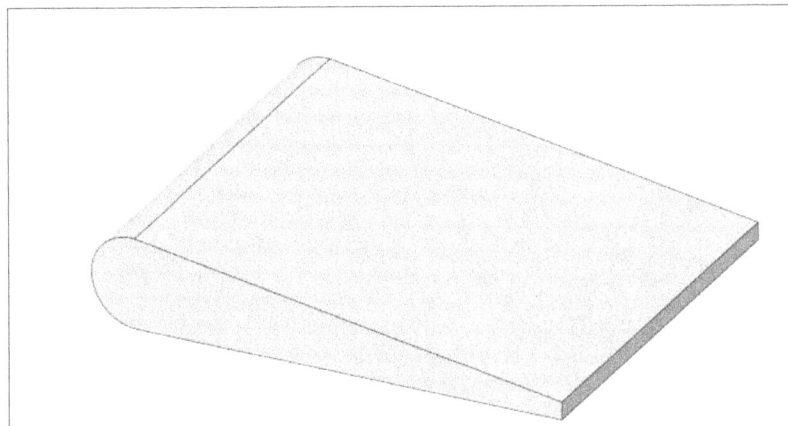

Figure-17. Model for external fluid flow analysis

- Switch to **EXPLORE** stage using the **Stage Selector** and click on the **EXTERNAL FLOW** tool from the **QUICK START** panel in the **SIMULATION** tab of the **Ribbon**. The **EXTERNAL FLOW HUD** toolbar will be displayed with preview of fluid domain; refer to Figure-18.

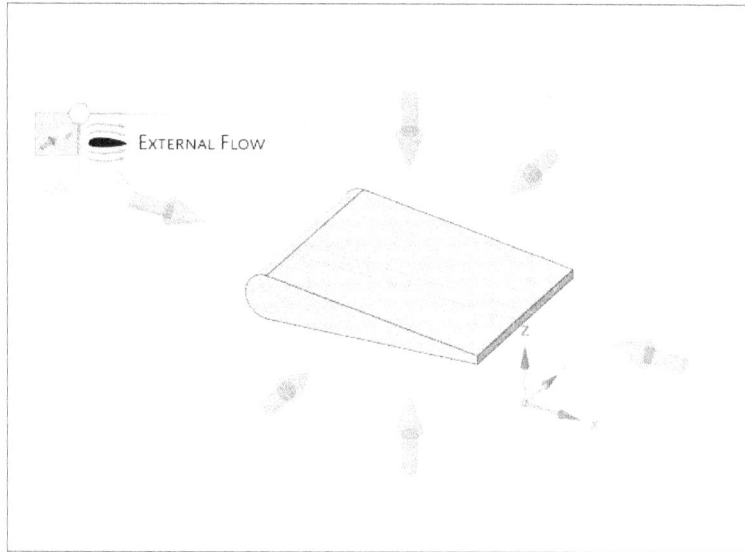

Figure-18. External flow domain

- Select desired arrow to define inlet. You will be asked to select ground plane for the domain.
- Select desired face for the ground plane; refer to Figure-19. The model will be displayed as shown in Figure-20 and parameters related to analysis will be added in the **Physics Tree**.

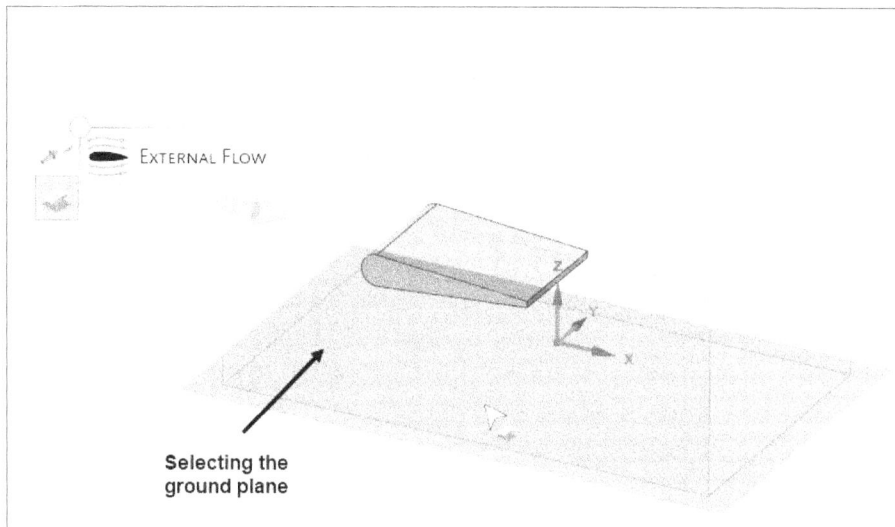

Figure-19. Selecting ground plane

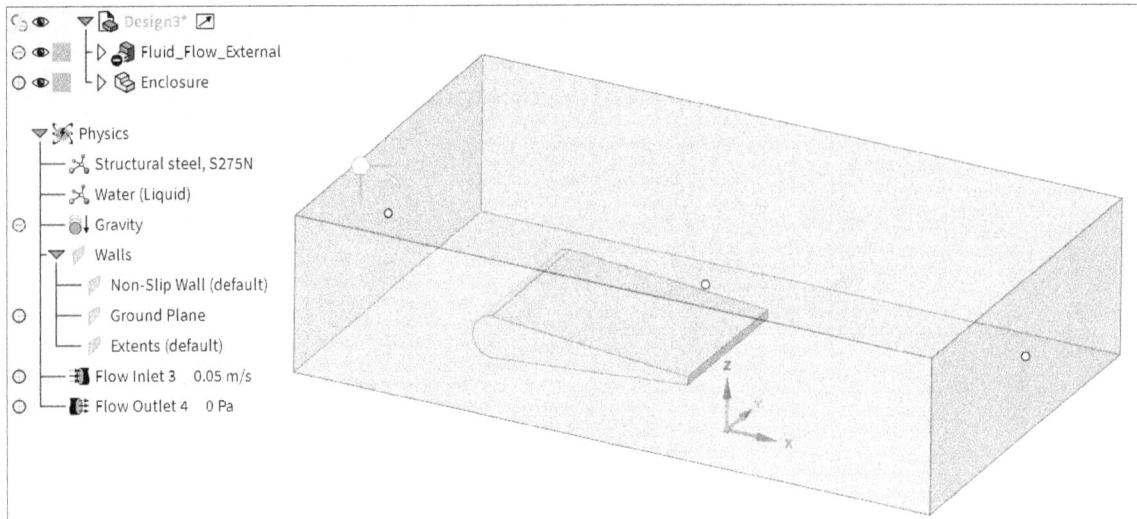

Figure-20. Domain for external fluid flow analysis

- Right-click on the **Water (Liquid)** node from the **Physics Tree** and select the **Edit** option from the shortcut menu. The **MATERIAL HUD** toolbar will be displayed.
- Select the **Air** option from the drop-down in right **HUD** toolbar; refer to Figure-21 and then press **ESC** twice to exit the tool.

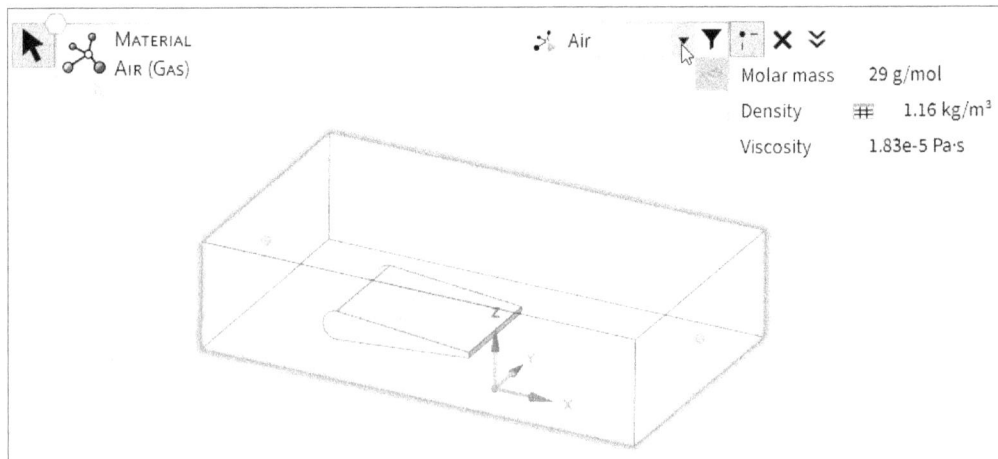

Figure-21. Selecting Air material

- Specify desired value in the **Flow Inlet** edit box to define the speed at which air is flowing around the model. We are using **800 km/h** speed for our analysis.
- Similarly, set the pressure at flow outlet to **1 atm**.
- After setting the parameters, click on the **Solve** button from the **Results Arc**. The results of analysis will be displayed; refer to Figure-22.

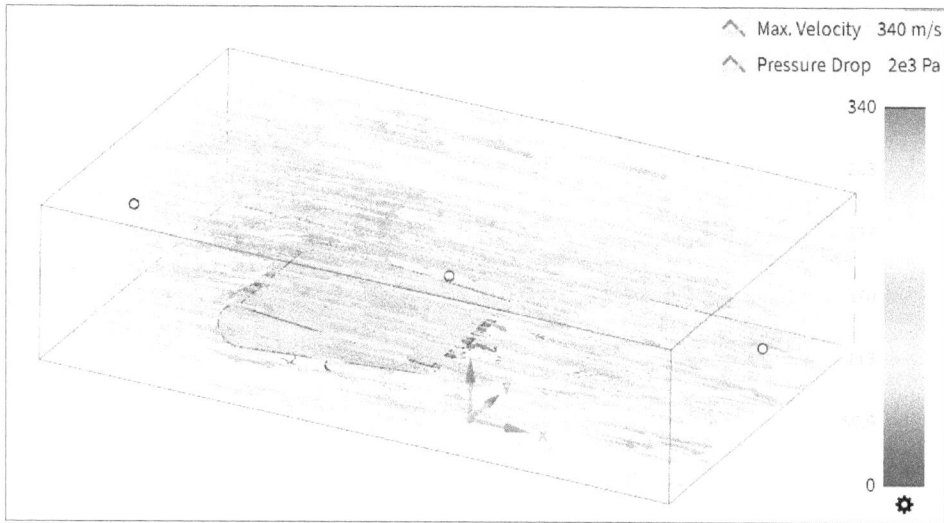

Figure-22. External flow analysis result

PERFORMING ELECTROMAGNETICS ANALYSIS

The Electromagnetics analysis is performed to check various details of electromagnetism in different applications. For example, it is important to analyze the area of magnetism around the conductor through which high current is passing in high power electronic circuits and similarly, you can understand the gain of antenna used in electronic devices based on electromagnetic properties of model. We will use the example of a GPS antenna to understand how electromagnetics analysis is setup in Ansys Discovery. Note that this analysis is available in Enterprises version of the software if you are commercial user of Ansys Discovery.

- Create/import the model for which you want to perform the analysis; refer to Figure-23. (Model is available in resource kit of this book).

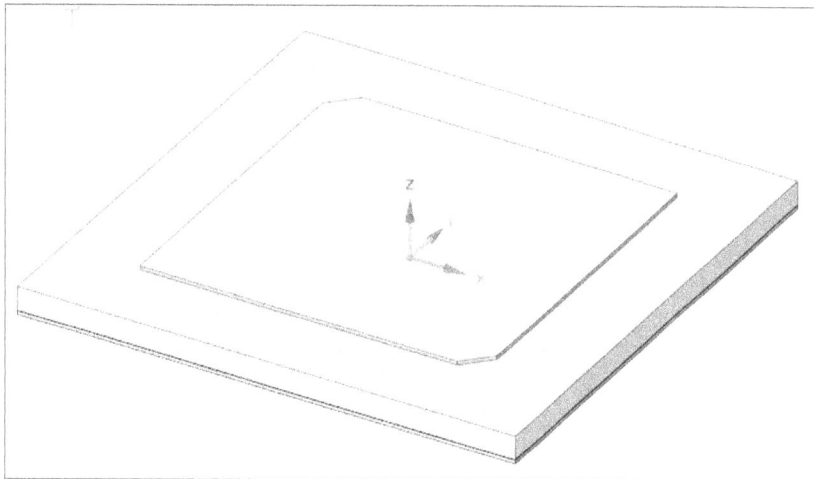

Figure-23. Model for electromagnetism

- Switch to **REFINE** stage from the **Stage Selector** to activate tools related to Electromagnetic analysis.

- Hide the **Substrate** component from the **Design Tree** so that you can clearly select edge of antenna in model; refer to Figure-24.

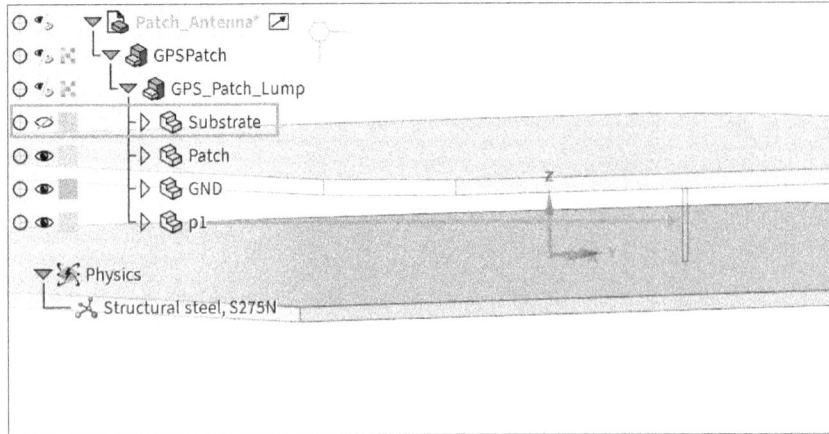

Figure-24. Showing antenna in model

- Click on the **Circuit Port** tool from the **Electromagnetics** drop-down in the **PHYSICS** panel of **SIMULATION** tab in the **Ribbon** and select the edge of antenna as shown in Figure-25. The **ELECTROMAGNETICS CIRCUIT PORT HUD** toolbar will be displayed; refer to Figure-26.

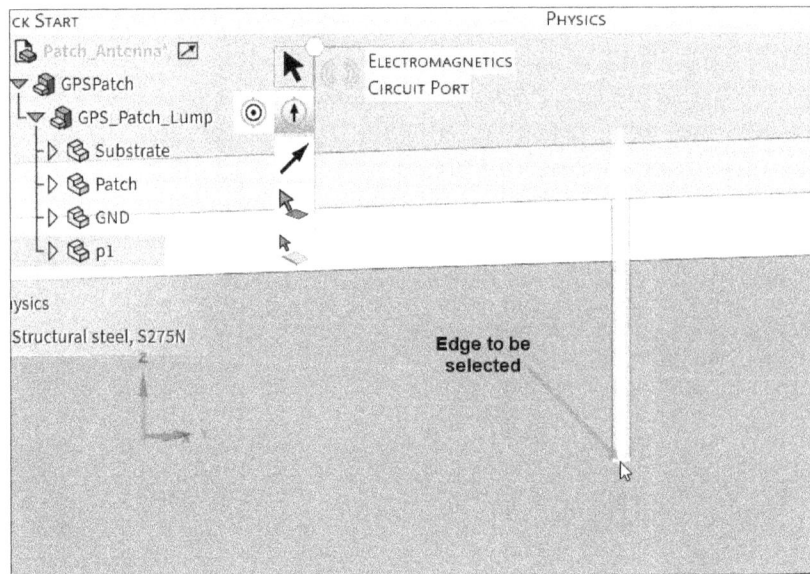

Figure-25. Edge selected for circuit port

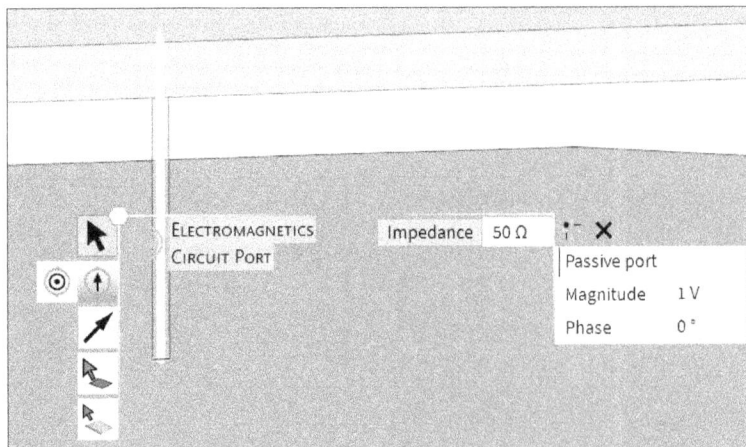

Figure-26. ELECTROMAGNETICS CIRCUIT PORT HUD toolbar

- Specify desired value of impedance in the edit box and press **ENTER** to apply the electromagnetic load. An electromagnetism analysis will be automatically set as analysis model; refer to Figure-27. Note that material of **Substrate** component should be PCB laminate and pure copper for other components.

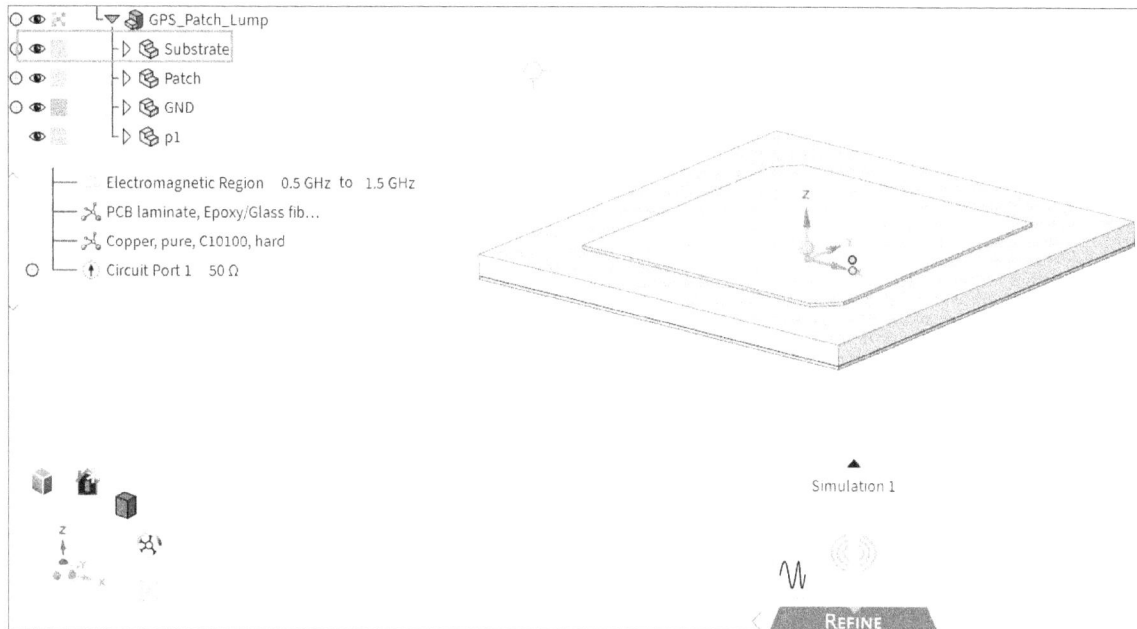

Figure-27. Model set for electromagnetism

- Set desired parameters in the **Electromagnetic Region** edit boxes in the **Physics Tree** to define range within which gain of antenna and far field is to be studied for interference.
- After setting desired parameters, click on the **Solve** button from the **Results Arc**. The result of analysis will be displayed with far field results; refer to Figure-28. Similarly, you can select **Near field pattern** result from the **Results Arc**.

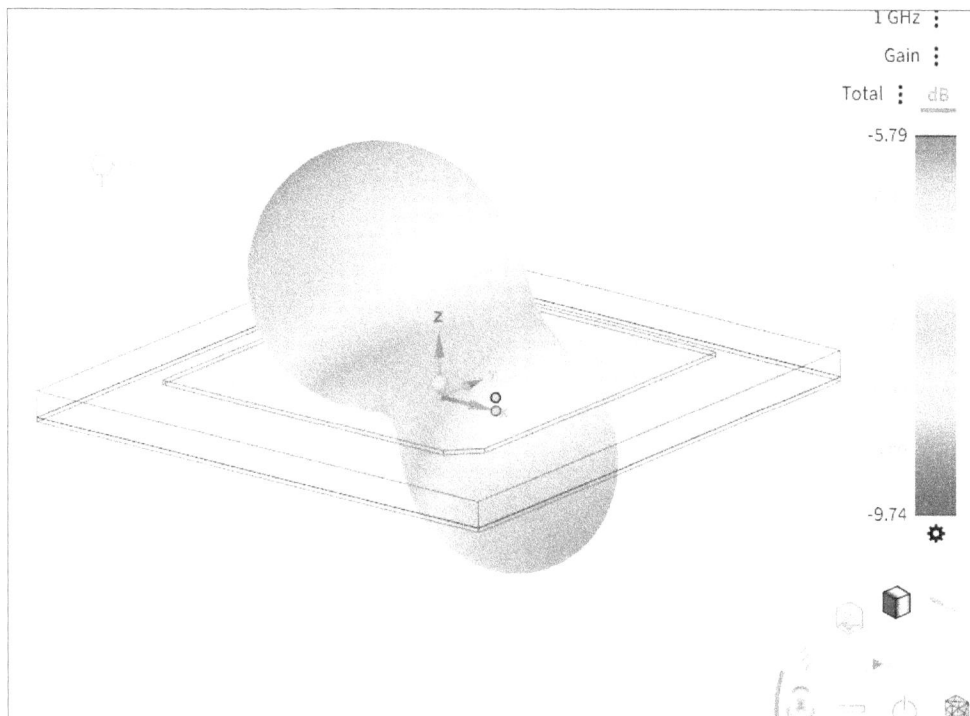

Figure-28. Far field analysis result of electromagnetism

PROBLEM 1

A metal sphere of diameter d = 35mm is initially at temperature Ti = 700 K. At t=0, the sphere is placed in a fluid environment that has properties of T∞ = 300 K and h = 50 W/m2-K. The properties of the steel are k = 35 W/m-K, ρ = 7500 kg/m3, and c = 550 J/kg-K. Find the surface temperature of the sphere after 500 seconds.

PROBLEM 2

A flanged pipe assembly; refer to Figure-29, made of plain carbon steel is subjected to both convective and conductive boundary conditions. Fluid inside the pipe is at a temperature of 130°C and has a convection coefficient of hi = 160 W/m²-K. Air on the outside of the pipe is at 20°C and has a convection coefficient of ho = 70 W/m²-K. The right and left ends of the pipe are at temperatures of 450°C and 80°C, respectively. There is a thermal resistance between the two flanges of 0.002 K-m²/W. Use thermal analysis to analyze the pipe under both steady state and transient conditions.

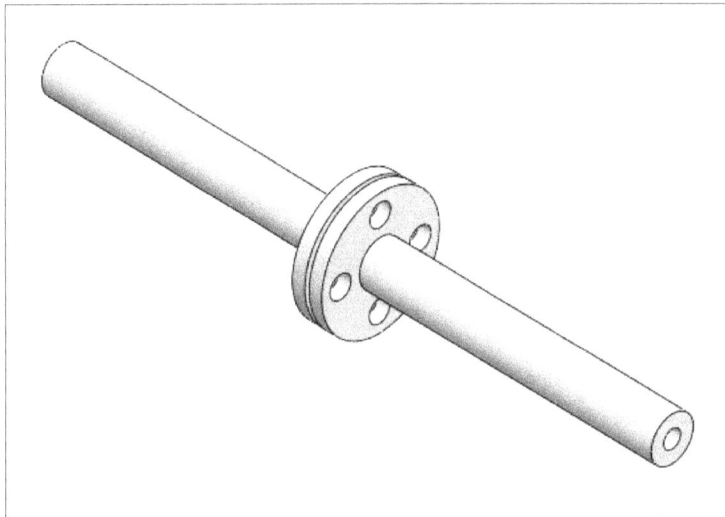

Figure-29. Flanged pipe assembly

PROBLEM 3

Perform fluid flow analysis and determine the temperature distribution in the heat exchanger based on conditions shown in Figure-30. (Model in resource kit)

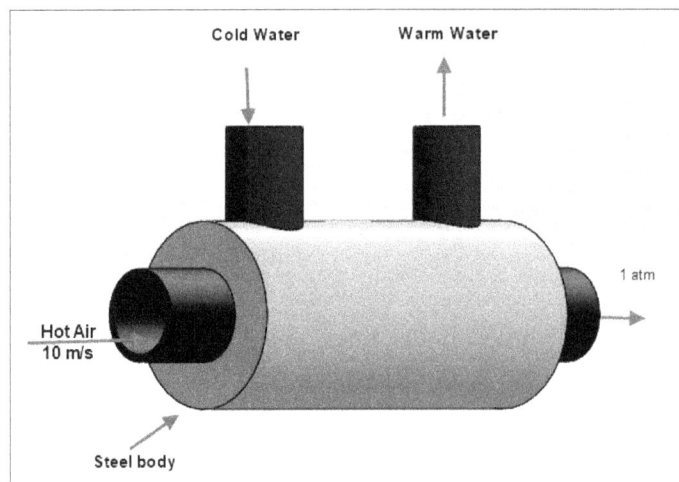

Figure-30. Practice 3

SELF-ASSESSMENT

Q1. What is thermal analysis used for?
A. Checking fluid flow in a pipe
B. Checking distribution of heat over a body due to applied thermal loads
C. Measuring electromagnetic waves
D. Calculating structural stress

Q2. Which of the following is NOT a mechanism of heat transfer?
A. Conduction
B. Convection
C. Radiation
D. Compression

Q3. In which heat transfer mechanism does heat flow without the need for an intervening medium?
A. Conduction
B. Convection
C. Radiation
D. Conduction and convection

Q4. What does thermal analysis output?
A. Electrical conductivity
B. Magnetic field strength
C. Temperature distribution
D. Stress-strain curves

Q5. In conduction, heat is transferred by:
A. Electromagnetic waves
B. Molecular interactions within a material
C. Bulk movement of fluid
D. Compression of gases

Q6. Which example best describes convection?
A. Heating a metal rod at one end and feeling warmth at the other end
B. Heat from the sun reaching the Earth
C. Boiling water in a pot
D. Electricity passing through a conductor

Q7. What is the best example of radiation heat transfer?
A. Heat traveling through metal
B. Heat from the Sun reaching the Earth
C. Heat transfer through boiling water
D. Air heated by a radiator

Q8. What is thermal gradient?
A. Rate of increase in temperature per unit depth in a material
B. The total temperature of a system
C. The resistance to heat flow
D. The ratio of heat absorbed to heat emitted

Q9. What is the thermal flux?
A. Heat loss per second
B. Rate of heat transfer per unit cross-sectional area
C. The total heat absorbed
D. Heat loss due to radiation only

Q10. What is bulk temperature?
A. The temperature at the core of a material
B. The surface temperature of a material
C. The temperature of a fluid flowing outside the material
D. The temperature measured at absolute zero

Q11. What does the emissivity of a material represent?
A. The amount of energy stored in the material
B. The ratio of energy radiated by the material to that radiated by a black body
C. The electrical conductivity of a material
D. The reflectivity of the material

Q12. What is the value of emissivity for a perfect black body?
A. 0
B. Less than 1
C. 1
D. More than 1

Q13. What is the Stefan-Boltzmann constant used for?
A. Defining the amount of heat conduction
B. Determining radiation energy from a black body
C. Calculating specific heat
D. Measuring convection heat transfer

Q14. What does thermal conductivity indicate?
A. The ability of a material to store heat
B. The ability of a material to conduct heat
C. The resistance of a material to heat flow
D. The total energy content of a system

Q15. What is specific heat?
A. Heat required per unit mass to raise temperature by one degree Celsius
B. Heat lost due to radiation
C. Heat required to convert solid to liquid
D. Heat required per unit area to transfer energy

Q16. In thermal analysis, why is a heat sink often used?
A. To store heat energy
B. To increase the temperature of the system
C. To dissipate heat and keep the temperature within limits
D. To prevent conduction

Q17. What is the assumed default convection rate between solid and air in open air conditions?
A. 1 W/m²·°C
B. 10 W/m²·°C
C. 100 W/m²·°C
D. 250 W/m²·°C

Q18. What is the main objective of performing fluid flow analysis?
A. To check the movement of solids in a system
B. To analyze how a fluid flows through a specified domain
C. To determine the weight of the fluid
D. To measure electromagnetic interference

Q19. When performing internal fluid flow analysis, what is needed?
A. An open domain for fluid movement
B. A closed region with defined inlet and outlet faces
C. A solid body without any openings
D. No boundary conditions

Q20. What is the purpose of external fluid flow analysis?
A. To analyze fluid movement inside a container
B. To determine pressure changes inside pipes
C. To check the aerodynamic behavior of a model
D. To measure heat conduction in a solid

Q21. In electromagnetics analysis, what is typically analyzed in high-power circuits?
A. The strength of the material
B. The area of magnetism around a conductor carrying high current
C. The amount of heat generated
D. The pressure of air around the circuit

Q22. What parameter is set in electromagnetics analysis to define the study range for interference?
A. Fluid velocity
B. Electromagnetic region
C. Radiation intensity
D. Heat transfer coefficient

FOR STUDENT NOTES

FOR STUDENT NOTES

Chapter 9

Ansys Discovery - Multi-physics Analysis

Topics Covered

The major topics covered in this chapter are:

- *Introduction*
- *Performing Multi-physics Analysis*

INTRODUCTION

Ansys Discovery is capable of performing multi-physics analysis. A multi-physics analysis is performed when you want to combine multiple analyses like fluid flow analysis, structural analysis, and thermal analysis. There are various examples of multi-physics analyses like temperature rise in wind turbine along with structural stress due to air flow. This example needs to be solved by combining fluid flow, thermal, and structural analyses. Similarly, an artificial heart valve should be analyzed using combination of fluid flow, structural analysis, and material deformation analysis. Fluid flow analysis is used to simulate blood flow through valve, structural analysis is performed to study stress distribution due to pressure change in valve, and fatigue analysis to check for life of valve.

PERFORMING MULTI-PHYSICS ANALYSIS

Find out the temperature distribution inside the pipe when hot water gets mixed with the cold water flowing in the pipe and heat applied to the surface of pipe by coil; refer to Figure-1.

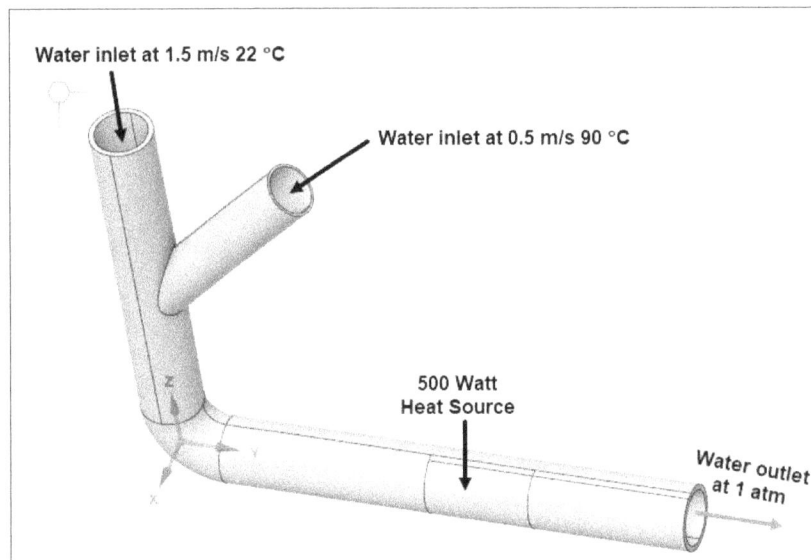

Figure-1. Model setup for multi-physics analysis

In case of this analysis, you will be combining fluid flow analysis with thermal analysis. The procedure to perform this analysis is given next.

- Start Ansys Discovery if not started yet.
- Open the **Heated water pipe** model from the Resource kit folder of the book.
- Switch to **EXPLORE** stage from the **Stage Selector**.
- Click on the **INTERNAL FLOW** tool from the **QUICK START** panel in the **SIMULATION** tab of the **Ribbon**. You will be asked to specify inlet face for fluid.
- Select the face as shown in Figure-2. You will be asked to select outlet face for fluid.
- Select the end face of model as shown in Figure-3. You will be asked to select another face as outlet but we want to add another inlet in the fluid domain. So, select the **INLET** button from **HUD** toolbar and select the face of opening as shown in Figure-4. The fluid volume will be created automatically and Fluid Flow study will be setup in graphics area; refer to Figure-5.

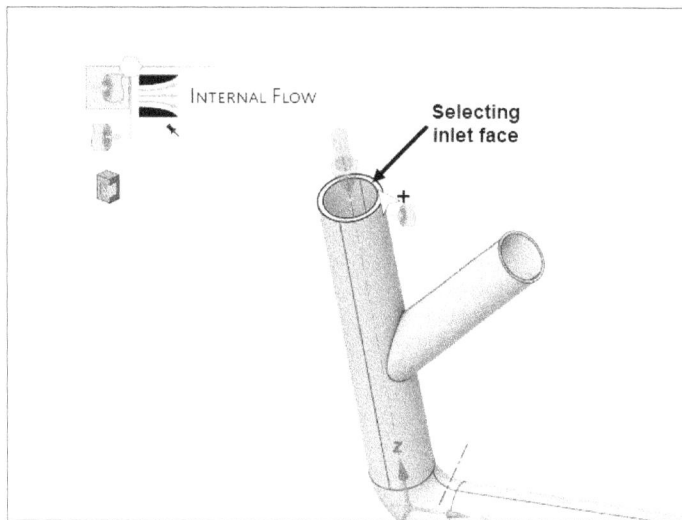

Figure-2. Selecting inlet face

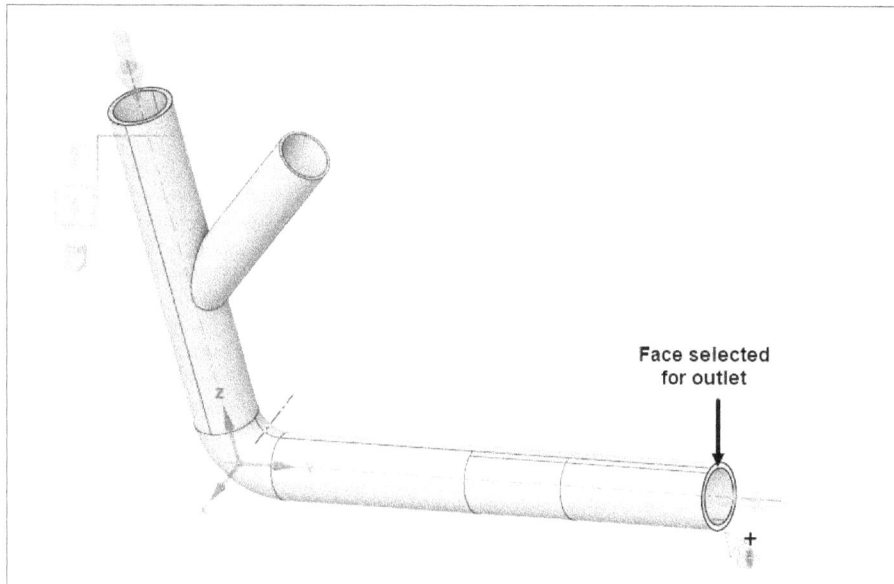

Figure-3. Selecting face for outlet of fluid

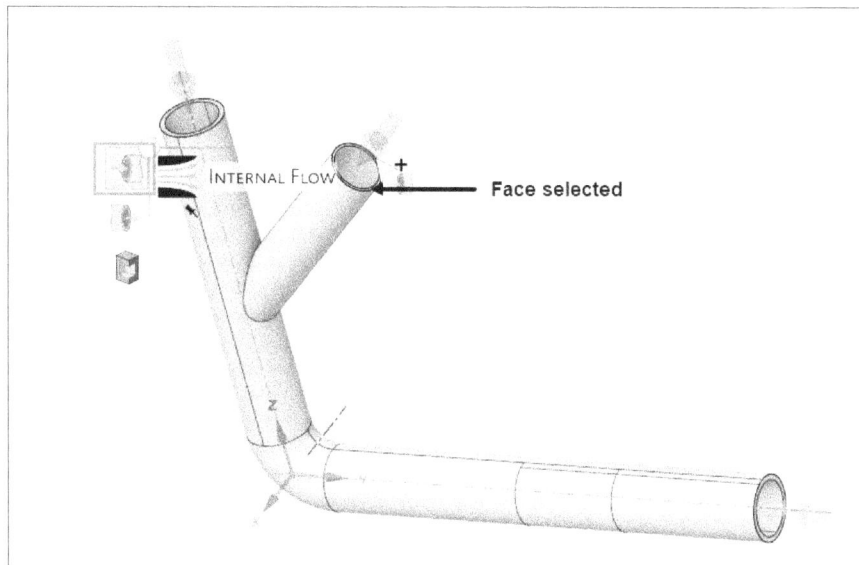

Figure-4. Face selected for another inlet

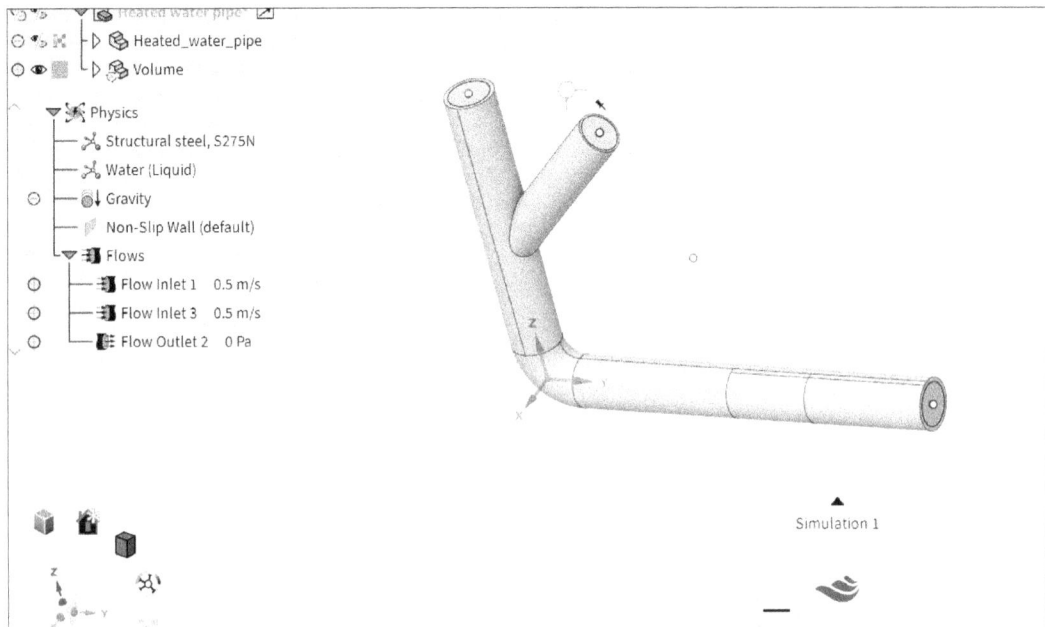

Figure-5. Fluid flow study setup

- Set the speed of flow inlets and pressure at outlet as per the parameters shown in Figure-6.

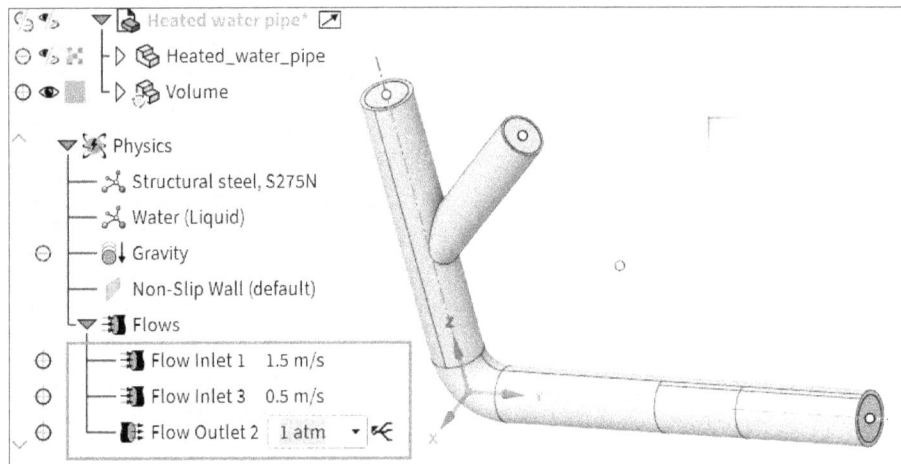

Figure-6. Setting flow rates

- Click on the **Heat** tool from the **Solid Thermal** drop-down in the **PHYSICS** panel of **SIMULATION** tab in the **Ribbon**. The **THERMAL HEAT HUD** toolbar will be displayed.
- Select the face of model as shown in Figure-7 and set the total heat value to **500 W**. Press **ESC** to exit the tool. The thermal study parameters will be added automatically; refer to Figure-8.

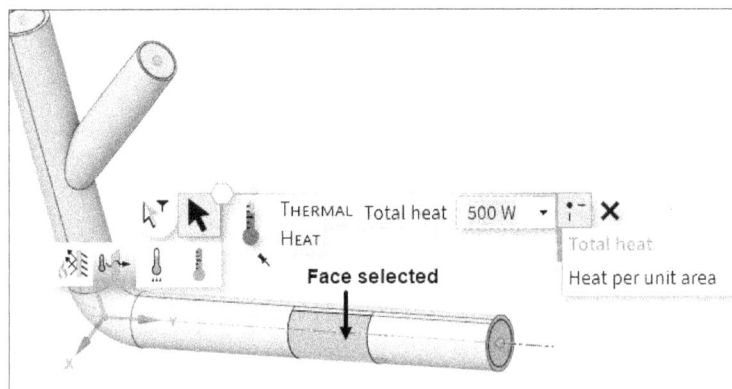

Figure-7. Applying heat parameter

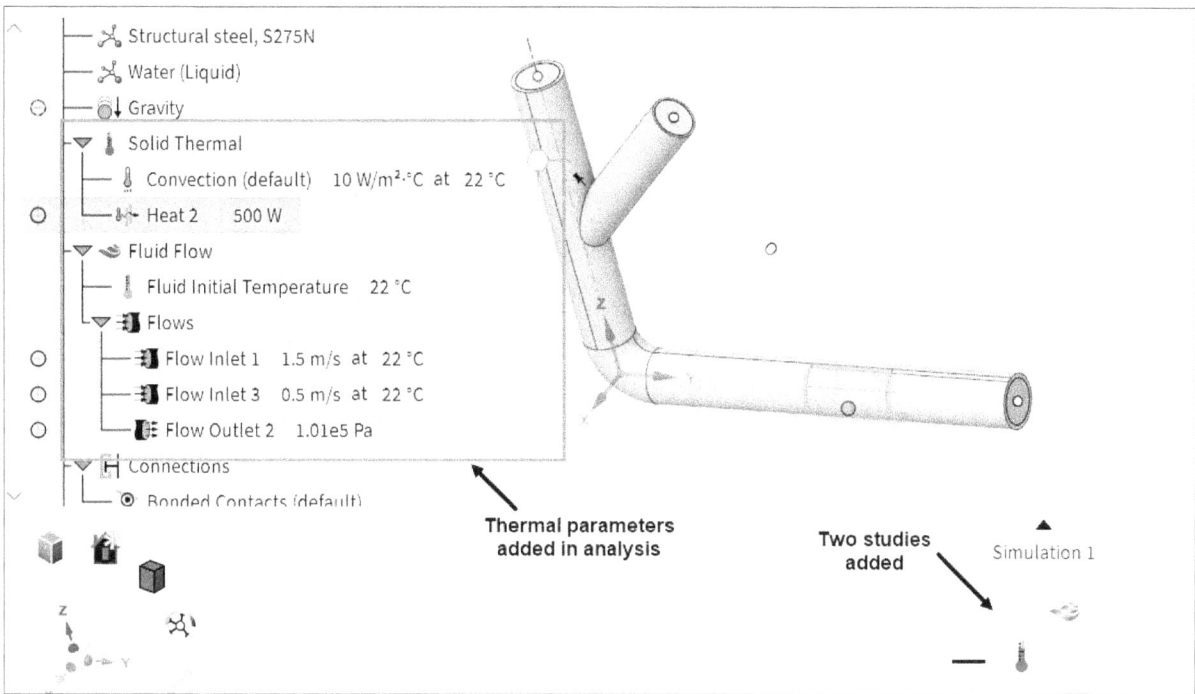

Figure-8. Thermal study parameters added

- After setting desired parameters, click on the **Solve** tool from the **Results Arc** to solve the analysis. The results will be displayed as shown in Figure-9.

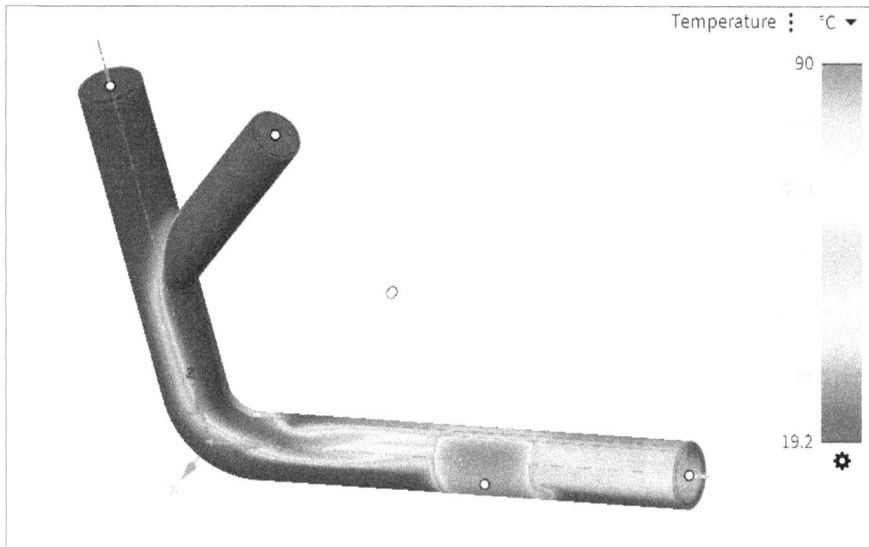

Figure-9. Result of multi-physics fluid flow analysis

SELF-ASSESSMENT

Q1. What type of analysis is Ansys Discovery capable of performing?
A. Single-physics analysis
B. Multi-physics analysis
C. Structural analysis only
D. Thermal analysis only

Q2. Which of the following is an example of a multi-physics analysis?
A. Calculating the stress in a bridge beam
B. Measuring airflow over an aircraft wing
C. Temperature rise in a wind turbine along with structural stress due to airflow
D. Finding the melting point of a metal

Q3. What type of analyses are required to study an artificial heart valve?
A. Only fluid flow analysis
B. Fluid flow, structural, and material deformation analysis
C. Only structural analysis
D. Only fatigue analysis

Q4. In the heated water pipe analysis, which two types of analysis are combined?
A. Structural and fatigue analysis
B. Fluid flow and structural analysis
C. Fluid flow and thermal analysis
D. Thermal and fatigue analysis

Q5. What is the first step to perform a multi-physics analysis in Ansys Discovery?
A. Open the Heated water pipe model
B. Select the inlet and outlet faces
C. Start Ansys Discovery
D. Set the total heat value

Q6. What should be done after opening the Heated water pipe model?
A. Switch to EXPLORE stage
B. Click on the Solve tool
C. Directly input temperature values
D. Start the external flow simulation

Q7. Which tool is used to define the fluid flow in the analysis?
A. EXTERNAL FLOW
B. INTERNAL FLOW
C. SOLID THERMAL
D. STRUCTURAL

Q8. What needs to be selected when prompted for the fluid inlet?
A. The face of the model
B. The edge of the pipe
C. The center of the model
D. Any random face

Q9. What additional step is taken before selecting the outlet face for the fluid?
A. Selecting another face as an additional inlet
B. Setting the temperature conditions
C. Applying structural stress
D. Adding a heat source

Q10. How is the thermal analysis applied to the model?
A. By selecting the Heat tool from the Solid Thermal drop-down
B. By setting structural stress first
C. By changing the material properties
D. By defining the fatigue parameters

Q11. What value of total heat is set for the thermal study in the example?
A. 200 W
B. 300 W
C. 500 W
D. 1000 W

Q12. Which tool is used to solve the analysis and display the results?
A. Heat tool
B. Solve tool
C. Fluid flow tool
D. Explore tool

FOR STUDENT NOTES

Index

OTHER BOOKS BY CADCAMCAE WORKS

Autodesk Revit 2025 Black Book
Autodesk Revit 2024 Black Book
Autodesk Revit 2023 Black Book
Autodesk Revit 2022 Black Book

Autodesk Inventor 2025 Black Book
Autodesk Inventor 2024 Black Book
Autodesk Inventor 2023 Black Book
Autodesk Inventor 2022 Black Book

Autodesk Fusion Black Book (V 2.0.21508)
Autodesk Fusion PCB Black Book (V 2.0.21528)

AutoCAD Electrical 2025 Black Book
AutoCAD Electrical 2024 Black Book
AutoCAD Electrical 2023 Black Book
AutoCAD Electrical 2022 Black Book

SolidWorks 2025 Black Book
SolidWorks 2024 Black Book
SolidWorks 2023 Black Book
SolidWorks 2022 Black Book

SolidWorks Simulation 2025 Black Book
SolidWorks Simulation 2024 Black Book
SolidWorks Simulation 2023 Black Book
SolidWorks Simulation 2022 Black Book

SolidWorks Flow Simulation 2025 Black Book
SolidWorks Flow Simulation 2024 Black Book
SolidWorks Flow Simulation 2023 Black Book

SolidWorks CAM 2025 Black Book
SolidWorks CAM 2024 Black Book
SolidWorks CAM 2023 Black Book
SolidWorks CAM 2022 Black Book

SolidWorks Electrical 2025 Black Book
SolidWorks Electrical 2024 Black Book
SolidWorks Electrical 2022 Black Book
SolidWorks Electrical 2021 Black Book

SolidWorks Workbook 2022

Mastercam 2023 for SolidWorks Black Book
Mastercam 2022 for SolidWorks Black Book
Mastercam 2017 for SolidWorks Black Book

Mastercam 2025 Black Book
Mastercam 2024 Black Book
Mastercam 2023 Black Book
Mastercam 2022 Black Book

Creo Parametric 11.0 Black Book
Creo Parametric 10.0 Black Book
Creo Parametric 9.0 Black Book
Creo Parametric 8.0 Black Book
Creo Parametric 7.0 Black Book

Creo Manufacturing 11.0 Black Book
Creo Manufacturing 10.0 Black Book
Creo Manufacturing 9.0 Black Book
Creo Manufacturing 4.0 Black Book

ETABS V22 Black Book
ETABS V21 Black Book
ETABS V20 Black Book
ETABS V19 Black Book

Basics of Autodesk Inventor Nastran 2025
Basics of Autodesk Inventor Nastran 2024
Basics of Autodesk Inventor Nastran 2022
Basics of Autodesk Inventor Nastran 2020

Autodesk CFD 2024 Black Book
Autodesk CFD 2023 Black Book
Autodesk CFD 2021 Black Book
Autodesk CFD 2018 Black Book

FreeCAD 1.0 Black Book
FreeCAD 0.21 Black Book
FreeCAD 0.20 Black Book
FreeCAD 0.19 Black Book

LibreCAD 2.2 Black Book

TinkerCAD Black Book

www.ingramcontent.com/pod-product-compliance
Lightning Source LLC
Chambersburg PA
CBHW081803200326

41597CB00023B/4124